STRUCTURE
AND EVOLUTION OF
THE STARS

By MARTIN SCHWARZSCHILD

DOVER PUBLICATIONS, INC.
NEW YORK

Published in Canada by General Publishing Company, Ltd., 30 Lesmill Road, Don Mills, Toronto, Ontario.
Published in the United Kingdom by Constable and Company, Ltd., 10 Orange Street, London W. C. 2.

This Dover edition, first published in 1965, is an unabridged republication of the work first published by Princeton University Press in 1958. This edition is published by special arrangement with Princeton University Press.

International Standard Book Number: 0-486-61479-4
Library of Congress Catalog Card Number: 65-25710

Manufactured in the United States of America

Dover Publications, Inc.
180 Varick Street
New York, N. Y. 10014

TO BARBARA

PREFACE

It may be folly to try to write a book on a subject where previous milestones have been set by books of men like Emden, Eddington, and Chandrasekhar. But since the writing of the latest of these milestones a rapid development has taken place in the theory of stellar structure. Under these circumstances I do hope one may not be thought presumptuous for attempting to write a book, for temporary use, which aims to summarize the present state of our subject and thus to help prepare the next developments.

And still I would have lacked the courage for this undertaking had it not been for three collaborators: Dr. Lyman Spitzer, who critically affected the over-all plan of the book and section by section added essential improvements; Mr. Richard Härm, who derived most of the numerical material and prepared with care the tables and figures; and my wife, who smoothed the English, checked understandability and consistency, and bolstered the writer's persistence. For their collaboration I am deeply grateful.

Martin Schwarzschild

Princeton University Observatory
November 1957

CONTENTS

INTRODUCTION

A little more than a decade ago research on the stellar interior underwent a profound change. The central cause of this change was the introduction of nuclear physics into astronomy. Nuclear physics has provided the theory of the stellar interior with the last—but not the least—of the fundamental physical processes which determine stellar structure and evolution. Thus a new and far-reaching development in this field of research became possible.

Simultaneously with this new theoretical development occurred an equally far-reaching upsurge in the relevant fields of observational astronomy, an upsurge due largely to the introduction of new spectrographic and photoelectric techniques. The combination of these developments suddenly opened up an unprecedentedly wide front of contact between observation and theory in the research field of the stellar interior, in striking contrast with the situation twenty-five years ago, when there was only one major point of contact, the mass-luminosity relation.

The rush of this development continues on a broad front. Under such circumstances it cannot be hoped that a technical book like this will remain up to date for long. Consequently, the purpose of this book cannot be to give an account of the theory of stellar structure and evolution that is even temporarily definitive. Instead, the purpose is to make the methods used and the results thus far obtained more easily available to all those who may press the next steps in this exciting part of today's astrophysics.

The eight chapters of this book are divided into three groups. In the first three chapters are assembled the fundamentals for a theory of the stellar interior, that is the relevant astronomical observations, the basic physical laws and processes, and the necessary mathematical techniques. The subsequent four chapters give the detailed construction of stellar models, starting with the initial main-sequence models, continuing with sequences of models for successive evolutionary phases, and ending with the final white-dwarf models. The last chapter contains a summary of the results and conclusions that can be drawn from the available model sequences and a description of our present tentative picture of stellar evolution.

To make the last chapter more generally useful, an attempt has been made to present the material of this chapter in a form understandable without the prerequisite of the preceding four more specialized chapters.

Most readers will not want to read all the sections of this book. A physicist, for example, might just glance through Chapter II to see the large variety of topics of his discipline that go into the theory of the stellar interior, and then read the last chapter to acquaint himself with the present status of this theory. An applied mathematician might just go through Chapter III to see what type of mathematical problem the stellar interior presents, and what techniques have thus far been employed. An observational astronomer might choose Chapter I to orient himself as to the observational data that may be of direct import to the theory of stellar evolution, and then follow this up with the last chapter to see the present results. Only an active student of the theory of the stellar interior may want to dive into the details of the model construction work described in Chapters IV to VII.

For a one-term graduate course on the topic of this book it might possibly be useful to divide up the available time as follows: 50 percent for the fundamentals described here in the first three chapters; then 20 percent for deriving as an example a particular model, such as one of those in §15, and even reconstructing such a model in full numerical detail; the following 10 percent for running through the evolution of a medium-weight star in a cursory manner, following §§16, 23, and 24; and, finally, the remaining 20 percent for a study of the white dwarfs, as given here in §§26 and 27, and a summary of the present picture of stellar evolution, according to §§28 to 30 (which, however, will soon have to be brought up to date by the addition of the latest publications in this field).

To keep the size of this book within reasonable bounds the topics to be covered had to be restricted to those most directly relevant to stellar structure and evolution, and several important topics of the general field of the stellar interior had to be omitted. The biggest omissions are the origin of the stars, structure and evolution of close double stars, physical variability such as pulsations, and the origin of the heavier elements. Some of these big topics here omitted have an up-to-date presentation in recent books. Some others may not be ripe yet for a full book presentation.

One further topic has been completely omitted from this book, the history of the science of the stellar interior. This omission is a consequence of the central aim to present here as concisely as possible the present status of the theory of stellar structure and evolution. Thus many investigations of the past, however fundamental and necessary at their time, are not described here if they have been followed up subsequently by a more direct or a more complete approach.

Consistent with this aim of presenting the present status, not the history, of our subject the text has been kept entirely free from any references (except in the headings of tables and figures, where references are given to the sources of the specific data shown), and all the references have been put together in a special section at the end of the book. This procedure, it is hoped, will permit the reader to concentrate more easily on the present theory. At the same time it entails one grave danger, that the reader may subconsciously assume that the author of this book is also the author of the major part of the investigations described. Nothing could be more false, not only with regard to the particular author, but also with regard to the circumstances under which the recent developments of our subject have occurred. These recent developments are characterized by the circumstance that they have not arisen mainly by the work of one man, but rather by the work of a whole group of men with a great variety in special ability and character. It is the results of their work which this book aims to present, work done sometimes in active cooperation, sometimes in gay competition, always pervaded by that deep delight that research brings when new fields open up.

STRUCTURE AND EVOLUTION
OF THE STARS

CHAPTER I
OBSERVATIONAL BASIS

1. Luminosities and Radii

If simple perfect laws uniquely rule the universe, should not pure thought be capable of uncovering this perfect set of laws without having to lean on the crutches of tediously assembled observations? True, the laws to be discovered may be perfect, but the human brain is not. Left on its own, it is prone to stray, as many past examples sadly prove. In fact, we have missed few chances to err until new data freshly gleaned from nature set us right again for the next steps. Thus pillars rather than crutches are the observations on which we base our theories; and for the theory of stellar evolution these pillars must be there before we can get far on the right track.

Only one chapter of this book will be devoted to all the relevant observations. The limitation is set by the plan of the book, not by the scope or value of the observations. This chapter can give but a slight hint of all the ingenuity and effort which had to be applied to wring these data securely from the near-mute starlight. That story could well fill another book. Here we can only summarize it.

The Hertzsprung-Russell Diagram

Soon after its first construction the Hertzsprung-Russell diagram was recognized as a powerful tool for the study of stellar evolution. This diagram represents the stars according to two readily observable characteristics, the visual absolute magnitude M_{vis} and the spectral type or in its place the color index B-V. Fig. 1.1 gives the Hertzsprung-Russell diagram for nearby stars with well-known distances and hence good absolute magnitudes. In this figure the points near the top represent luminous stars and those near the bottom faint ones: the more negative the magnitude, the greater the corresponding brightness; see Eq. (1.2). The points at the left represent blue stars with high surface temperatures and those at the right red stars with low surface temperatures: the more negative the color index, the more negative the blue magnitude B as compared with the visual magnitude V, that is the greater the brightness in the blue relative to that in the yellow. Points in the lower left-hand corner represent stars of small size (low luminosity in spite of high surface temperature indicates a small surface area), whereas points in the upper right-hand corner represent the stars with the biggest diameters.

We recognize in Fig. 1.1 the main sequence stretching from the upper left-hand corner to the lower right-hand corner, with its lower part sharply defined by the red dwarfs and its upper part occupied by luminous blue

1

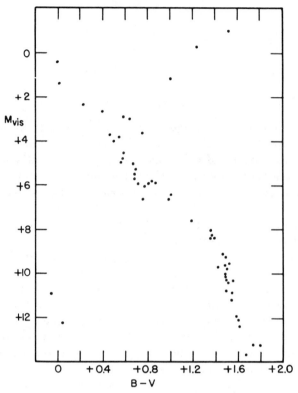

Fig. 1.1. Hertzsprung-Russell diagram of nearby
stars. (H. L. Johnson and W. W. Morgan, *Ap.J. 117,*
313, 1953).

stars sparsely represented in the small volume of the nearby stars. To
the right and above the main sequence are the red giants, spread over a
wide area, while in the lower left corner are the white dwarfs.

How can it help to sort all stars according to two characteristics
whose only advantage is the practical one of ready observability? In
their place, should we not consider the characteristics of a star that are
basic? We may assume—as we shall find right—that all characteristics
of a star, including those used in the Hertzsprung-Russell diagram, are
fixed completely by its mass, its initial composition, and its age. As a
star grows older it will run through the Hertzsprung-Russell diagram on
an evolutionary track fixed from the outset by its basic characteristics,
mass and composition. To derive such evolutionary tracks will be a
major aim of the theory of stellar evolution.

No easy clue to evolutionary tracks is provided by the Hertzsprung-
Russell diagram of Fig. 1.1. The random sample represented by the
nearby stars certainly includes a large range of masses, possibly some
variety in initial composition, and most likely a great mixture of stellar

ages. A great variety of evolutionary tracks therefore is involved in Fig. 1.1 and their disentanglement seems somewhat hopeless. This complication we may overcome if we exploit the infinitely neater sampling which stellar clusters provide.

Fig. 1.2 shows the Hertzsprung-Russell diagram for two galactic clusters observed with the highest accuracy. A picture quite different from Fig. 1.1 emerges. In each cluster the stars, with few exceptions, fall sharply along one line. (Most of the exceptions can be explained by unresolved double stars which give points lying too high in the Hertzsprung-Russell diagram by as much as 0.7 magnitudes.) We must conclude that the stars of any one cluster differ from one another in only one basic characteristic, and this must be the mass. That initial composition and age should be identical for all stars in one cluster hardly seems surprising in view of the common origin strongly suggested by the dynamical unity represented by a cluster.

While the stars of one cluster lie along one line in the Hertzsprung-Russell diagram, the lines of different clusters differ significantly from one another. The two examples of Fig. 1.2 are typical. The fainter stars in each cluster follow a line approximately parallel to the main sequence of the nearby stars, but the brighter stars turn upwards to the right from the main sequence and in some clusters—as for example in Praesepe —there are a few bright stars far to the right in the red giant region. The turnoff of the brighter stars from the main sequence to the right is a phenomenon common to all clusters. The spectral type or color index at

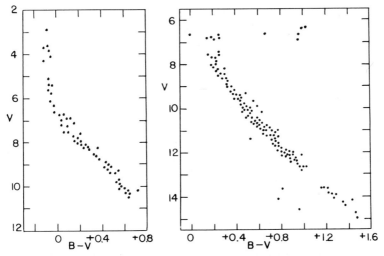

Fig. 1.2. Hertzsprung-Russell diagram of two typical galactic clusters. At the left the Pleiades (H. L. Johnson and W. W. Morgan, *Ap.J. 117*, 313, 1953), at the right Praesepe (H. L. Johnson, *Ap.J. 116*, 640, 1952). The fringe of stars just above the main sequence may consist of unrecognized binaries.

which this turnoff occurs, however, is different from cluster to cluster. Fig. 1.2 shows, for example, that the turnoff occurs in the Pleiades at a bluer color than in Praesepe. Thus one can arrange all clusters into a simple sequence according to the color index at the turnoff. We shall later recognize that by arranging the clusters in this sequence we have in reality arranged them in order of their age.

Let us look at two more Hertzsprung-Russell diagrams of galactic clusters, at opposite extremes of the age sequence. These extreme cases, shown in Fig. 1.3, present objects that are observationally much more difficult than the typical clusters shown in Fig. 1.2. Consequently the scatter in Fig. 1.3 is still uncomfortably large at present—not large enough, however, to obscure the main features seriously.

The upper graph of Fig. 1.3 represents an extremely young cluster. The turnoff occurs at a very blue color. Only the stars of intermediate brightness fall along a line approximately parallel to the main sequence. The fainter stars show a peculiar behavior not observed in the older clusters: they deviate to the right from the main sequence, for reasons quite different from those of ordinary red giants, as we shall see in § 19.

On the other extreme, the lower graph of Fig. 1.3 represents a very old galactic cluster. The turnoff occurs at a red color index and the brighter stars deviate greatly from the main sequence, reaching far into the red giant region.

In contrast to the extended age sequence of galactic clusters, the Hertzsprung-Russell diagrams of globular clusters, though different in a variety of detail from one another, seem much alike in their main features. A typical example is given in Fig. 1.4. Again the faintest stars follow a line parallel to, if not identical with, the main sequence. Again the brighter stars turn off the main sequence and occupy a complicated line through the red giant and supergiant region, as well as a horizontal branch which reaches from the bluest colors at the left all the way to the red giant branch. We find that the Hertzsprung-Russell diagram of a globular cluster (Fig. 1.4) is quite similar to that of an old galactic cluster (lower graph of Fig. 1.3) in a number of main features, and particularly in the reddish color index of the turnoff from the main sequence. Nevertheless several major differences exist, especially in the slope and extent of the red giant branch. We shall find an explanation of these differences in terms of the third basic characteristic of stars, their initial composition.

It will take all of the theory of stellar evolution described in the subsequent chapters, and more, to explain the observed Hertzsprung-Russell diagrams of stellar clusters. In § 29 we shall give a summary of the results as they stand now. In the present section we shall restrict ourselves to just a preparatory step: the transformation of the two observed quantities, magnitude and color index, into two quantities, luminosity and radius, more appropriate for the theoretical discussion.

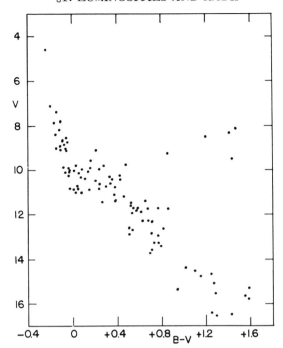

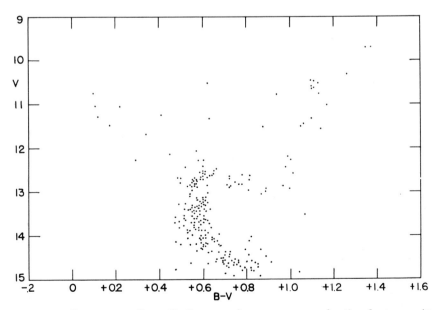

Fig. 1.3. Hertzsprung-Russell diagram of two extreme galactic clusters. At the top the very young NGC 2264 (M. Walker, *Ap.J.*, Supplement No. 23, 1956), at the bottom the very old M67 (H. L. Johnson and A. R. Sandage, *Ap.J. 121*, 616, 1955).

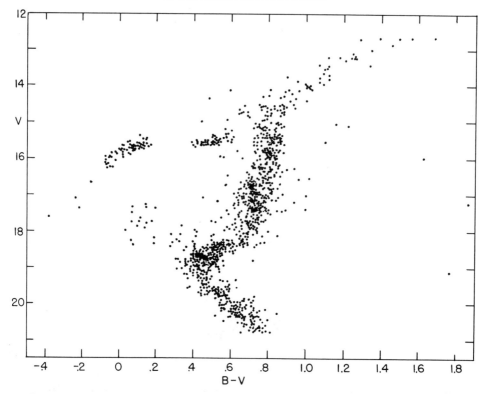

Fig. 1.4. Hertzsprung-Russell diagram of the globular cluster M3. (H. L. Johnson and A. R. Sandage, *Ap.J. 124*, 379, 1956)

Transformation to Luminosities and Radii

The first step in this transformation is the conversion of apparent magnitudes into absolute magnitudes. For this conversion distances are needed. For the nearby stars the distances are known with fair accuracy through the direct measurement of trigonometric parallaxes. Thus it is possible to draw the Hertzsprung-Russell diagram for the nearby stars directly in terms of the absolute magnitude, as in Fig. 1.1. Not so for the clusters.

All the stars within a cluster are at practically the same distance. Therefore if we plot the Hertzsprung-Russell diagram of an individual cluster in terms of the apparent magnitude, as we have done in Figs. 1.2 to 1.4, we can then see the relative positions of the stars in the diagram correctly and can determine unambiguously the slopes and forms of the lines occupied by the stars. But we cannot derive the absolute magnitude or luminosity of any star in a cluster without determining the cluster's distance.

No cluster is close enough to permit an accurate determination of its distance through the measurement of its trigonometric parallax. Only for one cluster, the Hyades, can the distance be determined securely by the "moving cluster method," that is by the measurement of the apparent contraction of the cluster caused by its motion away from us. Are we then restricted to the nearby stars and to the Hyades if we want to derive absolute magnitudes and luminosities? No, not if we permit ourselves one essential assumption.

Let us consider for this purpose only that segment in the Hertzsprung-Russell diagram of each galactic cluster which runs approximately parallel to the main sequence of the nearby stars. Then let us make the assumption that these segments assembled from the various galactic clusters all fall on one line in the Hertzsprung-Russell diagram, that they form one unique relation between the absolute magnitude and the color index. We shall call this line the "initial main sequence," for reasons which we will see in Chapter IV. Under this assumption of the existence of an initial main sequence common to all galactic clusters we can now easily derive the distances of all these clusters on the basis of the distance for the Hyades. All that is needed is to plot the Hertzsprung-Russell diagram for various clusters in terms of apparent magnitudes (Figs. 1.2 to 1.4) and—after correcting the colors for interstellar reddening by the three-color method—to superimpose them onto each other with such vertical shifts as will force the relevant segments into coincidence. For each cluster the vertical shift relative to the Hyades gives its distance. Thus we are set to transform the apparent magnitudes of all galactic cluster stars into absolute magnitudes.

To obtain the distance of a globular cluster we have three choices. We may assume that in the Hertzsprung-Russell diagram of the globular cluster the segment which runs essentially parallel to the main sequence coincides with the initial main sequence of the galactic clusters. Or we may assume that this segment coincides with the main sequence determined from the nearby red dwarfs. Or we may assume that the mean absolute magnitude of the short period variables occurring in globular clusters is identical with that of the nearby short-period variables, the distances of which can be estimated by a statistical investigation of their motions. None of these choices seems very safe. When applied to the typical globular cluster M 3, however, all three methods give results in fair agreement with the others, and the absolute magnitudes thus derived seem safe enough for our purposes.

The next necessary step is the transformation of the visual magnitude, which refers to the intensity in the limited visual wave length band, to the bolometric magnitude, which refers to the total intensity integrated over all wave lengths. This is done by applying the bolometric correction, B.C.,

$$M_{bol} = M_{vis} + B.C. \tag{1.1}$$

The bolometric correction depends on the surface temperature of the star. Since the color index also depends on the surface temperature, we can consider the bolometric correction as a function of the color index. This relation is given in Table 1.1. We see that at both ends of the table the bolometric correction is quite large, reflecting the fact that at very high and very low surface temperatures the visual magnitude measures only the tail end of the total intensity distribution. At these extremes the basic data for the bolometric corrections are still quite uncertain and errors in the bolometric correction of as much as 25 percent are not excluded.

After the bolometric absolute magnitude is obtained from Eq.(1.1), it can be finally transformed into the luminosity of the star, L, which is measured in ergs per second. This last step can be done most easily by referring to solar data with the help of the equation

$$\log \frac{L}{L_\odot} = \frac{1}{2.5} (M_{bol_\odot} - M_{bol})$$

(1.2)

with $L_\odot = 3.78 \times 10^{33} \frac{erg}{sec}$ and $M_{bol_\odot} = +4.63$

Having transformed the ordinate of the Hertzsprung-Russell diagram we turn to the abscissa. The color index—or the spectral type, whichever is given by the observations on hand—is a function of the surface temperature. More precisely, we shall here use the effective temperature, T_e, which is defined by the relation

$$L = \frac{ac}{4} T_e^4 \times 4\pi R^2$$

(1.3)

with $\log ac \, \pi = -3.148$.

According to this relation the effective temperature of a star is defined as the temperature of a black body which has the same surface brightness as the star. Eq.(1.3) can also be written in the following form which is more convenient for some applications

$$\log \frac{L}{L_\odot} = 4 \log \frac{T_e}{T_{e\odot}} + 2 \log \frac{R}{R_\odot}$$

(1.4)

with $T_{e\odot} = 5760°$ and $R_\odot = 6.95 \times 10^{10}$ cm.

The effective temperature as a function of the color index is given in the last column of Table 1.1, according to the best available data. These data are still afflicted by appreciable uncertainties and hence the T_e values may still require noticeable corrections, particularly at the top and the bottom of the table. Nevertheless, for consistency's sake, we shall use throughout this book the values of Table 1.1 as they stand.

With the help of Table 1.1 and Eqs.(1.1) and (1.2) we can transform the Hertzsprung-Russell diagram from the observed quantities M_{vis} and B-V to the theoretical quantities L and T_e. If in addition we want to derive the radii, this is easily done through Eq.(1.4).

TABLE 1.1.

The bolometric correction and the effective temperature as functions of the color index, B-V. (Bolometric correction from G. Kuiper, *Ap.J. 88,* 429, 1938; effective temperature from P. C. Keenan and W. W. Morgan in J. A. Hynek; ed., *Astrophysics,* McGraw-Hill, 1951; color from H. L. Johnson and W. W. Morgan, *Ap.J. 117,* 313, 1953, and H. L. Johnson and D. L. Harris, III, *Ap.J. 120,* 196, 1954.)

B-V	B.C.	$\log T_e$
−0.30	−4.1 :	4.65 :
−0.25	−3.0 :	4.46 :
−0.20	−2.3 :	4.32 :
−0.15	−1.7 :	4.21 :
−0.10	−1.25	4.13
−0.05	−0.95	4.08
0.00	−0.72	4.03
+0.05	−0.52	3.99
+0.1	−0.38	3.95
+0.2	−0.20	3.91
+0.3	−0.09	3.87
+0.4	−0.02	3.83
+0.5	0.00	3.80
+0.6	−0.03	3.76
+0.7	−0.09	3.73
+0.8	−0.18	3.69
+0.9	−0.29	3.66
+1.0	−0.42	3.63
+1.1	−0.57	3.61
+1.2	−0.74	3.58
+1.3	−0.92	3.56
+1.4	−1.11	3.55
+1.5	−1.3 :	3.53 :
+1.6	−1.5 :	3.52 :
+1.7	−1.7 :	3.52 :
+1.8	−1.9 :	3.51 :

The Hertzsprung-Russell diagram of the clusters is reproduced in the transformed form in Fig. 1.5. Each cluster is here represented not by its individual stars but by its mean line. The additional set of parallel diagonal lines in Fig. 1.5 refers to various values of the radii in accordance with Eq.(1.4).

Fig. 1.5 shows at once the key observational phenomena which the theory of stellar evolution has to explain. It shows the initial main sequence towards which all the clusters converge. It shows the turnoff from the main sequence, at different points for different clusters. And it shows the difference in form of the giant branch between the oldest galactic cluster and the globular cluster.

From Fig. 1.5 we may read the relation between L and T_e for the initial main sequence. These data are given in Table 1.2 (together with the corresponding values of the radius and of the related observational ꝗuantities), ready for comparison with the theoretical results of Chapter IV.

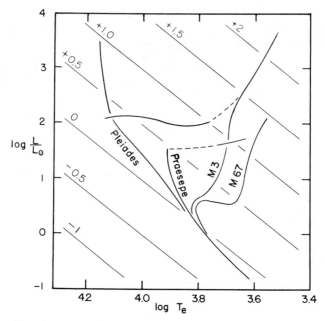

Fig. 1.5. Transformed Hertzsprung-Russell diagram for clusters. The diagonals represent lines of constant radius; the numbers attached to them give log $R/R_{\odot}$

TABLE 1.2.

Initial Main Sequence. (A. Sandage, *Ap. J. 125*, 435, 1957.)

M_{vis}	B-V	B.C.	log T_e	M_{bol}	$\log \dfrac{L}{L_{\odot}}$	$\log \dfrac{R}{R_{\odot}}$
−2.0	−0.24	−2.9	4.43	−4.9	+3.81	+0.57
−1.5	−0.22	−2.6	4.38	−4.1	+3.49	+0.51
−1.0	−0.19	−2.2	4.30	−3.2	+3.13	+0.49
−0.5	−0.16	−1.8	4.23	−2.3	+2.77	+0.45
0.0	−0.14	−1.61	4.19	−1.61	+2.50	+0.39
+0.5	−0.11	−1.34	4.15	−0.84	+2.19	+0.32
+1.0	−0.06	−1.01	4.09	−0.01	+1.86	+0.27
+1.5	0.00	−0.72	4.03	+0.78	+1.54	+0.23
+2.0	+0.08	−0.44	3.97	+1.56	+1.23	+0.20
+2.5	+0.17	−0.25	3.92	+2.25	+0.95	+0.16
+3.0	+0.28	−0.11	3.88	+2.89	+0.70	+0.11
+3.5	+0.39	−0.03	3.83	+3.47	+0.46	+0.09
+4.0	+0.47	−0.01	3.81	+3.99	+0.26	+0.03
+4.5	+0.55	−0.02	3.78	+4.48	+0.06	−0.01
+5.0	+0.63	−0.05	3.75	+4.95	−0.13	−0.04
+5.5	+0.72	−0.11	3.72	+5.39	−0.30	−0.07
+6.0	+0.82	−0.20	3.68	+5.80	−0.47	−0.07
+6.5	+0.93	−0.33	3.65	+6.17	−0.62	−0.09
+7.0	+1.04	−0.48	3.62	+6.52	−0.76	−0.10
+7.5	+1.16	−0.67	3.59	+6.83	−0.88	−0.10

Subdwarfs and White Dwarfs

In the lower left-hand section of the Hertzsprung-Russell diagram there exist two kinds of stars we still need to discuss, the subdwarfs and the white dwarfs.

Under the term "subdwarf" astronomers tend to include all the stars in the Hertzsprung-Russell diagram between the main sequence and the white dwarfs. This large area may well encompass two totally different types of stars, those closely related to main-sequence stars—stars in their youth, as we shall see—and those closely related to white dwarfs— stars nearing their death. In this book we shall use the term subdwarf only for the first class while we might call the second class incipient white dwarfs. The separation of these two distinct classes, though obviously of importance for the theory of stellar evolution, does not yet seem possible with any certainty on the basis of the available observational data.

If we want to plot the subdwarfs in the Hertzsprung-Russell diagram we encounter two practical difficulties. First, accurate trigonometric distances have been determined thus far for only a few subdwarfs. Second, the spectra of subdwarfs are characterized by an unusual weakness of most absorption lines. This affects seriously the spectral type estimates as well as the color measurements. In consequence a correction is needed for the relation between color index and effective temperature if this relation is to be applied to the subdwarfs. But so far this correction has been determined only in the roughest approximation.

With these uncertainties well in mind we may make bold to summarize the present observations as follows. The subdwarfs—in our restricted sense—fall in the Hertzsprung-Russell diagram in a band parallel to the main sequence and just below it. They do not seem to fall on a sharp line. Their average distance below the main sequence appears to be approximately one magnitude, that is the luminosity ratio of a subdwarf to an ordinary dwarf of equal effective temperature is given approximately by

$$\log \frac{L(\text{subdwarf})}{L(\text{dwarf})} \approx -0.4. \qquad (1.5)$$

We shall attempt to explain this ratio in §17 in terms of a composition difference.

The present observational data for the white dwarfs present the same difficulties as those for the subdwarfs. The number of white dwarfs for which trigonometric parallaxes have been measured is still limited; and, owing to the strong peculiarities in the white dwarf spectra, the appropriate relation between color and effective temperature is still rather uncertain. We shall have to approach these data therefore with some caution.

The Hertzsprung-Russell diagram of Fig. 1.6 represents the observational data for the white dwarfs as they stand at present. In spite of the uncertainties we have just mentioned, we may conclude from this figure

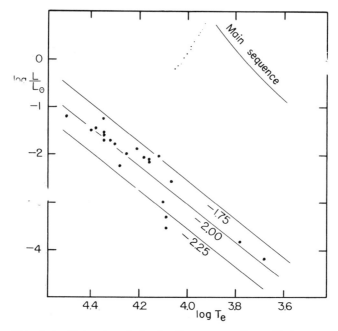

Fig. 1.6. White dwarfs in the Hertzsprung-Russell diagram (D. Harris, III, *Ap.J. 124*, 665, 1956; only the *U-V* colors were used for the white dwarfs, and the temperature scale of Table 1.1 was accordingly transformed with the help of the relation of *U-V* with *B-V* for main-sequence stars). The straight lines are loci of constant radii; they are labelled with the values of log $R/R_\odot$

with fair confidence that the white dwarfs occupy in the Hertzsprung-Russell diagram a strip running approximately parallel to the lines of constant R and that their radii amount to only about one hundredth that of the sun.

2. *Masses*

The wealth of data shown in the Hertzsprung-Russell diagram all refers to two stellar characteristics, the luminosity and the radius. These two characteristics gain their importance from the circumstance that they are observable for many stars. Hence they provide a multitude of clues and checks for the theory of stellar evolution. But they are not basic characteristics in the sense of being determined by the previous history of the star. Instead they are determined from moment to moment by the equilibrium conditions which we shall discuss in Chapter II. The basic characteristics of a star are rather its mass and its composition. The first of these basic characteristics is our topic for this section.

Only for one star, the sun, do we know the mass with high accuracy:

$$M_\odot = 1.985 \times 10^{33} \text{ g.}$$

Among all other stars masses can be determined directly only for certain special binaries. Earlier it had been hoped that the relativistic gravitational red shift could be used for the determination of stellar masses, at least for the heaviest stars. But unfortunately this hope had to be abandoned since the gravitational red shift was found inseparably entangled with effects caused by the motions of the atmospheric gases of these stars.

The restriction to binaries for the purpose of determining stellar masses raises one worrisome question: will the selected binaries be typical for all the stars? The answer appears to be yes if we consider only binaries with well-separated components, but no if we include close pairs. In a close pair the equilibrium and evolution of each component seems seriously affected by the existence of the other one. This phenomenon is a fascinating topic in itself. But for the purpose of this book we shall consider only wide pairs.

Spectroscopic Binaries

If for a distant binary the combined spectrum shows the absorption lines of both components so that both radial velocity curves can be measured, and if the binary is an eclipsing variable so that the inclination of the orbit can be ascertained, then the semi-major axes of the component orbits can be found. Knowing these axes, as well as the orbital period, we can compute the masses of the two components directly from Kepler's third law.

Furthermore, for these binaries the eclipse light curves give the radii in terms of the sum of the semi-major axes, which we know already. Finally, after estimating the effective temperatures from the spectral types, we can derive the luminosities and the absolute bolometric magnitudes with the help of Eqs.(1.3) and (1.2) respectively.

Many spectroscopic binaries have been observed. To select the best data we may apply the following rules. Let us exclude all binaries which do not have observed radial velocity curves of high accuracy. Let us consider only those binaries which are well separated, that is those in which the components are distinctly smaller than their "zero velocity surfaces," so that no serious interaction need be feared. Let us include only those binaries in which the components differ in luminosity by less than a factor 2—thus excluding many cases of uncertain interpretation and permitting the use of average values for the two components instead of the less certain individual values. Finally, let us for the moment restrict ourselves purely to main sequence pairs.

Under these rules the data now available shrink into the seventeen spectroscopic binaries listed in Table 2.1. Before we look into the data

TABLE 2.1.

Masses of main sequence spectroscopic binaries (Z. Kopal, *Trans. Int. Astr. Union*, Vol. IX, 1955). The values given are averages for the two components.

Binary	Sp. T.	M_{bol}	$\log \dfrac{M}{M_\odot}$
V478 Cyg	B0	−5.0	+1.16
Y Cyg	B0	−4.6	+1.24
AH Cep	B0	.−4.5	+1.18
V453 Cyg	B2	−4.2	+1.20
CW Cep	B3	−3.0	+1.00
AG Per	B4	−2.2	+0.68
U Oph	B6	−1.9	+0.70
σ Aql	B8	−1.3	+0.78
DI Her	B4	−1.3	+0.54
β Aur	A0	+0.1	+0.37
AR Aur	A0	+0.6	+0.39
RX Her	A0	+0.6	+0.30
WW Aur	A8	+1.8	+0.25
TX Her	A6	+1.9	+0.29
ZZ Boo	F0	+2.2	+0.24
VZ Hya	F6	+3.5	+0.06
YY Gem	M1	+7.8	−0.19

shown in this table let us first assemble the corresponding data from the visual binaries.

Visual Binaries

If a binary is sufficiently nearby so that its components can be resolved visually, if the relative orbit of the fainter component around the brighter one has been measured, and if the absolute orbit of the brighter component (or of the center of light) has been measured against a set of background stars, then the semi-major axes of the actual orbits of both components can be found in seconds of arc. If, furthermore, the trigonometric parallax has been determined, then these angular measures can be converted into linear ones. Thus again the determination of the semi-major axes together with the orbital period permits the computation of the masses of both components with the help of Kepler's third law.

Few visual binaries are sufficiently nearby to permit accurate determinations of their masses. The data for the five best determined pairs are given in Table 2.2, separately for each component. At the bottom of this table two additional pairs of lower observational accuracy are listed. In spite of their observational uncertainties, they are of special interest because they consist of very faint dwarfs. Throughout Table 2.2 only those pairs are included in which both components belong to the main sequence.

A compairson of Tables 2.1 and 2.2 shows that the spectroscopic binaries and the visual binaries supplement each other very well since the spectroscopic binaries contain predominantly stars heavier than the sun

TABLE 2.2.

Masses of main sequence visual binaries. (P. van de Kamp, *A. J. 59*, 447, 1954; K. Aa. Strand and R. G. Hall, *Ap.J. 120*, 322, 1954; S. L. Lippincott, *A.J. 60*, 379, 1955; P. van de Kamp and E. Flather, *A.J. 60*, 448, 1955.)

Binary	Sp. T.	M_{bol}	$\log \dfrac{M}{M_\odot}$
First class determinations:			
α Cen.A	G4	+ 4.4	+0.03
" B	K1	+ 5.6	−0.06
η Cass A	G0	+ 4.5	−0.03
" B	K5	+ 7.5	−0.24
ξ Boo A	G8	+ 5.4	−0.07
" B	K5	+ 6.7	−0.12
70 Oph A	K1	+ 5.6	−0.05
" B	K5	+ 6.8	−0.19
Kru 60 A	M4	+ 9.1	−0.57
" B	M6	+10.0	−0.77
Additional faint dwarfs:			
Fu 46 A	M3	+ 8.7	−0.51
" B	—	+ 8.8	−0.60
Ross 614 A	M6	+10.0	−0.85
" B	—	+12.	−1.10

while the visual binaries consist mostly of dwarfs less massive than the sun.

Mass-Luminosity Relation

All the data we have assembled above for main sequence stars are shown in Fig. 2.1, in which the luminosity is plotted against mass. A well-defined relation emerges. This is the mass-luminosity relation, which has been and still is one of the most powerful tools in the investigation of the stellar interior.

The rather moderate scatter in Fig. 2.1 around the mean relation and the large range in luminosity covered by the data are convincing witnesses to the growth in quality and quantity of the observations in this key field during the past decade. Fig. 2.1 clearly shows the compound character of the mass-luminosity relation, its steep slope in the middle section and its shallower slopes at both ends.

Do the data of Fig. 2.1 cover the entire range of stellar masses? Most certainly not. On the side of extremely large masses a few spectroscopic binaries are known, but they are not eclipsing variables and thus the orbital inclination can not be determined. These cases give only lower limits to the masses. Some of these lower limits are somewhat more than thirty solar masses. It seems probable, therefore, that stellar masses as high as 50 solar masses may occur. We seem to have no definite indica-

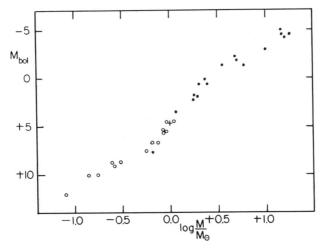

Fig. 2.1. Empirical mass-luminosity relation for main-sequence stars. Data from Tables 2.1 and 2.2. Dots represent spectroscopic binaries, circles visual binaries, and the cross the sun.

tions, however, that stellar masses can reach as high as 100 solar masses.

On the other extreme, of very small stellar masses, the available direct data are limited entirely by the great observational difficulties of determining the masses of very faint stars. Indirectly, however, the observed high frequency of stars with very low luminosities suggests that stars with extremely small masses, much smaller than the lowest listed in Table 2.2, may exist and may even be quite common.

Masses of Giants and White Dwarfs

All the observations we have thus far discussed in this section refer to main sequence stars. Regarding the masses of stars not belonging to the main sequence, the available accurate data are as yet depressingly few. In fact, they are all summarized in Table 2.3.

Let us turn first to the giants and subgiants in Table 2.3. If we plot these three stars in the mass-luminosity relation for main-sequence stars of Fig. 2.1, we find that they conform reasonably well to this relation. It would seem entirely unsafe, however, to conclude from these three stars, none of which lies extremely far from the main sequence in the Hertzsprung-Russell diagram, that all giants conform to the mass-luminosity relation of main-sequence stars. In fact, data on a number of less well measured binaries, as well as other indirect information, seem to indicate that the red giants and in particular the supergiants tend to deviate from the mass-luminosity relation of the main-sequence stars in the sense of being somewhat over-luminous for a given mass.

Finally, there are the white dwarfs, which are represented in Table 2.3 by three examples with well-determined masses. One glance at their bo-

TABLE 2.3

Masses of stars not on main sequence. (Capella: K. O. Wright, *Ap.J.* *119*, 471, 1954; Z Her: D. M. Popper, *Ap.J.* *124*, 196, 1956; ζ Her, Sirius and Procyon: P. van de Kamp, *A.J. 59*, 447, 1954; 40 Eri: D. M. Popper, *Ap.J. 120*, 316, 1954.)

Star	M_{bol}	$\log \dfrac{R}{R_{\odot}}$	$\log \dfrac{M}{M_{\odot}}$
Giant: $\overline{}$			
Capella AB	0.0	+1.0	+0.48
Subgiants:			
Z Her B	+ 3.5	+0.4	+0.04
ζ Her A	+ 2.9	+0.3	+0.03
White dwarfs:			
40 Eri B	+10.3	−1.8	−0.36
Sirius B	+11.2	—	−0.01
Procyon B	+12.6	−2.0	−0.19

lometric magnitudes shows that they are decidedly under-luminous in comparison with main-sequence stars of the same masses.

We will see in §26 that the theory for the white dwarfs gives a definite relation between mass and radius. In this connection it is unfortunate that for one of the three white dwarfs of Table 2.3, Sirius B, the effective temperature has not yet been determined with sufficient accuracy to compute the radius. The observational difficulty lies in the closeness of the much brighter component A. This complicating circumstance furthermore seems to have prevented thus far an accurate measurement of the gravitational relativistic red shift which, if available, could give the radius, as the mass is known from the binary orbit. At present therefore the available observations for the mass-radius relation of white dwarfs is restricted to just two stars, 40 Eri B and Procyon B, as shown by Table 2.3.

If you compare this survey of our present knowledge about stellar masses with similar surveys of two decades ago you may well be impressed by the strong progress that has been made regarding the main-sequence mass-luminosity relation. In contrast you may be struck by the unfortunately limited progress made both regarding the mass-radius relation of the white dwarfs and, worse, regarding the masses of the red giants and supergiants. Even one additional well-determined mass of a giant or supergiant would be an enormous help to the theory of stellar evolution.

3. Stellar Populations

The mass of a star is one of the basic characteristics determined by the preceding evolution. The age of a star is the changing parameter governing the evolution. If we could determine the age of various stars by methods not depending on the theory of stellar evolution, then we could apply strong direct tests to this theory. Unfortunately, however, such methods exist at present only for one star, the sun.

Geophysical evidence indicates that the age of the earth is about 4.5 billion years. To derive the age of the sun from the age of the earth we have to make an assumption regarding the origin of the solar system. Even though present ideas on this subject are still fairly divergent, it appears improbable that the origin of the planets occurred much before or much after the birth of the sun. We may therefore conclude that the sun and the earth are likely to be quite close in age, and hence that the sun is approximately 4.5 billion years old. For all other stars, as we shall see, we have to rely on the theory of stellar evolution for age determinations.

Even though at present observations do not permit the direct determination of stellar ages—except for the sun—they do permit another decisive operation: the classification of stars into a sequence of stellar populations. And it now appears that this classification in its main features amounts simply to sorting the stars in order of their age.

The art of classifying stars by populations is still in its infancy. A variety of unconnected observational data is being used, each by itself inaccurate and insufficient, but all put together sufficient to give a first outline of an age sequence among stars.

The three main fields of observation for population classification which we shall now discuss are: the brightest stars in stellar systems, the kinematical behavior of groups of stars, and spectroscopic peculiarities.

Brightest Stars in Stellar Systems

Perhaps the most striking observational phonomenon suggesting the classification of stars into populations is the wide difference in color between the most luminous stars of different stellar systems. We have already encountered a first example of this phenomenon in §1 when we discussed the Hertzsprung-Russell diagrams of various clusters. In most galactic clusters the brightest stars are blue supergiants while in globular clusters they are luminous red giants. The same differences are found in the much larger stellar systems, the galaxies. The galaxies which show spiral structure contain blue O and B stars as the most luminous members while in galaxies of the elliptical type the brightest stars are red. Could this observed contrast in color represent simply a contrast in the stars' ages? The answer is yes with great certainty.

For the blue supergiants we have two observations, both pointing towards extreme youth for these stars. First, blue supergiants are found only in or near clouds of interstellar gas and dust—an observation which suggests that these stars are still so young that they have not had time enough to move far from the matter out of which they sprang. Second, several galactic clusters or associations containing blue supergiants have been found to expand. Measurement of the expansion rate and of the present size gives ages for these clusters—and hence presumably for the member stars—of less than a hundred million years. Youthful indeed as compared with the sun.

On the other hand, for the luminous red giants in globular clusters and elliptical galaxies we do not have good direct observational evidence for their high age, though it may be significant that globular clusters and elliptical galaxies appear to be nearly completely void of interstellar gas and dust, and that the globular clusters are highly stable dynamical systems capable of a long life. The meagerness of direct observational evidence will not worry us for long, since we shall see in Chapter VIII that the theory of stellar evolution permits the determination of the age of globular clusters, and the result is an old age indeed.

Shall we then conclude that all stellar systems can be divided into just two kinds, the very young ones with blue supergiants, called Population I, and very old ones with luminous red giants, called Population II? Probably not. We shall encounter in the following paragraph indications for the necessity of finer population subdivisions. One such indication we have already seen in §1: the cluster M 67 appears in many ways to be of a type intermediate between the galactic and the globular clusters. We may well suspect then that here, as often, it is easier to recognize the extremes than to differentiate into finer subdivisions.

Kinematical Behavior

Early in the study of stellar populations—in fact before the term "stellar population" was invented—a mysterious but significant correlation was discovered. Stars of different astrophysical characteristics showed a different kinematical behavior within the galaxy. This correlation became clear at once when the stars in the solar neighborhood for which space velocities had been observed (relative to the local standard of rest) were divided into "high-velocity stars" with a space velocity (projected onto the galactic plane) in excess of 63 kilometers per second and "low-velocity stars" with space velocities below this limit. The high-velocity stars were found to contain many red giants, and when plotted in the Hertzsprung-Russell diagram gave an array fairly similar in the main features to that of the globular clusters. On the other hand, the low-velocity stars contained practically all the blue O and B stars and gave a Hertzsprung-Russell diagram strongly resembling that of the galactic clusters. This correlation was strikingly confirmed by the clusters themselves: the galactic clusters have on the average low space velocities while the globular clusters have very high ones.

The correlation between astrophysical characteristics of stars and their kinematical behavior has been substantiated in many details. For example, the smallest average space velocities were found for the O and

B stars, and this average velocity turned out practically indentical with that of the interstellar clouds. Thus once more the close relation between the young blue O and B stars and the interstellar matter out of which they were formed was confirmed.

On the other extreme, the highest average space velocities were found for the RR Lyrae variables and for the subdwarfs. This result for the RR Lyrae variables was not a surprise since such variables are quite common in globular clusters. The high average velocity found for the subwarfs introduces a new aspect, however; it suggests that since globular clusters also have very high velocities the main-sequence stars found in them may possibly be more closely related to the subdwarfs than to the ordinary dwarfs.

Though these extreme cases of very high and very low average space velocities provide the most striking examples of correlations, we must not neglect the equally important fact that all the most common types of stars, including the ordinary dwarfs, show an average space velocity intermediate between the two extremes we have discussed above. The analysis of observed frequency distributions of stellar velocities is always hazardous, but it appears fairly safe to conclude from the presently available material that the velocity distribution of the common types of stars can probably not be interpreted as a mixture of two extremes corresponding to the average velocities of the O and B stars on the one hand and to that of the subdwarfs and RR Lyrae variables on the other hand, but rather that the majority of the most common types of stars must actually belong to intermediate populations.

We have used thus far the average space velocity as a criterion for the kinematical behavior of a group of stars. We could also use the average distance from the galactic plane at which the type of stars in question is found. It is easily understandable that these two criteria should be closely related. If a star has a high velocity perpendicular to the galactic plane it can reach great distances from this plane before gravity will reverse its velocity. It is not surprising then that we find the low-velocity O and B stars to be restricted to the immediate neighborhood of the galactic plane, in the form of a "flat subsystem," whereas the high-velocity globular clusters turn out to occupy a large volume reaching to great distances from the galactic plane, in the form of a more or less "spherical subsystem." Accordingly, for any particular group of stars our choice between the average space velocity and the degree of flatness of the subsystem formed by it will be determined entirely by which one is easier and more accurately observable.

Even though on first sight the correlation between kinematical behavior and astrophysical characteristics of various stellar populations may seem mysterious, a possible explanation—at least a speculative one—is not lacking. Imagine that our galaxy at the time when it became an independent system well separated from nearby galaxies consisted of gases in a highly turbulent state. Any stars formed at this early time must have inherited the high random velocities of the gases out of which they contracted and cannot have noticeably altered their high average velocities since then. As time passed by, the turbulent velocities of the remaining gases must have decayed so that the stars born later inherited lesser average space velocities. The decay time of the turbulence in the interstellar medium can be estimated by order of magnitude from the initial velocities, say 100 km/sec., and the size of the galactic system, say 30,000 parsecs; the resulting value is approximately a third of a billion years. This suggests that all the high-velocity Population II stars must have formed in an interval of, say, half a billion years.

Eventually the turbulence in the interstellar matter must have reached rather low average velocities which could be more or less maintained by the stirring action caused by bright, hot stars heating the surrounding interstellar matter. This stirring action may have retarded, if not completely stopped, any further decay of the interstellar turbulent velocities. Such a circumstance would explain the relatively smaller differences in average velocity of the younger Population I stars. In fact, all Population I stars may well have been born with exactly the same low velocity dispersion, but the older ones among them may have had time enough to be speeded up by encounters with cloud complexes, which may have raised their velocities to the slightly higher values now observed.

Altogether, however uncertain the details of this speculative picture may be, it does suggest how the mysterious correlation between kinematical behavior and astrophysical characteristics could have arisen.

Be that as it may, for the purposes of this book we did not introduce the kinematical behavior of stars with the intent of studying the dynamics of the galaxy but rather with the hope of using the observable kinematical behavior to classify stars into a sequence of stellar populations. Can this be done? Yes, for groups of stars but not for individual stars. The kinematical characteristics are by their nature statistical quantities which do not apply to individual stars. In a group of Population II stars of high average velocities we should always expect a few individual stars with low space velocities, while on the other hand a group of Population I stars may well contain a small fraction of high-velocity stars belonging to the tail-end of the characteristic velocity distribution of this population. The kinematical characteristics can help us therefore only in the population classification of groups of stars but not of individual stars.

Spectroscopic Differences

When the two-dimensional classification of stellar spectra according to temperatures and absolute magnitudes was pushed to high accuracy it was noticed that certain individual stars could not be classified cleanly into the two-dimensional scheme. In particular, some high-velocity stars were found to show certain spectral peculiarities such as abnormal weakness of the CN bands and abnormal strength of the CH bands. It was immediately recognized that these peculiarities might possibly provide the means of a population classification of individual stars directly by astrophysical criteria.

Subsequent investigations on medium-dispersion spectra of stars in the spectral range from F5 to K5 considered not only the relative strength of the bands already mentioned but also the average strength of the fainter metal lines. With the help of these criteria it was possible not only to separate the small percentage of Population II stars in the solar neighborhood from the common Population I stars, but also to divide the latter into two halves, the "weak-line stars" and the "strong-line stars." It was even found possible to extend this delicate subclassification to much fainter stars with the help of objective prism spectra.

After a sufficient number of stars had been classified in this manner into subpopulations, the average space velocity could be computed for each subpopulation, and it was found that the weak-line stars have a larger average space velocity than the strong-line stars. If one adds to this the observation that the high-velocity subdwarfs as well as the giants in some globular clusters have unusually weak metal lines one recognizes that the subpopulations may be nothing but finer divisions in one continuous sequence of stellar populations.

Under these circumstances it appears clearly of importance to exploit the existence of these spectroscopic differences between stellar populations for an ever finer and more extensive population classification. The first photometric investigations in this direction have proved most encouraging. Measurements on F stars of certain parameters connected with the Balmer Jump, as well as photoelectric measurements of the strength of the G band in G and K stars have already proven themselves adequate for coarse population separations and appear entirely capable of being pushed to a much finer population classification. Only the future can tell to what extent this new field of fundamental astrophysical observations can be developed.

Sequence of Stellar Populations

In the preceding paragraphs we have surveyed a motley array of criteria for the classification of stars into stellar populations. The results thus far obtained from these criteria may be summarized in terms of a sequence of stellar populations as indicated in Table 3.1. In this table Population I has been divided into three subpopulations and Population II into two subpopulations. This amount of subdividing appears both justified and necessary for the observational data available at present.

None of the observable criteria, however, suggests that stars are in fact sharply divided into populations or subpopulations. On the contrary, all evidence seems to indicate that the sequence of stellar populations

TABLE 3.1

Sequence of stellar populations. (The velocity dispersion refers to the mean of the two components parallel to the galactic plane.)

Population	Typical Members	Velocity Dispersion	Subsystem
Young Pop. I	blue supergiants, galactic clusters	10 km/sec	flat
Intermed. Pop. I	"strong-line" stars ⎱ majority of	20	⎱
Old Pop. I	"weak-line" stars ⎰ nearby stars	30	⎰ intermed.
Mild Pop. II	majority of "high-velocity stars"	50	
Extreme Pop. II	bright red giants, globular clusters	130	spherical

is continuous and that any divisions are purely a matter of research convenience. Nothing appears to contradict the interpretation that in its main features the sequence of stellar populations is nothing but the sequence of stellar ages.

How far can we translate the sequence of stellar populations into actual stellar ages? We have already mentioned three relevant items. First, the expanding clusters indicate that at least a fair portion of the youngest subdivision of Population I is not older than a hundred million years. Second, for the sun we found from geophysical arguments an approximate age of 4.5 billion years. By a careful spectroscopic comparison of the sun with other G stars the sun has been classified as a weak-line star, that is an old Population I star. This suggests that the oldest Population I stars have ages probably not exceeding 5 billion years, approximately. Third, the argument regarding the decay of the turbulence in the galactic gases suggests that all Population II stars were born within an interval probably not exceeding half a billion years. Thus it seems likely that even the oldest stars in our galaxy are less than 6 billion years old.

We shall see in §29 to what extent the theory of stellar evolution confirms and strengthens this tentative scale of stellar ages.

4. Abundances of the Elements

Given the mass of a star and its age, we will find that its entire internal structure is determined when one more basic characteristic is fixed, the initial chemical composition.

The chemical composition of the stars is an important but unsafe subject which has been much tossed back and forth between the experts of the stellar interior and those of stellar atmospheres. To start with, thirty years ago, the two lightest elements, hydrogen and helium, were considered of so low abundance as to be ignorable in the stellar interior. When the very high abundance of hydrogen in stellar atmospheres was deduced from the relative strength of the Balmer lines, this predominance of hydrogen was suspected to be just a skin phenomenon. Only after it was realized that the introduction of a substantial amount of hydrogen into the stellar interior could remove a discrepancy between the observed and the computed mass-luminosity relations which had plagued the theoreticians until then, was hydrogen given first place among the abundant elements, a place it has held ever since.

Helium has had a similar fate. Completely ignored to begin with, its importance was realized only when the spectrographic analysis of some hot stars with strong helium lines indicated a high helium abundance in stellar atmospheres and when the discovery of the relevant nuclear processes suggested the possibility of a strong helium abundance in the stellar interior. Then helium rose to the second place among the abundant elements.

More than anything else it is this rise in our estimates of the abundance of hydrogen and helium relative to all heavier elements which has caused the substantial changes made during the past three decades in our temperature and density determinations for the stellar interior.

In this section we shall assemble the abundance determinations—as they now stand—derived from spectroscopic observations. The abundance determinations derived from the analysis of the stellar interior we shall discuss as we come to them at various points in the following chapters. The spectroscopic data refer of course directly only to the surface layers. There is as yet no evidence, however, to substantiate any process of stellar birth which could cause large and systematic differences between the composition of the stellar interior and that of the surface layers. Therefore it seems safe to assume that the spectroscopically-determined surface composition is practically identical with the initial composition of the star as a whole. Exceptions are those special cases which we shall mention later on in which the abundance changes caused by the nuclear burning in the deep interior seem to have reached the surface by convection.

Composition of the Sun

Perhaps the most striking feature of stellar spectra is their uniformity. If you exclude a few special and uncommon types of stars, you can classify all stellar spectra according to the coordinates of the Hertzsprung-Russell diagram, effective temperature and absolute magnitude, and you find that all stars of one effective temperature and one absolute magnitude show only very small variations in the relative strength of the absorption lines in their spectra. This uniformity of stellar spectra—at least for the majority of the nearby stars—indicates a fair degree of uniformity in composition. It makes it useful to analyze in great detail the spectrum of one star (the sun, for obvious practical reasons) and thus obtain the initial composition of most stars, at least in first approximation.

Table 4.1 contains the chemical composition of the sun according to recent spectroscopic analyses. Only the most abundant elements are listed in this table; the abundances of the rarer elements, though of great interest for the theory of the origin of the elements, have little direct influence on stellar structure and evolution.

Several of the values of Table 4.1 are still rather uncertain in spite of the great amount of work and ingenuity applied to this analysis. These uncertainties are largely caused by the circumstance that some of the most abundant elements, such as helium, are very poorly represented in the spectrum of the sun. As a matter of fact, the value given in Table 4.1 for the relative abundance of helium is based mainly on the spectroscopic analysis of some hotter stars which show the helium lines with greater strength. In view of these uncertainties we should not be surprised if later on several of the values listed will be found wrong by as much as a factor 2.

We may summarize the data of Table 4.1 in the following manner, as we shall see is convenient for the application to the stellar interior. Let us lump together all the elements heavier than helium, let us designate the abundance of hydrogen by X, that of helium by Y, and that of all the heavier elements by Z. Finally, let us give these three abundances in terms of weight fractions, so that X, for example, represents the weight of hydrogen contained in one gram of stellar matter. Under these definitions Table 4.1 gives for the sun

$$X = 0.73, \quad Y = 0.25, \quad Z = 0.02. \tag{4.1}$$

We again remember that Y as well as Z derived from spectroscopic observations may still be wrong by as much as a factor 2.

TABLE 4.1

Composition of Sun from Spectroscopic Observations. (L. H. Aller, *Astrophysics*, p. 327, 1953.)

Element	Atomic Weight	Abundance by number	Abundance by weight
H	1	1000.	1000
He	4	80.	320
C	12	0.1	1
N	14	0.2	3
O	16	0.5	8
Ne	20	0.5	10
Mg	24	0.06	1
Si	28	0.03	1
S	32	0.02	1
A	40	0.05	2
Fe	56	0.02	1

There is one further item we shall need from Table 4.1 when we come to the nuclear processes. This is the combined abundance of carbon and nitrogen. If we express this abundance as a fraction of the abundance of all the heavier elements, and if we again use weight fractions, we read from Table 4.1

$$X_{CN} = \frac{1}{7} Z. \tag{4.2}$$

We shall use this equation when we apply the carbon cycle to the stellar interior.

One final item regarding the solar composition needs mentioning, the extremely low abundance of the light elements lithium, beryllium, and boron. These low abundances do not surprise us, since we are used to them from the chemical analysis of the earth's crust. As a matter of fact, the abundance of beryllium and boron seems to be the same in sun and earth as accurately as this can be determined at present. Lithium, however, appears to be rarer in the sun than in the earth by a factor 10. Can this extra low abundance of lithium in the sun be explained by nuclear processes? We shall return to this question in §§10 and 23.

Composition Differences between Populations

Up till now we have stressed the great uniformity in the composition of the stars. Now we shall discuss a difference in composition which has a significant influence on stellar evolution, as we shall see in §24. This composition difference appears when different stellar populations are compared. We have already encountered in §3 the first qualitative indications of this difference.

If the purpose of a spectroscopic investigation is the classification of stars into the population sequence, one may limit oneself to the use of just a few criteria. For these criteria one may use spectral features, however complex in structure and cause, which show significant differences between the populations and which can be recognized on spectrograms of sufficiently low dispersion so that the classification is not limited just to the brighter stars. On the other hand, if a spectrophotometric investigation has the purpose of determining the actual composition differences between populations then many individual lines have to be measured on high-dispersion spectrograms so that curves of growth can be constructed and actual abundance ratios be determined. Only a very few such spectrophotometric investigations have as yet been carried out.

If stars of different populations are to be compared for the purpose of determining composition differences, it is advantageous to select stars as similar as possible in effective temperature and absolute magnitude so as to minimize the effects of differences in these two basic parameters. Accordingly, two types of stars appear particularly useful for population comparisons, the F dwarfs and the G and K giants, since both these types are well represented among the brighter stars of various populations.

For these two types of stars spectrophotometric comparisons have been carried out between intermediate Population I stars (represented by nearby low-velocity stars) and mild Population II stars (represented by nearby high-velocity stars). The main result of these measurements is this: the abundance of all the heavier elements appears to be smaller in Population II stars as compared with Population I stars by about a factor 3. It does not yet appear certain from these investigations to what degree the various heavier elements share the same differences between stellar populations. Indeed, there are some indications that the elements of the second period, represented mainly by carbon, nitrogen, and oxygen, differ less from population to population than the metals. Here we shall ignore this more delicate differentiation and as before consider all elements heavier than helium in one lump.

In another high-dispersion spectrophotometric investigation F type subdwarfs belonging to the extreme Population II were compared with ordinary dwarfs of Population I. Again it was found that the abundance of the heavier elements is lower in Population II than in Population I—in this more extreme comparison by approximately a factor 10.

Finally there is the observed phenomenon which we have already represented in Table 3.1, that the majority of the nearby stars may be divided into strong-line stars (intermediate Population I) and weak-line stars (old Population I). We may interpret this phenomenon as indicating that the abundance of the heavier elements which provide the majority of the lines is somewhat lower in stars of old Population I than in stars of intermediate Population I. In this case the abundance difference is probably less than a factor 2, as one may estimate from the delicacy of the observed differences.

These various observational results are summarized in Table 4.2 in terms of Z, the abundance of the heavier elements by weight fraction. The dependence of the abundance of the heavier elements on stellar population, that is on stellar age, as shown by Table 4.2 has a significant influence on stellar evolution, as we shall see. We shall discuss a

TABLE 4.2

Approximate Composition Differences between Stellar Populations.

Population	Abundance of heavier elements Z
Young Pop. I	0.04
Intermed. Pop. I	0.03
Old Pop. I	0.02
Mild Pop. II	0.01
Extreme Pop. II	0.003

process which may possibly have caused this dependence in the very last section of this book. Here we shall leave these data as they stand, but emphasize once more their great uncertainty at present; large changes in the numerical values listed in Table 4.2 may still be expected.

One essential question which remains completely unanswered by the present spectroscopic data is: do stars of different stellar populations differ in their initial helium abundance? It does not seem likely that this question will be answered soon by direct spectroscopic observations. The reason is that the helium lines show up well only in the spectra of hot stars, and hot stars occur only in Population I, not in the older Population II. (Exceptional hot stars in Population II, like those of the horizontal branch in the Hertzsprung-Russell diagram of globular clusters, will be identified later with very advanced evolutionary phases in which the helium abundance even in the atmospheres may well be much higher than the initial helium abundance.) There exists, however, another possibility of answering the important question about the initial helium abundance of Population II through the analysis of the interior of the subdwarfs—as we shall see in §17.

Composition Anomalies

The composition of the vast majority of stars may be represented in fair approximation by the composition of the sun given in Table 4.1, appropriately modified for the population differences in accordance with Table 4.2. There exist, however, a few groups of uncommon stars which show in their spectra peculiarities strongly suggesting composition anomalies. We should at least mention these groups, if for no other reason than to avoid being accused of oversimplification.

On the one hand, there are the magnetic A stars and the metallic-line A stars. The magnetic A stars show strong lines of certain normally rare elements. It appears plausible that the increased abundance of these elements might be caused by nuclear processes in the surface layers made possible by electromagnetic acceleration of atoms. It is not known as yet whether either the magnetic fields themselves or the composition anomalies have a significant influence on the internal structure and evolution of these stars.

Regarding the metallic-line A stars, it does not seem certain at present whether their striking spectral peculiarities are caused by composition anomalies. Instead, the analysis of these peculiarities suggests that they might be caused by a difference in the atmospheric structure. The basic reason, however, for such a structure difference does not seem clear as yet.

On the other hand, there are the carbon stars and the S stars. Both these groups belong to the red giants. In the carbon stars the abundance of carbon is greater than normal by a large factor, whereas in the S stars a group of heavy elements—including the unstable technitium—shows an unusually high abundance. It seems significant that these two types of composition anomalies occur only in the red giants and not in the red dwarfs. We shall see that evolution has gone a long way in the red giants, but has not had sufficient time to cause significant changes in the slowly evolving red dwarfs. This circumstance suggests that the abnormal composition of the carbon stars and the S stars is more likely a consequence of stellar evolution than of abnormal initial composition.

Thus, altogether, none of these special groups of common stars appears to indicate a definite difference in the initial composition. For the purposes of this book we shall therefore ignore them, except that we must remember that eventually the theory of advanced stellar evolution will have to explain the phenomenon of the carbon stars and the S stars.

This brings us to the end of our survey of the observations relevant to our subject. If after reading this chapter you feel that the observational data now available are somewhat slim for our purposes, compare them with the data available ten years ago and you will be encouraged by their enormous growth.

CHAPTER II
PHYSICAL STATE OF THE STELLAR INTERIOR

5. *Hydrostatic and Thermal Equilibrium*

In Chapter I we have surveyed a large body of observational data. Now, which of these data relate directly to the internal structure of an individual star? Even for the best observed stars we find just four basic items: the mass, the luminosity, the radius, and the composition of the outer layers. Can these four items suffice for us to derive uniquely the internal structure? They could not if it were not for one additional observed item: the constancy of stars over long time intervals.

Intensive astrophysical observations over the past fifty years have revealed no changes for the vast majority of stars. Even variables like Cepheids and novae appear remarkably free of sizeable long-term changes if one averages their short-term variations. Such observations, however, have little weight since fifty years is a minute span compared with the lifetime of a star. The two following features are of more weight. First, geologists have found fossil algae more than a billion years old. They estimate that the temperature of the earth then cannot have differed from what it is now by more than 20°C to permit the flourishing of these algae. Thus, the luminosity of the sun must have been reasonably constant for a billion years. Second, the pulsation periods of classical Cepheid variables can be measured with striking accuracy. Over the past fifty years they have been found to change, but at such a slow rate that it will take of the order of a million years to change the period by a large fraction. Thus, the internal structure, which determines the pulsation period of these stars, must be reasonably constant over at least a million years. It is this constancy of the stars which makes a definite analysis of the stellar interior possible. It asserts that the stellar interior must be in perfect equilibrium and thus limits the necessary analysis to equilibrium configurations.

We shall assemble in this chapter the complete set of equilibrium conditions which must be fulfilled throughout a star. We shall ignore for the time being such perturbations as rotation, pulsation, tidal distortion, and large-scale magnetic fields. In consequence we may assume spherical symmetry from the outset.

Condition of Hydrostatic Equilibrium

The first condition which has to be fulfilled throughout the stellar interior is that of hydrostatic equilibrium: all the forces acting on any small volume within the star must compensate each other exactly, since a non-vanishing net force would cause motions and hence changes in the structure. The only forces we need to consider here are the gravitational force, which is directed inwards, and the pressure force, which is directed outwards.

Let us consider a small cylindrical volume at a distance r from the star's center, with its axis pointing toward the center, and with a cross-section ds and a length dr. The pressure force acting on this volume will be

$$-\frac{dP}{dr}\, ds\, dr$$

where P is the pressure, which will be a monotonic decreasing function of the distance r from the center. The gravitational force acting on the same volume will be given by the mass of the volume times the gravitational acceleration, i.e.

$$\rho\, ds\, dr\, \frac{GM_r}{r^2}.$$

Here ρ is the density and G is the constant of gravitation. We have expressed the gravitational acceleration in terms of M_r, which designates the mass in a sphere with the radius r and which can be expressed in terms of the density run by

$$M_r = \int_0^r \rho 4\pi r^2 dr. \tag{5.1}$$

Setting the two opposing forces equal to each other we obtain the hydrostatic equilibrium condition

$$\frac{dP}{dr} = -\rho\, \frac{GM_r}{r^2}. \tag{5.2}$$

Eqs.(5.1) and (5.2) are the first two of the basic equilibrium equations which govern stellar structure. By themselves they are clearly insufficient to determine uniquely the runs of pressure, density, and mass distribution throughout the interior. But they do permit us to gain some insight immediately into the order of magnitude of pressures and temperatures which we will encounter. Let us apply the hydrostatic condition (5.2) to a point in the sun midway between center and surface. Let us take for the density there—in the roughest approximation—the known mean density of the sun, for M_r one-half the solar mass, and for r one-half the solar radius. Further, on the left-hand side of Eq.(5.2) let us take for dr the radius of the sun and for dP the difference between the central pressure and the surface pressure. If we then trust that the surface pressure will be quite ignorable compared with the pressure at the center, the only unknown in Eq.(5.2) is the central pressure, for which we get

$$P_{c\odot} \approx 2\bar{\rho}_\odot \frac{GM_\odot}{R_\odot} = 6 \times 10^{15}.$$

(Here, as throughout this book, c.g.s. units are used.)

From this estimate of the pressure we can derive directly an estimate of the temperature if we permit ourselves to use the equation of state of an ideal gas, which we will find to hold closely in most stars. This equation can be put into the form

$$P = \frac{k}{m} \rho T \tag{5.3}$$

where T is the temperature, k is the Boltzmann constant, and m is the mean molecular weight. For m we shall use half the proton mass since we saw that hydrogen is the most abundant element, and for ionized hydrogen one proton and one free electron act like two particles of a mean mass of half a proton. Let us apply this equation again to the median point in the sun. If we take the pressure there to be about half the pressure in the center, we find for the temperature

$$\bar{T}_\odot \approx 10^7.$$

Thus we find for a typical temperature in the stellar interior 10 million degrees.

These estimates set at once the scene in which we have to work. We have to work with gases too hot to contain any chemical compounds and hot enough to be highly ionized. We do not need to consider the complex physics of solids and fluids. It is this lucky limitation of our working scene that permits us now to investigate stellar structure and evolution with happy confidence.

Before we decide on stern obedience to the hydrostatic condition (5.2), we should establish what punishment would follow non-obedience. Assume that somewhere in a star the gravitational acceleration is not quite counterbalanced by the pressure force, with a fraction f left unbalanced. The stellar material will then actually be accelerated by the amount

$$\frac{d^2r}{dt^2} = f \frac{GM_r}{r^2}.$$

We may solve this equation for the dt in which the unbalance causes a noticeable displacement dr. If we use as this noticeable displacement the fraction f of the radius and if we use the same values for the median point in the sun as before, we obtain

$$dt \approx \left(2G \frac{M_\odot}{R_\odot^3} \right)^{-\frac{1}{2}} = 10^3 \text{ sec} = \frac{1}{4} \text{ hr.}$$

There is no doubt that punishment would follow swiftly for any disobedience of the hydrostatic law.

The Energy Stores of a Star

To insure hydrostatic equilibrium is not sufficient to insure a constant star. Thermal equilibrium must also be considered. Perfect thermal equilibrium is obtained within a system if all parts of the system have reached the same temperature and if no further flow of energy between the parts occurs. Such perfect thermal equilibrium cannot hold within a star. We have already seen that the interior temperatures are of the order of 10 million degrees, whereas the surfaced layers are observed to have temperatures of the order of only several thousand degrees. Furthermore, we see a flow of energy coming through the surface—as measured by the star's luminosity. The existence of this flux in itself presents a deviation from perfect thermal equilibrium.

What kind of thermal equilibrium then holds within a star? To answer this question we must first find the energy store which feeds the flux that we see coming through the surface. There are three types of energy to be considered: thermal energy E_T, gravitational energy E_G, and nuclear energy E_N. The first two types of energy can each be represented by an integral over the star:

$$E_T = \int_0^R \left(+ \frac{3}{2} \frac{k}{m} T \right) \rho 4\pi r^2 dr = + \overline{\frac{3}{2} \frac{k}{m} T} \times M \approx + 5 \times 10^{48}, \quad (5.4)$$

$$E_G = \int_0^R \left(- \frac{GM_r}{r} \right) \rho 4\pi r^2 dr = - \overline{\frac{GM_r}{r}} \times M \approx - 4 \times 10^{48}. \quad (5.5)$$

Here the parentheses in Eq.(5.4) contain the thermal energy of an ideal, monatomic gas for one gram of stellar matter, while the parentheses in Eq.(5.5) contain the energy needed to move one gram of stellar matter from its position in the star to infinity after all layers further out in the star have already been so removed. The numerical values given in these equations give the order of magnitude of the two energies derived from the above rough figures for the sun.

It is no accident that the orders of magnitude of these two energies are equal. That this should be so follows directly from hydrostatic equilibrium. Multiply Eq.(5.2) by $4\pi r^3$ and integrate over the star. Using integration by parts on the left-hand side, you find

$$\int_0^R 3P \times 4\pi r^2 dr = \int_0^R \rho \frac{GM_r}{r} \times 4\pi r^2 dr.$$

The left-hand integral is twice the thermal energy, as can be seen by comparing Eqs.(5.3) and (5.4). The right-hand integral is directly the negative gravitational energy. Hence, we obtain

$$2E_T = -E_G. \tag{5.6}$$

This relation is equivalent to the virial theorem of classical dynamics. We shall not be dismayed that the relation (5.6) is not fulfilled exactly by the numbers of Eqs.(5.4) and (5.5) since these numbers were but order-of-magnitude estimates.

Let us now follow a contracting star. It decreases steadily in gravitational energy. Exactly one half of this energy decrease will be compensated by an increase in thermal energy, according to the relation (5.6). The other half will be lost by radiation from the surface. Thus, the net energy store available for radiation from the surface is just equal to the thermal energy. How long then will this store cover the radiative surface losses? Our numerical estimates for the sun give us for this time, which is often called the Kelvin contraction time,

$$\frac{E_{T\odot}}{L_\odot} \approx 10^{15} \text{ sec} = 3 \times 10^7 \text{ yrs.}$$

This time is short, even compared with the interval since the first occurrence of algae on the earth. Similarly, we find for the Cepheids, which are much more luminous than the sun, a Kelvin contraction time of the order of one third of a million years—again short compared with the time indicated by the constancy of their pulsation periods, though in this case not by a very large margin. Hence we conclude that the thermal and gravitational energies of a star are not sufficient to cover the surface losses for the whole life of a star, though they may play an important role in short, critical phases of stellar evolution. We shall return to this problem later on, and now we turn to the third type of energy in a star, nuclear energy.

Nuclear processes release energy which comes from the mass equivalent of the nuclei involved. One might therefore suppose that the total available nuclear energy of a star is c^2M. This, however, is an overestimate since this amount of energy would be released only if complete nuclei were annihilated. It appears now fairly certain that annihilation of nuclei will not occur at the relatively moderate temperatures encountered in stars. We have therefore to consider only such nuclear processes as transmute one chemical element into another. The energies released in such processes are equivalent to the relevant mass defects, which are much smaller than the masses of the whole nuclei. The largest possible mass defect is available when hydrogen is transmuted into iron. In this process the mass defect released amounts to eight thousandths of the nuclear masses involved.

Does the actual store of nuclear energy of a star approach this theoretical maximum? Yes, very closely. We have already seen that the spectroscopic evidence strongly indicates that most stars consist mainly of hydrogen, the most advantageous fuel for a star to start with. And as to the end product, it makes little difference whether or not the transmutations lead all the way to iron; the transmutation of hydrogen just into helium liberates a mass defect of seven thousandths. Hence the theoretical maximum gives a close approximation to the actual nuclear energy store of a star. Thus we obtain for the sun

$$E_{N\odot} = 0.008 \ c^2 M \approx 10^{52},$$

which is over a thousand times more than the thermal and gravitational energies. In the case of the sun this store will cover the losses by radiation from the surface over the time interval

$$\frac{E_{N\odot}}{L_\odot} \approx 3 \times 10^{18} \ \text{sec} = 10^{11} \ \text{yrs},$$

which is amply long.

Condition of Thermal Equilibrium

We conclude from the preceding discussion that the energy loss at the surface as measured by the luminosity of the star is compensated by the energy release from nuclear processes throughout the stellar interior. This condition may be expressed by the equation

$$L = \int_0^R \varepsilon \rho 4\pi r^2 dr \tag{5.7}$$

where ε is the energy released from nuclear processes per gram per second. This nuclear energy production ε will depend on temperature, density, and composition. We will discuss it in detail in §10.

Do we have to be as careful of the minute-by-minute fulfillment of the energy balance condition (5.7) as we have to be with the hydrostatic condition (5.2)? No, not at all. If we were to turn off the nuclear energy sources within the sun it would continue to shine, feeding on its gravitational energy store. If we turned on the nuclear sources again after, say, a million years—that is long before a Kelvin contraction time had elapsed —the sun would not have been seriously upset by the interference. Over such times the gravitational and thermal energy store acts like an effective buffer. However, if we consider averages over times longer than the Kelvin contraction time, then the energy balance condition (5.7) must be fulfilled exactly.

Eq.(5.7) ensures energy balance for the star as a whole. But the same type of balance must hold for any section within the star. Energy gains in one section and energy losses in another section would lead in time to a change of the temperature structure of the interior and would thus make the star inconstant. Let us consider a spherical shell with radius r and unit thickness. The energy balance for this shell may be written as

$$\frac{dL_r}{dr} = \varepsilon\rho 4\pi r^2 \qquad (5.8)$$

where L_r is the energy flux through the sphere with radius r. The left-hand side of this equation represents the net loss to the shell caused by the excess of the flux leaving the shell through the outer surface over the flux entering the shell through the inner surface. The right-hand side represents the energy produced within the shell by nuclear processes. Eq.(5.8) represents the third of the basic equilibrium conditions which must hold throughout the stellar interior.

Eq.(5.8) has to be modified in the short but critical phases of stellar evolution in which the changes in the internal structure are so fast that variations in the two minor types of stellar energy—thermal and gravitational—play an important role. In these phases we cannot expect that the flux will carry out of a volume just as much energy per second as the nuclear reactions produce inside the volume, as expressed in Eq.(5.8). Instead we should expect that the energy loss by the flux, the energy gain by nuclear reactions, and the work done by the pressure together will determine the rate of change of the internal energy of the volume. The internal energy of one gram of an ideal gas is given by $\frac{3}{2}\frac{k}{m}T$. The work done by the pressure is $-PdV$, where the specific volume V can be replaced by its reciprocal, the density. The nuclear energy released per gram per second is by definition ε. The net loss of energy by the flux, for a complete spherical shell of unit thickness, is dL_r/dr, which has to be divided by the mass of the shell $4\pi r^2\rho$ to get the loss per gram. Hence, altogether we have

$$\frac{d}{dt}\left(\frac{3}{2}\frac{k}{m}T\right) = +\frac{P}{\rho^2}\frac{d\rho}{dt} + \varepsilon - \frac{1}{4\pi r^2\rho}\frac{dL_r}{dr}. \qquad (5.9)$$

With the help of the equation of state (5.3) this may be brought into the more convenient form

$$\frac{dL_r}{dr} = 4\pi r^2\rho\left[\varepsilon - \frac{3}{2}\rho^{\frac{2}{3}}\frac{d}{dt}\left(\frac{P}{\rho^{\frac{5}{3}}}\right)\right]. \qquad (5.10)$$

This equation takes the place of Eq.(5.8) in those phases in which the evolutionary changes are unusually fast. It is identical with Eq.(5.8) in the normal phases in which the changes are so slow that the time-derivative term in Eq.(5.10) can be ignored.

This far we have only considered the condition which the energy flux must fulfill to balance the energy sources. Physically however, the actual flux is determined by the mechanism which provides the energy transport. This may be conduction, convection, or radiation. For all three of these mechanisms it is essentially the temperature gradient which determines the energy flux. Therefore we have to consider next these mechanisms in detail in order to determine the condition on the temperature gradient that will produce a flux which just fulfills the energy balance condition (5.8) or (5.10).

Conduction turns out to be far too slow to contribute seriously to the energy transport through the stellar interior. Because of the high gas densities encountered there the free path lengths for the ions and electrons are so short compared with stellar radii that conduction may be quite neglected. There is one exception. We will find that the gas is degenerate in the white dwarfs as well as in the cores of red giants. In a highly degenerate gas the electrons may have very long free paths, which lead to very effective electron conduction. We shall return to this special mechanism of heat transport in §§25 and 27. In the following two sections we shall consider in detail the two transport mechanisms which dominate in the stellar interior, namely radiation and convection.

6. Radiative Energy Transport

If the stellar interior were isothermal, the radiation intensity would be isotropic and a net flux of radiation in any particular direction could not exist. In fact, however, there does exist a radial temperature gradient throughout the interior. In consequence, if you look from any given point inside a star towards the center you receive radiation coming from a slightly hotter region than at the given point, and if you look towards the surface you receive radiation coming from a slightly cooler region. The radiation directed outwards, which comes from the hotter region, will have a somewhat higher intensity than the radiation directed inwards, which comes from the slightly cooler region. Thus, there will be a net flux of radiation outwards.

How large will this radiation flux be? This will depend on the opacity of the gases. If the opacity is low, you will be able to see, from a given point, far inwards to much hotter regions and far outwards to much cooler regions; the anisotropy of the radiation at a given point will be large and the net flux outwards great. Let us represent the opacity by the absorption coefficient per gram, $\varkappa$, defined so that

$$\varkappa \rho dl$$

gives the fraction of the energy of the beam which is lost by absorption over the distance dl. We will see in §9 that the absorption coefficient in the stellar interior is generally of the order of magnitude of unity—and in no case very much smaller. If we once again use the mean density of the sun as representative, we find that $\varkappa\rho$ is of the order of unity, and we see that in the stellar interior a distance dl of as little as a centimeter is sufficient to absorb a large fraction of the intensity of a beam. In fact, a thickness of several centimeters of these gases will be effectively completely opaque. You will therefore not see very far, inwards or outwards, from any given point in the interior. The temperature differences over this small region will be of the order of only a thousandth of a degree since the temperature drop over the entire stellar radius must be of the order of 10,000,000 degrees. The radiation field therefore must be very nearly isotropic, and we surely would neglect this minute degree of anisotropy if the flux it causes were not the essential link between the nuclear processes at the center and the radiative losses at the surface.

The General Radiation Transfer Equation

Let us now derive the exact relation between the radiation flux and the temperature gradient. For this purpose we describe the radiation field by the function $I(r, \theta)$, which is defined to be the intensity (in ergs per square centimeter per second per unit solid angle) of the radiation at a distance r from the center and in a direction inclined by the angle θ to the radius vector. Consider now the gains and losses which the radiation within the solid angle $d\omega$ receives per second in a cylinder with cross section ds, length dl, and location as shown in Fig. 6.1. There will be a gain by radiation entering at the bottom surface

$$+ I(r, \theta)\, d\omega\, ds$$

and there will be a corresponding loss at the top surface

$$- I(r + dr, \theta + d\theta)\, d\omega\, ds.$$

In the last expression an increment occurs in the first argument since the top of the cylinder is further from the center than the bottom, and an increment occurs in the second argument since the curvature introduced

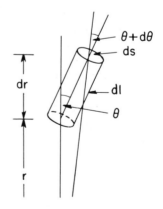

Fig. 6.1. Diagram for the derivation of the radiation transfer equation.

by spherical symmetry causes the vertical at the top of the cylinder to be somewhat inclined to that at the bottom. Next, we have the loss by absorption over the length of the cylinder

$$- I\,d\omega ds \times \varkappa\rho\,dl.$$

Finally, there is the gain by the emission of the gases in the cylinder. Let j represent the total energy emitted by one gram per second isotropically in all directions. The emission of all the matter in the cylinder in the directions contained within the solid angle $d\omega$ is then

$$+ j\rho\,ds\,dl\,\frac{d\omega}{4\pi}.$$

If we now apply the condition of thermal equilibrium specifically to the radiation field, we must require that the gains of the radiation exactly balance its losses, that the sum of the above four terms must be zero. This condition, together with the two geometrical relations

$$dl = dr/\cos\,\theta,\ d\theta = -\,dl\,\sin\,\theta/r,$$

gives us

$$\frac{\partial I}{\partial r}\cos\,\theta - \frac{\partial I}{\partial \theta}\frac{\sin\,\theta}{r} + \varkappa\rho I - \frac{1}{4\pi}j\rho = 0. \tag{6.1}$$

This is the basic equation of radiative transfer, which must hold at every point within a star.

Radiative Equilibrium in the Stellar Interior

The solution of Eq.(6.1) for the radiation field presents one of the main problems in the theory of stellar atmospheres. For the stellar interior the solution is greatly simplified by the fact that here the radiation field is so very nearly isotropic. Instead of working with the function I, which represents the distribution of the radiation over all directions, let us consider the first three moments of this distribution function:

$$\text{Energy density}\qquad E(r) = \frac{1}{c}\int I\,d\omega, \tag{6.2}$$

$$\text{Radiation flux}\qquad H(r) = \int I\,\cos\theta\,d\omega, \tag{6.3}$$

$$\text{Radiation pressure}\ P_R(r) = \frac{1}{c}\int I\,\cos^2\theta\,d\omega. \tag{6.4}$$

The physical meaning of the first two moments is clear. The dynamical meaning of the third moment we will discuss in §8. For these moments we can obtain differential equations by forming the corresponding moments of the basic equation (6.1), i.e. by multiplying this equation by powers of $\cos\theta$ and integrating over all directions. The first two moments of Eq.(6.1) (multiplication by 1 and $\cos\theta$ respectively) are

$$\frac{dH}{dr} + \frac{2}{r} H + c\varkappa\rho E - j\rho = 0, \tag{6.5}$$

$$\frac{dP_R}{dr} + \frac{1}{r}(3P_R - E) + \frac{\varkappa\rho}{c} H = 0. \tag{6.6}$$

We have obtained only two equations for the three functions E, H, and P_R. This insufficiency cannot be removed by deriving one more equation corresponding to the next higher moment of the basic equation (6.1) because this new equation would involve a fourth function representing the fourth moment of I. In fact, such an insufficiency is nearly always encountered if the method of moments is used to replace a partial differential equation, such as Eq.(6.1), by a set of ordinary differential equations, such as Eqs.(6.5) and (6.6). To make the entire procedure definite, one must find an additional, approximate relation between the moments considered. In our case, such an additional relation is easily derived because of the near isotropy of the radiation field in the stellar interior.

Let us represent the radiation field at a given point by the series

$$I = I_0 + I_1 \cos\theta + I_2 \cos^2\theta + \ldots \tag{6.7}$$

Let us determine the rate of convergence of this series by introducing it into the basic equation (6.1). We may for the moment ignore the second term of this equation, which arises from the curvature, since this term complicates matters but does not affect the rate of convergence. Finally, let us assemble terms with the same power of $\cos\theta$. This gives us the recursion formula

$$\frac{d I_{n-1}}{dr} + \varkappa\rho I_n = 0.$$

This formula holds for $n > 0$; the corresponding formula for $n = 0$, which involves the j term, does not have to be considered here. Since we want to use the recursion formula only for order of magnitude estimates, we may replace in it dI_{n-1} by I_{n-1} and dr by R. With these replacements we obtain

$$\left| \frac{I_n}{I_{n-1}} \right| \approx \frac{1}{\varkappa\rho R} \approx 10^{-10}.$$

Hence it is clear that the series (6.7) converges extremely rapidly in the stellar interior. We may therefore restrict ourselves to the first two terms in the series (6.7); we could not restrict ourselves to the first term only since then we would have fallen back to an isotropic radiation field without any net flux. If we introduce the first two terms of the series (6.7) into Eqs.(6.2) to (6.4) for the three moments, we obtain

$$E = \frac{4\pi}{c} I_0, \ H = \frac{4\pi}{3} I_1, \ P_R = \frac{4\pi}{3c} I_0,$$

from which follows

$$P_R = \frac{1}{3} E. \tag{6.8}$$

The relative error of this relation will be of the order of I_2/I_0, which in turn is of the order of 10^{-20}. No higher accuracy is needed. Eq.(6.8) is the additional relation which, together with Eqs.(6.5) and (6.6), completes the necessary set of three equations for the three moments.

With the help of the relation (6.8) we may now simplify the two differential equations (6.5) and (6.6). First, we may replace the flux per square centimeter, H, by the flux through an entire sphere, L_r, by using the geometrical relation

$$L_r = 4\pi r^2 H.$$

Next, we may introduce the appropriate expression for the emission j. The emission consists of two parts. The first contribution arises from the normal thermal emission, which according to Kirchhoff's law is proportional to the absorption coefficient $\varkappa$, to the Stefan-Boltzmann constant a, to the velocity of light c, and to the fourth power of the temperature of the emitting gases. The second contribution arises from the nuclear processes and is equal to the nuclear energy production per gram per second, ε. The full expression for the emission is therefore

$$j = \varkappa a c T^4 + \varepsilon.$$

Finally, we may eliminate ε with the help of the basic thermal equilibrium condition (5.8). After these several substitutions our equation (6.5) simplifies into

$$E = a T^4. \tag{6.9}$$

This relation between the energy density of the radiation and the temperature of the matter is exactly that which is obtained for perfect thermal equilibrium. Hence, we find that this relation is not affected by the small anisotropy of the radiation field in the stellar interior.

After having reduced Eq.(6.5) to a simple form, we still have to do the same for Eq.(6.6). The relations (6.8) and (6.9) give for the radiation pressure

$$P_R = \frac{a}{3} T^4. \tag{6.10}$$

If this is introduced into Eq.(6.6), it takes the form

$$H = -\frac{1}{3\varkappa\rho} \frac{d}{dr}(a\,c\,T^4).$$ (6.11)

Replacing H by L_r as before we obtain finally

$$L_r = -4\pi r^2 \frac{4ac}{3} \frac{T^3}{\varkappa\rho} \frac{dT}{dr}.$$ (6.12)

This is our fourth basic equilibrium condition. It fixes the value of the net flux of radiation as a function of the temperature gradient and of the opacity of the gases through which the radiation flows.

The Magnitude of Stellar Luminosities and the Mass-Luminosity Relation

We may use this radiative equilibrium condition (6.12) for a preliminary order-of-magnitude estimate just as we have used the hydrostatic equilibrium conditions in the preceding section. Let us apply Eq.(6.12) to a point in the sun midway between center and surface. There we may take for r half the radius of the sun, for T ten million degrees in accordance with our previous estimates, for the reciprocal of $\varkappa\rho$ one centimeter, and for the differentials the corresponding differences. With these crude estimates we obtain from Eq.(6.12)

$$L \approx 6 \times 10^{35}.$$

Our result exceeds the luminosity of the sun by a factor of about a hundred. The cause of this discrepancy will turn out to be mainly the circumstance that our temperature estimate is somewhat too high; in §16 we shall find from a detailed solution of the basic equations for the sun that the temperature at the median point is nearer three million degrees than ten million degrees. But rather than fret about the fact that we have missed the observed luminosity of the sun by two powers of ten, we should emphasize that even these extremely rough estimates, which take no account of the detailed structure of the basic differential equations, have already given us a luminosity within the general range of observed stellar luminosities.

In the same spirit of roughest approximation we may use the radiative equilibrium condition to answer right now another question: how does the luminosity of a star depend on its mass? The necessary sequence goes as follows. A representative density within a star will vary with the mass and radius of the star according to

$$\rho \propto \frac{M}{R^3}.$$

If we introduce this proportionality into the condition of hydrostatic equilibrium (5.2), and if we replace the differentials as before by the corresponding differences, we find for the dependence of a representative pressure on the mass and radius

$$P \propto \frac{M^2}{R^4}.$$

If we introduce these two proportionalities into the equation of state for an ideal gas (5.3), we obtain for the temperature

$$T \propto \frac{M}{R}.$$

Now we may introduce the proportionalities just found for P and T into the radiative equilibrium equation (6.12), in which we may replace the differentials as before with the corresponding differences and in which we may assume the absorption coefficient $\varkappa$ to be substantially constant. Thus, we find

$$L \propto M^3.$$

The dependence on the radius has cancelled out and we have derived the theoretical mass-luminosity relation in its crudest form. Our result indicates that the luminosity of a star should increase with the third power of its mass. The observations which we discussed in §2 gave on the average much the same power of the mass, through with significant variations in the low, intermediate and high mass ranges, as shown by Fig. 2.1. But again, detailed comparison with observations is not the aim of these preliminary estimates. Their usefulness lies rather in illustrating the physical implications of the basic equilibrium conditions.

Stability of Thermal Equilibrium

Our last estimates point out that the luminosity of a star is not determined by the rate of energy generation by nuclear processes—for which no estimates have entered our discussions this far—but by the radiative equilibrium condition (6.12). We may summarize the physical reasons for this circumstance as follows. The pressure force must counteract gravity according to the hydrostatic equilibrium condition (5.2). If the internal pressure is to be high enough for this purpose, the internal temperature must have certain relatively high values according to the equation of state (5.3). The temperature gradient from the high internal temperatures to the low surface temperatures will cause a net radiation flux according to Eq.(6.12). The strength of this radiation flux will be fixed by the radiative equilibrium condition (6.12) irrespective of whether or not the energy loss caused by the radiation flux is compensated by the nuclear energy production in the interior. If the total nuclear energy generation is smaller than the energy loss by radiation from the surface, the star suffers a net energy loss. The only way this loss can be made up is from the gravitational energy by contraction.

According to Eq.(5.6), only one half of the gravitational energy released during a contraction is available for radiation losses from the surface; the other half must automatically go into an increase of the thermal energy. During the contraction, therefore, the internal temperatures will rise and in consequence the rate of nuclear energy production will increase. The contraction will stop when the over-all rate of nuclear energy release is equal to the radiative surface losses, that is to the luminosity of the star. Thus the star has the ability of balancing its nuclear energy source and its radiation loss. It achieves this balance, however, not by adjusting its luminosity but rather by adjusting its nuclear energy sources through an appropriate contraction or expansion.

There exist special circumstances in which a star is not able to balance its nuclear energy source with its radiation loss by a moderate expansion or contraction. This occurs when the internal densities are so high that the equation of state (5.3) for an ideal gas does not hold. We shall discuss such special circumstances in §§25 and 27.

We have derived in this section the basic relation (6.12) between the energy flux and the temperature gradient for the case that radiation provides the energy transport. We now turn to the case where transport by convection is essential.

7. Convective Energy Transport

Let us assume that in a certain layer of a given star radiative equilibrium holds as discussed in the preceding section. If this equilibrium is stable against perturbation, then no convective mass motions can persist and convective energy transport does not occur. If, however, the radiative equilibrium is found to be unstable, mass motions and convective energy transport may be the consequence. Clearly, we should determine the conditions under which radiative equilibrium is unstable before we investigate convection.

Stability Condition for Radiative Equilibrium

Consider the following particular perturbation. Take a volume element of small size within a star. Displace this element of matter upwards by the distance dr. Let the element expand adiabatically until the pressure inside the element is in balance with the pressure of the surroundings. Release the element and determine whether it will start moving downwards towards its original position or whether it will continue to move upwards. In the first case the radiative equilibrium is stable in the layer in question, whereas in the second case radiative equilibrium is unstable and convective motions will persist.

To follow this perturbation in more detail we may use the nomenclature indicated in Fig. 7.1; quantities referring to the inside of the element are

Fig. 7.1. Perturbation of radiative layer to test for stability against convection.

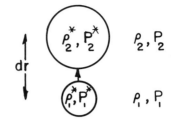

designated by an asterisk whereas quantities referring to the surrounding unperturbed medium have no asterisk, the subscript 1 refers to the original position of the element whereas the subscript 2 refers to the higher position to which the element is raised. Before the start of the perturbation the element considered does not differ from the surroundings, so that

$$\rho_1{}^* = \rho_1 \text{ and } P_1{}^* = P_1.$$

After the perturbation the pressure is again in equilibrium with the surroundings, but the density within the element may now differ from that of the surroundings since the internal density is determined by the adiabatic expansion of the element. Thus we have

$$P_2{}^* = P_2 \text{ and } \rho_2{}^* = \rho_1{}^* \left(\frac{P_2{}^*}{P_1{}^*}\right)^{\frac{1}{\gamma}}$$

where γ is the ratio of the specific heats c_p/c_v and has the value 5/3 for a highly ionized gas. The pressure force exerted on that volume which the element occupies after its displacement and expansion has not been altered by the perturbation. The gravitational force on the same volume, however, will be altered if the density within the element after the perturbation differs from that in the surroundings. Specifically, if the internal density is larger than the surrounding one, then the gravitational force is increased and the element will experience a net force downwards so that it will start moving back towards its original position. Hence, under the condition

$$\rho_2{}^* > \rho_2$$

any perturbation will immediately be damped out and the layer will be completely stable.

This stability condition may be transformed into a more convenient form. First, the asterisked quantities may be eliminated with the help of the four relations above so that the stability condition is expressed entirely in terms of quantities referring to the unperturbed layer. Second, the quantities at the higher position (subscript 2) can be expressed by Taylor series in terms of the quantities and their derivatives in the lower position (subscript 1). If finally the limit for small displacements dr is considered—so that higher than linear terms can be ignored—one obtains

$$-\frac{1}{\gamma}\frac{1}{P}\frac{dP}{dr} < -\frac{1}{\rho}\frac{d\rho}{dr}. \tag{7.1}$$

This inequality is an exact and general form of the condition for stability against convective motions of any layer in a star.

For the case that the equation of state (5.3) of an ideal gas holds, the stability condition can be brought into a still more practical form. Let us further specialize to the case of a layer which is homogeneous in chemical composition so that the molecular weight m is constant. Now by logarithmic differentiation of Eq. (5.3) we can express the density gradient in terms of the pressure and temperature gradients and eliminate it from the stability condition (7.1). We thus obtain

$$-\left(1-\frac{1}{\gamma}\right)\frac{T}{P}\frac{dP}{dr} > -\frac{dT}{dr}. \tag{7.2}$$

Since the pressure gradient and the temperature gradient are both always negative, both sides of this equation contain positive quantities. The right-hand quantity is the absolute amount of the actual temperature gradient in the layer. The left-hand quantity is often called the "adiabatic temperature gradient" since this quantity would represent the temperature gradient in the layer if the pressure and temperature followed the adiabatic relation throughout the layer. Condition (7.2) amounts to saying that for the layer to be stable the actual temperature gradient must be lower, in absolute amount, than the adiabatic temperature gradient.

The stability conditions (7.1) and (7.2) cannot be applied without special considerations to layers with inhomogeneous composition. We shall encounter in §20 a typical example of such considerations.

If a stable model of a star is to be constructed, the stability condition (7.2) must be checked at every layer within the model. That is, the pressure gradient has to be computed from the hydrostatic equilibrium condition (5.2), the temperature gradient has to be derived from the radiative equilibrium conditions (6.12), and both gradients have to be inserted into the stability condition (7.2). If this condition is found fulfilled, then the layer is stable and radiative equilibrium applies exactly as discussed in the preceding section. But what happens if the stability condition (7.2) is not fulfilled? This is the question we have now to consider in full detail. This problem is of appreciable consequence for stellar models. In stellar cores very high fluxes per square centimeter are often encountered and in stellar envelopes occasionally very high opacities occur; according to Eq. (6.12), both these circumstances lead to high—and therefore unstable—radiative temperature gradients.

Convective Equilibrium

Let us then consider a layer in which the stability condition (7.1) or (7.2) is not fulfilled. A perturbed element which is displaced upwards will have an internal density lower than the surrounding density. It experiences therefore a net force upwards and will in consequence continue its upward motion. Similarly, an element displaced downwards will be too heavy compared with its surroundings and thus continue its motion downwards. Hence, upon the slightest perturbation convective motions will break out throughout the unstable layer. What thermal consequences will these motions have? An upwards moving element has, as we said, a density lower than that of the surroundings. Since it has adjusted its internal pressure by expansion to equal the surrounding pressure, its temperature must exceed the surrounding temperature according to the equation of state (5.3); the element carries an excess of thermal energy upwards. Similarly, a downwards moving element which has an excess density and hence a deficient temperature carries a deficient thermal energy downwards. Both the upwards and downwards moving elements contribute to a convective energy transport upwards.

This additional energy flux has the following effect on the structure of the unstable layer. Assume for the moment that the layer was in precarious radiative equilibrium with the radiative flux carrying out the energy produced by the nuclear processes. Now, because of the instability convective motions break out throughout the layer. The convective flux will transport thermal energy from the lower levels to the upper levels of the layer; the temperature at the lower, hotter levels will decrease while the temperature at the upper, cooler levels will increase. Thus the temperature gradient will be reduced by the convection. The lowering of the temperature gradient will immediately reduce the radiation flux, according to Eq. (6.12). It will also reduce the convective flux since a reduction of the excess of the actual temperature gradient in the layer over the adiabatic temperature gradient causes a reduction in the average temperature excesses and deficiencies of the moving elements, thereby lessening the convective transport of thermal energy. The lowering of the temperature gradient by convection will continue until the radiation flux and the convective flux added together have reached a value which fulfills exactly the basic thermal equilibrium condition (5.8). At this stage radiation and convection together produce an energy flux which carries outwards exactly the amount of energy produced further in by the nuclear processes and no further temperature changes will occur at any level. Thus instability of radiative equilibrium in a layer of a star leads to another state of equilibrium—"convective equilibrium"—in which convective motions occur throughout the layer.

Energy Transport by Convection

We now have to derive the relation between the temperature gradient and the total energy flux in this state of convective equilibrium. For this purpose we have to consider in detail the transport of thermal energy by the moving elements. The temperature excess of a rising element over its surroundings is given by the difference between the adiabatic temperature change within the element and the actual temperature change in the surroundings from the starting point of the motion to its end point. If the element has risen by the distance dr, its temperature excess will therefore be

$$dT = \left(1 - \frac{1}{\gamma}\right) \frac{T}{P} \frac{dP}{dr} \times dr - \frac{dT}{dr} \times dr = \Delta\nabla\, T \times dr.$$

Here the symbol $\Delta\nabla T$ is defined by

$$\Delta\nabla T = \left(1 - \frac{1}{\gamma}\right) \frac{T}{P} \frac{dP}{dr} - \frac{dT}{dr} \tag{7.3}$$

and represents the excess of the actual temperature gradient—in absolute amount—over the adiabatic temperature gradient. If we multiply the temperature excess by $c_p\rho$ we obtain the excess of thermal energy per cubic centimeter. If we further multiply by the velocity of the element, v, we obtain the energy flux per square centimeter per second

$$H = \Delta\nabla T \; dr \; c_p\rho \; v. \tag{7.4}$$

Exactly the same equation holds for downwards moving elements since the change in sign of dr compensates the change in sign of v. In fact Eq. (7.4) represents the average flux produced by the convective motions if dr is taken to stand for the average displacement (that is the vertical distance from the level at which the element had the same internal temperature as the average surroundings) and if v is taken to stand for the average vertical velocity of all elements at one level.

Eq. (7.4) already represents the needed relation between the convective energy transport and the temperature gradient. It is, however, not yet in a useful form since there occurs in it the convection velocity v which first has to be determined by the following dynamical consideration. The deficiency of density of a rising element over its surroundings can be computed, much like the temperature excess, by the following formula

$$d\rho = -\frac{1}{\gamma} \frac{\rho}{P} \frac{dP}{dr} \times dr + \frac{d\rho}{dr} \times dr = \frac{\rho}{T} \Delta\nabla T \times dr.$$

If this density deficiency is multiplied by the gravitational acceleration, the deficiency in gravitational force, the excess force upwards, is obtained. Since this force acts only at the end of the displacement, while at the beginning of the displacement the excess force is zero, the average excess force is obtained by multiplying by 1/2. By multiplying this average excess force by the distance dr, we obtain the work done by the excess force on the element. It is this work which produces the kinetic energy of the element. Thus we obtain

$$\frac{1}{2} \rho \, v^2 = \frac{\rho}{T} \, \Delta \nabla T \; dr \, \frac{GM_r}{r^2} \frac{1}{2} \; dr. \tag{7.5}$$

Since both sides of Eq.(7.5) are quadratic in v and dr it holds, like Eq. (7.4), for downwards moving elements as well as upwards moving ones. Hence we may take Eq. (7.5) to represent the average over all elements at one level if we again take v and dr to stand for the appropriate averages. Eq.(7.5) gives the convection velocity as a function of the temperature gradient. It may be used to eliminate the convection velocity from Eq. (7.4) for the flux. At this time we may introduce the average mixing length l which is to represent the average vertical distance which an element moves before it dissolves into its surroundings. In terms of this mixing length we can represent the average distance by which an element has moved at an arbitrary moment by

$$\overline{dr} = \frac{1}{2} \, l. \tag{7.6}$$

Thus we obtain from Eqs. (7.4) and (7.5)

$$H = c_p \rho \left(\frac{GM_r}{Tr^2} \right)^{\frac{1}{2}} (\Delta \nabla T)^{\frac{3}{2}} \frac{l^2}{4}. \tag{7.7}$$

Eq. (7.7) represents our final relation between the convective energy flux and the temperature gradient. It involves one great uncertainty, the value of the mixing length appropriate for convection in stars. In laboratory experiments the effective mixing length is usually found comparable to the linear size of the volume in which convection is observed. Correspondingly, we might therefore set the mixing length in our situation equal to the depth of an unstable layer. This could, however, be a gross overestimate of the mixing length for unstable layers in which the density drops by a large factor from the bottom level to the top level—as is the case when convective instability occurs in a layer close to the surface. For the following estimates we shall use

$$l \approx \frac{1}{10} \, R.$$

We shall see immediately that the uncertainty in this value for the mixing length has little consequence for convection zones in the deep interior of a star. It does introduce, however, a noticeable uncertainty if convective instability occurs in the layers just below the photosphere of a star—as is often the case—and thus introduces in some cases a great uncertainty in the structure and the extent of the outer parts of a stellar model. In fact, for all but the most advanced phases of stellar evolution the lack of a definite theory for the effective mixing length appears to be the most serious hole in our present knowledge of the basic physical laws which govern the structure of the stellar interior.

Adiabatic Approximation for Temperature Gradient

To obtain the complete relation between the total energy flux and the temperature gradient we may write

$$H = H_{\text{Radiative}} + H_{\text{Convective}} \tag{7.8}$$

If we introduce here for the radiative flux the value given by Eq. (6.11) or Eq. (6.12) and for the convective flux the value given by Eq. (7.7), we have obtained a relation which we can solve for the temperature gradient. This solution is greatly simplified by the following order-of-magnitude estimate. Let us use again our numerical estimates for the median point in the sun. Let us furthermore use for the convective flux its upper limit, which is the total flux. If we introduce these values into Eq. (7.7) we obtain for the temperature gradient excess

$$\Delta \nabla T \approx 2 \times 10^{-10}.$$

This value should be compared with the temperature gradient itself, which we may estimate to be

$$\left| \frac{dT}{dr} \right| \approx \frac{T_c}{R} \approx 3 \times 10^{-4}.$$

We see that the excess of the actual temperature gradient over the adiabatic temperature gradient is only about one millionth of the temperature gradient itself. Within our accuracy requirements it is therefore entirely permissible to ignore the temperature gradient excess completely and, in a convective layer, to set the temperature gradient equal to the adiabatic temperature gradient. Thus according to Eq. (7.3) we have

$$\frac{dT}{dr} = \left(1 - \frac{1}{\gamma}\right) \frac{T}{P} \frac{dP}{dr}. \tag{7.9}$$

Because of the large margin of safety, this simple equation holds with satisfactory accuracy for the stellar interior even if our estimate of the mixing length should turn out wrong by a very large factor. It is only if we approach the photosphere, where the density and the mixing length must be much smaller, that we cannot expect Eq. (7.9) to remain a good approximation. In that case we have to work explicitly with Eq. (7.7) with its uncomfortable uncertainty in l.

Character of Convection in Stellar Interior

With the help of our earlier numerical estimates we may gain a somewhat clearer picture of the motions which occur in a convective layer in the stellar interior. For the average temperature excess or deficiency within a moving element over the surroundings we find

$$\overline{dT} = \Delta\nabla T \ \overline{dr} \approx 1°K.$$

This is indeed a small fluctuation compared with a mean temperature of several million degrees. The average velocity of the moving elements can be computed from Eq. (7.5), which gives

$$v \approx 3 \times 10^3 = 0.03 \ \frac{km}{sec}.$$

Again, these velocities are very small compared with the thermal velocities which amount to several hundred km per sec in the stellar interior. Since the convective velocities are smaller than the thermal velocities by about four powers of ten, the hydrodynamic effects of the convective motions must be smaller than the gas pressure force by about eight powers of ten. This circumstance is particularly important because it justified our tacit assumption that the convective motions do not disturb the hydrostatic equilibrium.

With the help of our estimate for the convective velocities we may compute the Reynolds number in the usual way and we find a value very much larger than the critical value for the Reynolds number. This is, of course, a direct consequence of the very large linear scale of the motions in the stellar interior. It shows that the convection will not occur in orderly, semi-stable patterns—such as in Benard cells—but rather in a chaotic, turbulent manner. The slowness of these turbulent motions may be emphasized once more by computing the average lifetime of a turbulent element

$$t \approx \frac{l}{v} \approx 2 \times 10^6 \ sec = 20 \ days.$$

This time scale is long from the point of view of turbulence but is extremely short compared with the time scale of stellar evolution. This has the consequence that a convective layer must be extremely well mixed. Thus when nuclear transmutations change the composition in the hottest parts of a convective region these changes become apparent by turbulent mixing in every part of the convective region in a very short time.

We may summarize our picture of the convective motions as follows. The motions in a convective layer in the stellar interior are turbulent in character and so slow that they have no hydrodynamic effect. The convective motions are highly efficient in transporting energy because of the high content in thermal energy of the gases in the stellar interior. The turbulent mixing is so fast that a convective region is practically homogeneous in composition at all times.

From the point of view of the construction of stellar models we may extract from this section the following recipe. In every layer of a stellar model compute the pressure gradient from the hydrostatic equilibrium equation (5.2) and the temperature gradient from the radiative equilibrium equation (6.12). Introduce these two gradients into the stability condition (7.2). If you find that this condition is fulfilled, convection does not occur and the temperature gradient computed from the radiative equilibrium condition (6.12) is the correct one. If you find that the stability condition (7.2) is not fulfilled, convection does occur and Eq. (6.12) cannot be used. Instead, use Eq. (7.9) which gives the actual temperature gradient in a convective layer with satisfactory accuracy.

8. Equation of State

In the three preceding sections we have assembled the equilibrium equations which must be fulfilled throughout the stellar interior. If we were now to start the building of theoretical stars, we would soon find ourselves floundering for lack of certain additional physical data. The equilibrium equations contain the pressure, the density, and the temperature; we shall need the equation of state which relates these three variables. The opacity is a decisive factor in the equation of radiative equilibrium; we shall need to know the opacity as a function of temperature and density. The basic equation of thermal equilibrium is useless without knowledge of the rate of energy production by nuclear processes for various conditions of temperature and density. Hence, we still have to postpone the attack on our main problem, the construction of theoretical stars of various masses and ages, and to continue patiently the assemblage of all the physical data relevant to the stellar interior.

The three additional relations we still need—the equation of state and the equations for the opacity and the energy generation—we may call the "gas characteristics relations." They differ in two respects from the equilibrium equations which we have already discussed. First, on the mathematical side, they are explicit relations while the equilibrium conditions take the form of differential equations. And second, whereas we managed to write down the equilibrium conditions in a general form without specific regard to the composition of the stellar matter, we shall find that the gas characteristics relations will depend explicitly on the composition. We shall now concentrate on the equation of state and take up the discussion of the opacity and the energy generation in the next two sections.

Ideal Gas

The equation of state of an ideal gas—which we have already used in §5 for numerical estimates—may be written in the form

$$P = NkT.$$

Here N represents the number of free particles per cubic centimeter. This simple equation of state holds with high accuracy for gases of low density. Its application to stellar atmospheres where the density is of the order of 10^{-7} g/cm³ appears therefore amply justified. Its application to the stellar interior, however, where the gas densities are very high, would not be justified if the temperature were not also very high and the matter therefore nearly completely ionized. Under these circumstances the effective volume of an individual particle is not the volume of a complete atom but rather a volume of very much smaller dimensions, and the deviations from the ideal gas equation caused by the particle volume are negligibly small.

If we now want to relate the number of free particles, N, to the density, we have to introduce the composition of the gases explicitly for the first time. Let X stand for that fraction of the matter, by weight, which consists of hydrogen, Y that fraction which consists of helium, and Z the remaining fraction consisting of all the heavier elements. By definition, the three fractions must fulfill the relation

$$X + Y + Z = 1.$$

With this form of designating the composition we can tabulate as follows the number of atoms and the number of corresponding electrons per cubic centimeter (where H stands for the mass of a proton and A for the average atomic weight of the heavier elements):

Element:	Hydrogen	Helium	Heavier	
No. of atoms:	$\dfrac{X\rho}{H}$	$\dfrac{Y\rho}{4H}$	$\left[\dfrac{Z\rho}{AH}\right]$	(8.1)
No. of electrons:	$\dfrac{X\rho}{H}$	$2\dfrac{Y\rho}{4H}$	$\dfrac{1}{2}A\dfrac{Z\rho}{AH}$	

The first item in the last column is put in brackets since it is generally negligibly small; the abundance of the heavier elements, Z, is of the order of a percent as we have seen in §4, and the average atomic weight for the heavier elements, A, is at least as high as sixteen. In the second item of the last column we set the number of electrons per atom equal to $\frac{1}{2}A$, which is exactly true, or nearly so, for most of the heavier elements. Since in a completely ionized gas all the electrons are free, we can obtain the total number of particles per cubic centimeter, N, by a simple summation of all the items in the enumeration (8.1):

$$N = \left(2X + \frac{3}{4}Y + \frac{1}{2}Z\right)\frac{\rho}{H}.$$

With this expression for the number of free particles per cubic centimeter
we obtain for the equation of state

$$P = \frac{1}{\mu}\frac{k}{H}\,\rho\,T \quad \text{with} \quad \frac{1}{\mu} = 2X + \frac{3}{4}\,Y + \frac{1}{2}\,Z \qquad (8.2)$$

where we have introduced the molecular weight μ measured in proton
masses.

Eq.(8.2) for the molecular weight does not hold in the outermost layers
of a star where the ionization is not complete. The limit to which Eq.(8.2)
is applicable may be determined by computing the degree of ionization of
the most abundant elements with the help of the Saha equation for various temperatures and densities. The results of such computations are
shown in Fig. 8.1, a temperature-density diagram for the ranges of interest for the stellar interior. In the lower left-hand portion of the diagram are shown three curves. The lowest of these curves represents the
conditions of temperature and density at which hydrogen is just half
ionized. The middle curve represents the condition where half of the
helium is singly ionized, and the highest curve the condition where half
of the helium is doubly ionized (the exact positions of these three curves

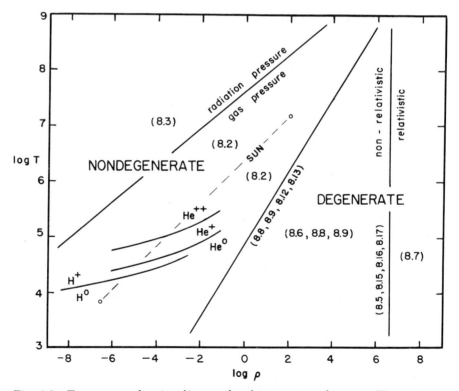

Fig. 8.1. Temperature-density diagram for the equation of state. The numbers
in parentheses refer to the relevant equations representing the equation of state.

depend somewhat on the composition; in Fig. 8.1 they are based on the limiting case of virtually pure hydrogen). The dashed line in Fig. 8.1 gives for comparison the approximate run of temperature and density in the sun from the photosphere (lower left circle) to the center (upper right circle). The intersection of this solar line with the three ionization curves shows that in the sun hydrogen and helium are practically completely ionized even at a temperature only a little above a hundred thousand degrees. Since, furthermore, slightly incomplete ionization of the heavier elements has little effect because of their low abundance, we may conclude that Eq.(8.2) for the molecular weight holds in all but the outermost layers of the sun and similar stars.

Radiation Pressure

At this point it may be useful to correct an omission which we have made in our discussion of hydrostatic equilibrium in §5. There we had assumed that the only force which counterbalances gravity is the gas pressure. In fact, however, we should add to this the force exerted on matter by radiation. Whenever an atom emits a photon it loses momentum and when it absorbs a photon it gains momentum. Since an atom emits isotropically over all directions no resultant momentum can accrue when one averages vectorially over many emission processes. The absorption processes, however, are not isotropically distributed since the radiation has a net flux outwards.

The net energy transport per square centimeter per second is H and therefore the momentum transport of the radiation per square centimeter per second is H/c. Of this momentum transport only the fraction $\varkappa\rho$ is absorbed by the matter over a vertical distance of one centimeter. The momentum absorbed by the matter per cubic centimeter per second, that is the force exerted by the radiation on the matter, is then

$$F_{Rad} = \varkappa\rho\, H\, \frac{1}{c}.$$

This equation for the radiation force may be transformed into a more convenient form with the help of Eq.(6.11) for radiative equilibrium, giving

$$F_{Rad} = -\frac{d}{dr}\left(\frac{a}{3}\, T^4\right) = -\frac{dP_R}{dr}.$$

Thus we see that the quantity which we have introduced in §6 as the "radiation pressure" acts in fact as a pressure in the sense that its gradient represents a force. We can therefore take the radiation force properly into account by simply adding the radiation pressure to the gas pressure, thereby obtaining for the total pressure

$$P = \frac{1}{\mu}\frac{k}{H}\rho T + \frac{a}{3}\, T^4. \tag{8.3}$$

Under what conditions is the radiation pressure of importance? We have drawn in the temperature-density diagram, Fig. 8.1, a line—in the upper left-hand portion—which represents the condition at which the radiation pressure exactly equals the gas pressure (this line depends somewhat on the molecular weight and hence on the composition; we have here again used pure hydrogen). To the left of and above this line radiation pressure dominates the gas pressure whereas to the right of and below the line the reverse is true. We see that throughout the sun radiation pressure is of no importance. We will find the same to be true in the majority of the stars. Only in fairly special circumstances, particularly in the heaviest stars, has radiation pressure to be taken into account.

Degeneracy

Eq.(8.3) would be our last word about the equation of state if it were not for the complicating phenomenon of degeneracy. Degeneracy occurs at high densities. It is not caused, however, by the gas density approaching nuclear densities, that is by the nuclei starting to "touch." Nuclear densities are of the order of 10^{12} g/cm^3, which is more than a factor a thousand higher than the highest densities which we will encounter. Rather, degeneracy is a direct consequence of the exclusion principle.

Consider for the moment only the free electrons in the gas. The electrons contained in a small volume $dx\, dy\, dz$ have, under conditions of low density, a velocity distribution given by Maxwell's law. To visualize this three-dimensional distribution function in momentum space we may consider a cross-section through it, i.e. we may look at the distribution of the x component of the momentum, p_x, for just those electrons which have y and z momentum components near zero. This one-dimensional cross-section has, according to Maxwell's law, the form of an error curve of which the dispersion is determined entirely by the temperature (see Fig. 8.2, curve a). If we put twice as many free electrons into our small space volume, without however changing the temperature, the distribution in momentum space is changed as shown by curve b in Fig. 8.2; the number density is doubled at every point in momentum space, but the dispersion of the distribution is not altered.

Such an increase in the number density in momentum space is not possible indefinitely; the exclusion principle sets a ceiling. Take a small volume $dp_x\, dp_y\, dp_z\, dx\, dy\, dz$ in six-dimensional phase space and divide it into cells of the size h^3 where h is the Plank constant. The exclusion principle states that at most two electrons—differing from each other in their spin—may be contained in each cell. We have therefore for the number density of electrons in phase space the upper limit

$$n_E\, dp_x\, dp_y\, dp_z\, dx\, dy\, dz \leq \frac{2}{h^3}\, dp_x\, dp_y\, dp_z\, dx\, dy\, dz.$$

Now, if we put still more free electrons into our small space volume than shown in curve b of Fig. 8.2, the maximum of the distribution function

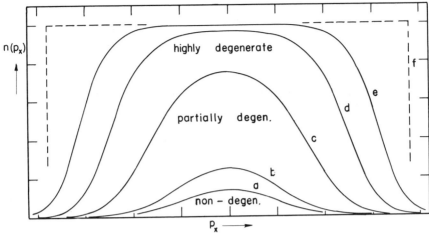

Fig. 8.2. Cross-section of momentum distribution for electrons.

will soon approach the ceiling; most of the cells at the lower momenta will be filled and the majority of the additional electrons have to be put into cells further out, at larger momenta, that is at larger energies. The distribution function will be deformed by this process as shown by curves c, d, and e of Fig. 8.2. Finally, if we put still more electrons in the same space volume, the distribution function will approach the form shown in curve f of Fig. 8.2: all cells for momenta smaller than a threshold momentum, p_0, are completely occupied, virtually no electrons are found at momenta larger than the threshold, the original distribution function which was governed by the temperature is completely obliterated, the new distribution function is entirely determined by the ceiling set by the exclusion principle and by the threshold value, that is the electron gas is "completely degenerate."

Complete Degeneracy

Let us derive the equation of state for the completely degenerate case. For this purpose we have to determine both the density and the pressure from the momentum distribution by the appropriate integrals over momentum space. A one-dimensional cross-section across the momentum distribution is shown in curve f of Fig. 8.2. The corresponding three-dimensional distribution function corresponds simply to a sphere with radius p_0; within this sphere the number density is equal to the upper limit given by the exclusion principle, while outside the sphere the number density is zero. The first integral, that for the number of electrons per cubic centimeter, takes correspondingly the simple form

$$N_E = \int\!\!\!\int\!\!\!\int_{|p|=0}^{|p|=p_0} \frac{2}{h^3}\, dp_x\, dp_y\, dp_z = \frac{2}{h^3}\frac{4\pi}{3}\, p_0^{\,3}.$$

If we want to relate the number of electrons to the density, we have to add up the items listed in the last line of the enumeration (8.1):

$$N_E = \frac{1}{\mu_E} \frac{\rho}{H} \quad \text{with} \quad \frac{1}{\mu_E} = X + \frac{1}{2} Y + \frac{1}{2} Z = \frac{1}{2} (1 + X). \qquad (8.4)$$

If we eliminate N_E between the last two equations, we obtain for the density as a function of the threshold p_0

$$\frac{1}{\mu_E} \rho = \frac{8\pi}{3} \frac{H}{h^3} p_0^{\,3}. \qquad (8.5)$$

The second integral, that for the electron pressure, may be written as

$$P_E = \int\!\!\int\!\!\int_{|p| = 0}^{|p| = p_0} p_x v_x \frac{2}{h^3} dp_x dp_y dp_z.$$

Here we have expressed the pressure as the rate of transport of momentum, as the momentum of p_x of a cubic centimeter transported through a square centimeter at the rate given by the corresponding velocity v_x. That we here have singled out the x direction is of no consequence since the momentum distribution is spherical. To evaluate this integral we have to distinguish two cases. In the non-relativistic case the velocity and the momentum are related by

$$v_x = \frac{p_x}{m}$$

where m is the mass of the electron. Introducing this into the pressure integral, we obtain by integration

$$P_E = \frac{8\pi}{15 m h^3} p_0^{\,5}.$$

For the case that most of the electrons have relativistic velocities, the total velocity is equal to the light velocity so that an individual velocity component is given by

$$v_x = c \frac{p_x}{|p|}.$$

If we introduce this relation into the pressure integral and perform the integration we have

$$P_E = \frac{2\pi c}{3 h^3} p_0^{\,4}.$$

Thus we have obtained for both cases a relation between the electron pressure and the threshold momentum.

If we express the threshold momentum in terms of the density with the help of Eq.(8.5), we arrive at the equation of state for a completely degenerate electron gas in the non-relativistic case

$$P_E = K_1 \times \left(\frac{1}{\mu_E} \rho \right)^{\frac{5}{3}}$$

(8.6)

$$\text{with } K_1 = \frac{h^2}{20m\,H} \left(\frac{3}{\pi H} \right)^{\frac{2}{3}} = 9.91 \times 10^{12}$$

and in the relativistic case

$$P_E = K_2 \times \left(\frac{1}{\mu_E} \rho \right)^{\frac{4}{3}}$$

(8.7)

$$\text{with } K_2 = \frac{hc}{8H} \left(\frac{3}{\pi H} \right)^{\frac{1}{3}} = 1.231 \times 10^{15}.$$

We still have to consider the atoms since their pressure P_A has to be added to the electron pressure to obtain the total pressure

$$P = P_A + P_E.$$

(8.8)

Are there conditions in stars where the atoms also are degenerate? In non-degenerate thermal equilibrium the atoms have on the average the same energy per particle as the electrons. Because of their higher mass their average momentum must therefore be larger by the square root of the mass ratio. In momentum space they are therefore scattered over a volume larger by the three-halves power of the mass ratio than that volume occupied by the electrons. Hence, the atoms have available nearly a hundred thousand times more cells in momentum space than the electrons and degeneracy of the atoms would occur only at densities a hundred thousand times higher than for degeneracy of the electrons. Such extreme densities will occur in the stellar interior only in the rarest of circumstances, if at all. We may therefore continue to use for the atoms the non-degenerate equation of state

$$P_A = \frac{1}{\mu_A} \frac{k}{H} \rho T \text{ with } \frac{1}{\mu_A} = X + \frac{1}{4} Y$$

(8.9)

where the appropriate molecular weight has been found by adding the items in the first line of the enumeration (8.1). The combination of Eqs.(8.6) and (8.9) gives the equation of state for nonrelativistic electron degeneracy while the combination of Eqs.(8.7) and (8.9) gives the equation of state for relativistic electron degeneracy.

Degeneracy Limits

To know when to apply which equation of state we must now determine two boundaries in the temperature-density diagram, first the boundary between the non-degenerate region and the degenerate region, and second the boundary which divides the latter region into its non-relativistic and relativistic parts.

We define the first boundary as that line where the non-degenerate and degenerate formulae give the same electron pressure. Thus, according to Eq.(8.6), the condition for this boundary is

$$\frac{1}{\mu_E} \frac{k}{H} \rho \, T = K_1 \left(\frac{1}{\mu_E} \rho\right)^{\frac{5}{3}},$$

which gives for the density as a function of the temperature along this line

$$\frac{1}{\mu_E} \rho = \left(\frac{kT}{HK_1}\right)^{\frac{3}{2}} = 2.40 \times 10^{-8} \times T^{\frac{3}{2}}. \tag{8.10}$$

The boundary, given by this equation, is represented in the temperature-density diagram Fig. 8.1 by the heavy diagonal line (the position of this line depends somewhat on the molecular weight, i.e. the composition; since degeneracy usually occurs in the stellar interior in regions in which the hydrogen has already been exhausted, we have used here a composition of pure helium). We see that conditions all through the sun fall into the non-degenerate region. On the other hand, we will find that conditions in the interior of white dwarfs fall well within the degenerate region—as is already suggested by the high mean densities of white dwarfs. Degeneracy will occur also during certain evolutionary phases of red giants, as we shall see in §24.

The second boundary, which divides the degenerate region into its non-relativistic and relativistic parts, may be defined as the line where Eqs.(8.6) and (8.7) for the two cases of degeneracy give the same electron pressure. Accordingly we have at this boundary

$$\frac{1}{\mu_E} \rho = \left(\frac{K_2}{K_1}\right)^3 = 1.916 \times 10^6. \tag{8.11}$$

This boundary is represented in Fig. 8.1 by the vertical line at the right (here again we have used a composition of pure helium). Clearly, very high densities have to be reached for relativistic degeneracy to occur. So far such high densities have been encountered only in the case of white dwarfs.

Partial Degeneracy

In the temperature-density diagram in the vicinity of either of the two boundaries just discussed the equation of state is obviously not accurately represented by any of the simple forms we have derived. Near the first boundary the momentum distribution takes the distorted form shown in curve c of Fig. 8.2, which is intermediate between the error curve characteristic for non-degenerate conditions and the square form characteristic for degeneracy. Near the second boundary part of the electrons will have relativistic velocities and part non-relativistic ones. In both boundary regions the exact equation of state can be derived by the same principles we have employed thus far. We shall not give these derivations here but rather only report the results.

In the neighborhood of the first boundary, which separates the non-degenerate region from the degenerate region, that is in the region of partial degeneracy, the exact equation of state can be written in the parametric form

$$\frac{1}{\mu_E} \rho = \frac{4\pi}{h^3} (2mkT)^{\frac{3}{2}} H F_{\frac{1}{2}}(\psi) \tag{8.12}$$

$$P_E = \frac{8\pi}{3h^3} (2mkT)^{\frac{3}{2}} kT F_{\frac{3}{2}}(\psi) \tag{8.13}$$

where the two new functions are defined by

$$F_{\frac{1}{2}}(\psi) = \int_0^\infty \frac{u^{\frac{1}{2}}}{e^{u-\psi}+1}\,du \quad \text{and} \quad F_{\frac{3}{2}}(\psi) = \int_0^\infty \frac{u^{\frac{3}{2}}}{e^{u-\psi}+1}\,du. \tag{8.14}$$

For large negative values of ψ Eqs.(8.12) and (8.13) go over into the equation of state for an ideal electron gas, as can be seen from the limiting behavior of the two F functions,

$$\text{for } \psi << 0: F_{\frac{1}{2}}(\psi) \approx e^\psi, F_{\frac{3}{2}}(\psi) \approx \frac{3}{2} e^\psi.$$

On the other hand, for large positive values of ψ these equations go over into the degenerate equation of state (8.6), as follows from the limiting behavior of the two F functions,

$$\text{for } \psi >> 0: F_{\frac{1}{2}}(\psi) \approx \frac{2}{3} \psi^{\frac{3}{2}}, F_{\frac{3}{2}}(\psi) \approx \frac{2}{5} \psi^{\frac{5}{2}}.$$

Semi-Relativistic Degeneracy

In the neighborhood of the second boundary, which separates the non-relativistic degenerate region from the relativistic degenerate region, the electron pressure is accurately given by

$$P_E = \frac{\pi m^4 c^5}{3h^3} f(x) \tag{8.15}$$

with $f(x) = x(2x^2 - 3)(x^2 + 1)^{\frac{1}{2}} + 3 \text{ arc sinh } x$ (8.16)

$$\text{and} \qquad x = \frac{p_0}{mc}. \qquad\qquad (8.17)$$

These equations together with Eq.(8.5) give the equation of state in parametric form. For smaller densities this set of equations goes over into Eq.(8.6) for non-relativistic degeneracy since we have

$$\text{for } p_0 << mc, \text{ i.e. } x << 1: f(x) \approx \frac{8}{5} x^5.$$

For higher densities it goes over into Eq.(8.7) for relativistic degeneracy since we have

$$\text{for } p_0 >> mc, \text{ i.e. } x >> 1: f(x) \approx 2 x^4.$$

Summary

By now we have collected for each relevant portion of the temperature-density diagram the equations which accurately represent the equation of state. In Fig. 8.1 we have indicated in each region the relevant equations by numbers in parentheses. All these equations together determine the pressure as a function of temperature and density uniquely and continuously throughout that portion of the temperature-density diagram relevant to stars. The fact that the equation of state cannot be given by one equation for this entire region is a bother in practice but not a difficulty in principle.

9. Opacity

The radiative opacity of the matter which makes up a star is one of the most essential factors in the structure of the stellar interior. For the astronomer who tries to reconstruct the stellar interior the opacity is by far the most bothersome factor in the entire theory. The opacity is caused by a multitude of atomic processes involving many elements and many stages of ionization. The sum of all these processes can not be represented by simple formulae in most cases, at least not with high accuracy over large regions of the temperature-density diagram.

This troublesome situation has been enormously alleviated in recent years by the publication of extensive tables which are based on detailed computations and which give the absorption coefficient for many different compositions and for a large number of points in the temperature-density diagram. Because of the existence of these tables it does not appear useful to describe the necessary computations here in full detail. Instead, in this section we shall first enumerate the main atomic processes which cause the opacity. Next we shall give the principle steps and equations used in the computation of the absorption coefficient, and finally we shall discuss the commonly used approximate equations which represent the opacity with rough accuracy.

Atomic Absorption Coefficients

The three atomic processes which contribute most to the opacity are, first, photoionization, the bound-free transition of an atom which absorbs a photon and in the process frees a formerly bound electron; second, free-free transitions in which an electron while passing an atom absorbs a photon and jumps from one free orbit to another more energetic one; and third, Thompson scattering of free electrons.

Consider first the bound-free transitions. Since hydrogen and helium are practically completely ionized in the interior, bound-free transitions have to be considered only for the heavier elements. Because even the heavier elements are very highly—though not completely—ionized in the interior the hydrogenic approximation for the atomic transition probabilities is applicable with good accuracy. In this approximation the absorption coefficient for one bound electron in one atom is given by

$$a_{bf} = \frac{64\pi^4 m\,e^{10}}{3\sqrt{3}\,c\,h^6} \frac{Z^{1^4}}{n^5} \frac{g}{\nu^3} \tag{9.1}$$

Here Z^1 is the effective charge of the ion considered, n is the principal quantum number of the electron considered, ν is the frequency of the light absorbed, and g is a non-dimensional factor called the Gaunt factor, which is of the order of unity and varies only slowly with n and ν.

Eq.(9.1) can hold only for frequencies for which the corresponding photon energies exceed the ionization energy χ_n, i.e.

$$h\nu > \chi_n = \frac{2\pi^2 m\,e^4}{h^2} \frac{Z^{1^2}}{n^2} \tag{9.2}$$

Hence, a specific bound-free transition causes an absorption continuum which is zero at low frequency, which jumps abruptly to its maximum at the critical frequency, and which tapers off at the higher frequencies proportionally to the reciprocal of the cube of the frequency. The absorption coefficient of a specific ion as a function of frequency will consist of the sum of several bound-free continua corresponding to several values of the principal quantum number n—with substantial contributions usually only from the few lowest n values. The absorption coefficient of a particular element will consist of one or two such series of continua corresponding to the one or two stages of ionization of the element which prevail at a given temperature and density. Altogether, therefore, the absorption coefficient of a mixture of the heavier elements will consist of an appreciable number of individual bound-free continua and correspondingly will tend to have a rather irregular behavior with frequency as well as with temperature and density.

Next consider the free-free transitions. The absorption coefficient for one atom and one free electron per cubic centimeter can be written in the hydrogenic approximation as

$$a_{ff} = \frac{4\pi e^6}{3\sqrt{3}\,c\,h\,m^2}\,\frac{Z^{1^2}}{v}\,\frac{g_{ff}}{v^3} \qquad (9.3)$$

Here v is the velocity of the electron and g_{ff} is again the Gaunt factor. This absorption coefficient varies smoothly from high values at low frequencies to low values at high frequencies, and there are no restrictions in the frequency range since photons of all energies can be absorbed by free-free transitions. In consequence, summation over the relevant ionization stages and elements produces an over-all free-free absorption coefficient which—except for minor variations caused by the Gaunt factor—is well behaved as a function of frequency, temperature, and density.

The last and simplest of the three absorption mechanisms is electron scattering. The absorption coefficient per electron is given by

$$\sigma_E = \frac{8\pi e^4}{3\,c^4\,m^2} \qquad (9.4)$$

This absorption coefficient is independent of frequency.

The Over-all Absorption Coefficient

How do we compute now from the three basic atomic absorption coefficients the over-all absorption coefficient of a cubic centimeter of stellar matter? To start again with the bound-free transitions, we have to multiply the atomic absorption coefficient a_{bf} by the number of atoms per cubic centimeter of atomic weight A, which is given by $X_A\rho/AH$ if X_A represents the abundance (by weight) of this element. We have further to multiply by the number of electrons, $N_{A,n}$, per atom which are bound in the nth state. We may compute this number with the help of the Saha equation as a function of the temperature and the number of free electrons per cubic centimeter. Finally, we have to sum over all elements and their various states of ionization and also over the various contributing values of n. Thus we obtain for the over-all absorption coefficient from bound-free transitions

$$\varkappa_{bf}(\nu) \times \rho = \sum_{A,n} a_{bf} \times \frac{X_A\rho}{AH} \times N_{A,n} \qquad (9.5)$$

A similar procedure is applicable to the free-free transitions. Again we multiply the atomic absorption coefficient a_{ff} by the number of atoms per cubic centimeter. Next we have to multiply by the number of free electrons per cubic centimeter N_E. Since the atomic absorption coefficient for free-free transitions depends on the velocity of the electron we must here use the number of electrons as a function of velocity, which is given by the Maxwell distribution, and then integrate over all velocities. Finally, we again sum over all relevant elements. Thus we obtain for the over-all absorption coefficient by free-free transitions

$$\varkappa_{ff}(\nu) \times \rho = \sum_A \int_\nu a_{ff} \times \frac{X_A \rho}{AH} \times N_E(\nu) d\nu \qquad (9.6)$$

For electron scattering we obtain the absorption coefficient per cubic centimeter simply by multiplying the atomic absorption coefficient σ_E by the number of electrons per cubic centimeter, thus obtaining

$$\varkappa_E \times \rho = \sigma_E \times N_E \qquad (9.7)$$

Now we add the contributions from the three atomic processes

$$\varkappa(\nu) = \varkappa_{bf}(\nu) + \varkappa_{ff}(\nu) + \varkappa_E \qquad (9.8)$$

and thus obtain the total absorption coefficient as a function of frequency.

The actual computations of the absorption coefficient are often trickier than our simple summary may have implied. For accurate computations it is necessary, for example, to take into account the perturbations by neighboring particles, which alter both the atomic transition probabilities and the ionization equation. Here we shall not enter upon these complicating features but shall rather continue with our survey of the principal steps.

The Rosseland Mean

Eq.(9.8) does not yet give the absorption coefficient in the form in which we need it. This equation presents the coefficient in its full frequency dependence, while in our discussion of radiative equilibrium in §6 we have used in every layer only one absorption coefficient, which clearly must stand for an appropriate mean over all frequencies. How is this mean to be taken?

If we consider radiative equilibrium not for the total radiation as we did in §6, but for monochromatic radiation, we find that all the derivations of §6 hold for monochromatic radiation up to Eq.(6.8). In particular, we obtain from Eqs.(6.6) and (6.8)

$$\frac{1}{3}\frac{dE(\nu)}{dr} + \varkappa^*(\nu)\frac{\rho}{c}H(\nu) = 0 \qquad (9.9)$$

The monochromatic energy density $E(\nu)$ and the monochromatic net flux $H(\nu)$ give the total energy density E and the total net flux H—to which we want to return as soon as we can—by straight integration over all frequencies:

$$E = \int_0^\infty E(\nu)d\nu \quad \text{and} \quad H = \int_0^\infty H(\nu)d\nu \qquad (9.10)$$

The absorption coefficient occurring in Eq.(9.9) has been given an asterisk to indicate the following correction which we have to add here. In §6 we had completely ignored stimulated emission. It can be shown that this as well as the relations between the three Einstein coefficients for emission and absorption are accurately taken into account if one uses a reduced absorption coefficient defined by

$$\varkappa^*(\nu) = \varkappa(\nu) \times (1 - e^{-h\nu/kT}) \qquad (9.11)$$

With this correction the absorption coefficient given by Eq.(9.8) can be introduced into Eq.(9.9). But before we can average this equation over all frequencies we must consider the frequency dependence of the monochromatic energy density, $E(\nu)$. The energy density is determined according to Eq.(6.2) by integrating the intensity over all directions. The outwards radiation passing a given point arises from a region slightly further in and at a somewhat higher temperature. Its intensity therefore is a little higher than the black body intensity corresponding to the temperature at the given point. This excess is very nearly exactly compensated by tne corresponding deficiency in the inward radiation. Therefore to obtain the energy density to a very high degree of accuracy we may use the black body formula

$$E(\nu) = \frac{4\pi}{c} B(\nu, T) = \frac{4\pi}{c} \times \frac{2h\nu^3}{c^2} \times (e^{h\nu/kT} - 1)^{-1} \qquad (9.12)$$

We see that in the stellar interior the frequency dependence of the energy density is not affected by the frequency dependence of the absorption coefficient.

The same thing can not be said at all for the flux, $H(\nu)$. For a frequency at which the absorption coefficient is high the outwards radiation passing a given point will arise from a region only just a little further in and hence only just a little hotter than the given point, and similarly the inwards-going radiation will arise from a nearby point only a very little cooler than the given point. Hence, the excess of the outwards radiation over the inwards radiation, that is the net flux, will be small. The reverse will be true for frequencies at which the matter is particularly transparent. Here the radiations will come from relatively more distant points with larger temperature differences and hence the net flux will be large. We may say that the net flux seeks those frequencies at which it experiences the least absorption. Thus no simple formula, corresponding to Eq.(9.12), can be given for the frequency dependence of the net flux.

In consequence it would not be useful to average Eq.(9.9) over all frequencies as it stands, but rather we shall first divide this equation by the absorption coefficient and then integrate over all frequencies, obtaining

$$H = -\frac{c}{3\rho}\int_0^\infty \frac{1}{\varkappa^*(\nu)}\frac{dE(\nu)}{dr}\,d\nu = \frac{c}{3\rho}\frac{dE}{dr}\times\frac{\displaystyle\int_0^\infty \frac{1}{\varkappa^*(\nu)}\frac{dE(\nu)}{dr}\,d\nu}{\displaystyle\int_0^\infty \frac{dE(\nu)}{dr}\,d\nu}$$

This equation goes over into our original Eq.(6.11) for radiative equilibrium if we define the absorption coefficient by the following means

$$\frac{1}{\varkappa} = \frac{\displaystyle\int_0^\infty \frac{1}{\varkappa^*(\nu)}\frac{dE(\nu)}{dT}\frac{dT}{dr}\,d\nu}{\displaystyle\int_0^\infty \frac{dE(\nu)}{dT}\frac{dT}{dr}\,d\nu} = \frac{\displaystyle\int_0^\infty \frac{1}{\varkappa(\nu)}(1 - e^{-h\nu/kT})^{-1}\frac{dB(\nu,T)}{dT}\,d\nu}{\displaystyle\int_0^\infty \frac{dB(\nu,T)}{dT}\,d\nu} \qquad (9.13)$$

which is called the Rosseland mean. Thus we have completed our description of how one derives the mean absorption coefficient from the atomic data.

Kramers' Law for Bound-Free Transitions

We shall now drop all considerations of accuracy and derive some very approximate formulae for the absorption coefficient. For bound-free transitions we may obtain a rough estimate of the absorption coefficient from Eq.(9.5) as follows. First we have to determine from the ionization equation the average number of electrons $N_{A,n}$ per atom which are bound in the nth state. For the case of high ionization, the Saha equation gives

$$N_{A,n} = N_E n^2 \frac{h^3}{2(2\pi m k T)^{\frac{3}{2}}} e^{+X_n/kT} \qquad (9.14)$$

The first factor, N_E, represents the number of free electrons per cubic centimeter, which is given by Eq.(8.4), and the second factor, n^2, arises from the statistical weight of the nth state. Next we introduce this equation as well as Eq.(9.1) for the atomic absorption coefficient into Eq.(9.5) and obtain

$$\varkappa_{bf} = \frac{2}{3}\sqrt{\frac{2\pi}{3}}\frac{e^6 h^2}{c H^2 m^{1.5}k^{3.5}}\cdot\frac{Z^{12}}{A}\cdot\left[\frac{1}{n}\frac{X_n}{kT}e^{+X_n/kT}\cdot\left(\frac{kT}{h\nu}\right)^3\cdot g\right]\cdot$$

$$Z(1 + X)\frac{\rho}{T^{3.5}} \qquad (9.15)$$

Here we have assembled the various factors in the following way. Two of the powers of Z^1 from Eq.(9.1) we have combined with the A from the denominator of Eq.(9.5). The ratio of these two quantities does not vary greatly from element to element if one considers only the more abundant of the heavier elements and we may use as a representative value

$$\frac{Z^{1^2}}{A} = 6$$

The remaining two powers of Z^1 from Eq.(9.1) we have replaced by the ionization potential, χ_n, which is given by Eq.(9.2). We have multiplied and divided kT in such a way that in Eq.(9.15) both the ionization potential and $h\nu$ occur always divided by kT. The summation over the elements was performed by simply summing over the abundances X_A, which gives Z, since hydrogen and helium do not contribute noticeably to the bound-free transitions. The summation over the various states was ignored since the lowest state with $n = 1$ usually provides the main contribution.

Consider now the factors in the bracket of Eq.(9.15). At any given temperature and density those particular elements and ionization stages which contribute most to the absorption coefficient will be those having ionization energies of the order of kT; those with much lower ionization potentials have too few bound electrons to be effective—as is the case for hydrogen and helium throughout the interior—while those with much higher ionization potentials require energies too high for the average photon at the given temperature. We may therefore take χ_n/kT to be of the order of one—at least as long as there exist some ionization potentials in the element mixture considered which are high enough for this condition to hold; and in reality the metals provide high enough ionization potentials for all temperatures and densities for which the bound-free absorption coefficient is of importance. The ratio $h\nu/kT$ likewise can be taken to be of the order of one since this gives the frequency just of the order of magnitude which occurs with highest weight in the Rosseland mean of Eq.(9.13). Hence all the factors in the bracket of Eq.(9.15) are of the order of unity. Therefore after introducing all these numerical factors, we may write Eq.(9.15) in the form

$$\chi_{bf} = 4.34 \times 10^{25} \times \frac{\bar{g}}{t} \times Z(1 + X) \frac{\rho}{T^{3.5}} \tag{9.16}$$

This is Kramers' Law of Opacity. It gives in a simple, though only very approximate, form the dependence of the bound-free absorption coefficient on temperature, density, and composition.

The factor $\bar{g}$ in Eq.(9.16) represents the mean Gaunt factor, and t represents an appropriate mean for the other factors in the bracket of Eq.(9.15); both these correction factors are of the order of unity as we have seen. We may also take t to include all the other modifying phenomena which we have not included in this very rough derivation. In particular it should include the following correction, which tends to reduce the bound-free absorption coefficient. For low temperatures the ionization equation (9.14) which holds only for high degrees of ionization, leads to an overestimate of the number of bound electrons per atom since this number can at most rise to the number of sub-states contained in the nth principal state. When this number is reached the guillotine falls on any further increase of the number of bound electrons and the absorption coefficient falls below the value otherwise expected. The factor t which takes care of this important correction in Kramers' Law (9.16) has therefore been named the guillotine factor.

Eq.(9.16) may be used from another point of view. Assume that this equation represents the exact bound-free absorption coefficient. Then this equation may be considered to define the guillotine factor, which in a strict sense now is a function of temperature, density, and composition. In fact, Eq.(9.16) has frequently been used in this sense; the results of detailed and accurate computations of the absorption coefficient have been given not in tables of the absorption coefficient itself but in tables of the corresponding guillotine factor—or, more precisely, in tables of the guillotine factor divided by the mean Gaunt factor. For the limited range of conditions occurring within a particular star one can often represent the guillotine factor with fair, though not high, accuracy by

$$\frac{t}{g} = \left(\frac{t}{g}\right)_0 [(1 + X)\rho]^\alpha \qquad (9.17)$$

Here the coefficient with zero subscript and the exponent α are two available constants which should be adjusted so that the interpolation formula (9.17) represents the accurate tabular value of the guillotine factor—or of the absorption coefficient itself—as well as possible. The numerical values for the coefficient usually range between one and ten and those for the exponent between zero and one-half. If Kramers' Law (9.16) is used not with a constant, average guillotine factor but with a varying guillotine factor as given by Eq.(9.17), it is often referred to as Modified Kramers' Law.

Kramers' Law for Free-Free Transitions

We continue the derivations of simple, approximate formulae for the absorption coefficient and turn now to the free-free transitions. In this case we have to consider only hydrogen and helium since for a mixture of the heavier elements the free-free transitions are fairly negligible compared with the bound-free transitions.

We proceed by introducing the atomic absorption coefficient of Eq.(9.3) into Eq.(9.6). Again the factor Z^2/A occurs; for hydrogen and helium this factor is exactly one. The integration over the electron velocities prescribed by Eq.(9.6) results in two factors. The first factor arises from the reciprocal velocity occurring in Eq.(9.3). Averaging over the Maxwellian distribution gives

$$\overline{\frac{1}{v}} = \sqrt{\frac{2m}{\pi k T}}$$

The second factor is the total number of free electrons per cubic centimeter, which again can be replaced with the help of Eq.(8.4). The frequency dependence of the atomic absorption coefficient of Eq.(9.3) can be averaged—accurately except for the Gaunt factor—with the help of Eq.(9.13) for the Rosseland mean. The result is

$$\overline{\nu^3} = 196.5 \left(\frac{kT}{h}\right)^3$$

This rather large value for the mean frequency reflects the fact that the free-free transitions give a much higher transparency at high frequencies than at low ones and that it is transparency, that is the reciprocal of the opacity, rather than opacity which is averaged in the Rosseland mean. Finally, the summing over the abundances X_A in Eq.(9.6) gives here $X + Y$. When these various factors are introduced into Eq.(9.6) this is transformed into

$$\chi_{ff} = \frac{2}{3} \sqrt{\frac{2\pi}{3}} \frac{e^6 h^2}{c H^2 m^{1.5} k^{3.5}} \frac{1}{196.5} \times [g_{ff}] \times (X + Y)(1 + X) \frac{\rho}{T^{3.5}} \quad (9.18)$$

$$= 3.68 \times 10^{22} \times \overline{g}_{ff} \times (X + Y)(1 + X) \frac{\rho}{T^{3.5}}$$

Thus we see that—in this rough approximation—the free-free absorption coefficient of hydrogen and helium has the same dependence on temperature and density as the bound-free absorption coefficient for the heavier elements. However, the numerical coefficient is very much smaller. Only when the abundance of the heavier elements is small will the absorption by hydrogen and helium play an important role. If we take the average Gaunt factor of Eq.(9.16) to be about equal to that of Eq.(9.18), and if we use a typical guillotine factor of ten, we find that the bound-free absorption by the heavier elements will dominate the free-free absorption of hydrogen and helium as long as the abundance of the heavier elements, Z, is of the order of two percent or larger—as is probably the case for the young Population I stars. However, free-free transitions of hydrogen and helium will dominate the absorption processes of the heavier elements if the abundance of the latter is of the order of one half of one percent or less—as may well be the case for the extreme Population II stars.

Electron Scattering

Turning finally to the last of our three atomic processes, we can obtain an explicit equation for the absorption coefficient arising from electron scattering extremely simply. If we introduce into Eq.(9.7) the atomic absorption coefficient from Eq.(9.4) and the expression for the number of electrons from Eq. (8.4), we are finished since stimulated emission does not occur for electron scattering and since the evaluation of the Rosseland mean can be skipped for a frequency-independent absorption coefficient. Thus we obtain for electron scattering, without approximations,

$$\varkappa_E = \frac{4\pi}{3} \frac{e^4}{c^4 H m^2} \times (1 + X) = 0.20 \times (1 + X) \qquad (9.19)$$

Temperature-Density Diagram for the Opacity.

At what temperatures and densities is electron scattering more important than the other sources of opacity? Electron scattering will dominate where Kramers' Law gives an unusually small absorption coefficient. This occurs according to Eq.(9.16) at low densities or at high temperatures. In the temperature-density diagram one may define the demarcation between the two opacity sources as given by that line where Eqs.(9.16) and (9.19) give the same absorption coefficient. On this line we have

$$\rho = 4.4 \times 10^{-27} \times \frac{t/\overline{g}}{Z} \times T^{3.5}$$

This line is shown in Fig. 9.1 for the representative values $Z = 0.01$ and $t/\overline{g} = 3$ (if we had used the corresponding demarcation between electron scattering and free-free absorption by hydrogen and helium, the line in Fig. 9.1 would have to be shifted just a little to the right. We see that conditions throughout the sun, which are indicated approximately by the dashed line in Fig. 9.1, fall into the region dominated by Kramers' Law. We shall find, however, that electron scattering plays an important role in the cores of the heavier stars, both on the main sequence and among the red giants.

To complete our discussion of the temperature-density diagram shown in Fig. 9.1 we may remark—as we have already mentioned at the end of §5—that under conditions of degeneracy conduction by the free electrons is highly effective so that energy transport by conduction is much more important than energy transport by radiation. We have therefore copied the demarcation line between degenerate and non-degenerate conditions from Fig. 8.1 on to Fig. 9.1. Thus the temperature-density diagram is

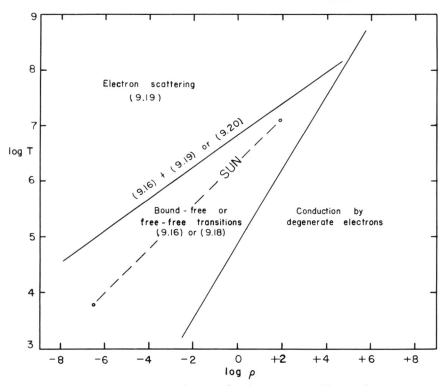

Fig. 9.1. Temperature-density diagram for the opacity. The numbers in parentheses refer to the relevant equations representing the absorption coefficient.

divided into three regions, electron scattering at the upper left, bound-
free and free-free absorption in the middle, and electron conduction to
the right.

Mixed Opacities

To return to the approximate equations for the absorption coefficient,
we still have to discuss the appropriate formulae for the transition
region around the demarcation line between the electron scattering
region and the region of bound-free and free-free transitions.   In this
transition region we may be tempted simply to add the two relevant
absorption coefficients—either Eqs.(9.16) and (9.19) or Eqs.(9.18) and
(9.19). This procedure is not strictly correct since according to Eq.(9.8)
we should add the monochromatic absorption coefficient before we take
the frequency average of the Rosseland mean by Eq.(9.13). The simple
addition of the average absorption coefficients will, however, represent
the true total absorption coefficient with tolerable accuracy in many
cases.

For the particular case of negligible abundance of the heavier elements
it has recently been shown that the absorption coefficient in the transi-

tion region is well represented by

$$\varkappa_{ff+E} = 1.6 \times 10^{11} \times (1 + X) \times \left(\frac{\rho}{T^{3.5}}\right)^{\frac{1}{2}} \qquad (9.20)$$

This formula represents the absorption coefficient with an accuracy of about ten percent over the range

$$4 \times 10^{-24} < \frac{\rho}{T^{3.5}} < 10^{-22}$$

This range of applicability contains only a rather narrow strip of the transition zone. It is, however, of particular importance since it includes much of the interior conditions in sub-dwarfs.

When in the subsequent chapters we make extensive use of Eqs.(9.16) to (9.20) for the absorption coefficient, even though we have here greatly emphasized their approximate nature, we do so in the understanding that their use is only a poor—but often efficient—substitute for the use of the detailed tables now available. We will consider the use of these formulae justified only if they are checked with and adjusted to the tabular values for every sample star.

10. Nuclear Reactions

Nuclear physics has provided the last missing link in the chain of basic phenomena which determine the structure and evolution of stars. Our knowledge of the nuclear processes in the stellar interior is still fascinatingly surrounded by the thrill of the new. But not only for such subjective reasons must we ascribe exceptional importance to the nuclear processes; it is they which provide the immense energies needed to maintain the stars as luminous bodies for their long life, it is they which provide the one-way transmutations that cause the stars to evolve.

Nuclear Processes at Stellar Temperatures

Over two hundred stable isotopes of all the elements are known. Each of these isotopes might undergo nuclear reactions with any other one. Can we hope to master this complex cookery sufficiently for our stellar purposes at this early date in the art of nuclear physics? Against this doubt we may gain confidence by the following happy circumstance. At a temperature of ten million degrees, typical for the stellar interior, the thermal energy of a particle is on the average only

$$E_{\text{Thermal}} = \frac{3}{2} k\, T \approx 2 \times 10^{-9} \approx 1 \text{ kev}$$

This is a small energy indeed if one compares it with the potential barrier between two nuclei caused by the Coulomb repulsion. If we take the distance over which nuclear forces are effective to be of the order 10^{-13} cm and if we consider only particles with the lowest charges, we obtain for the Coulomb potential

$$E_{\text{Coulomb}} = \frac{Z_1' \, Z_2' \, e^2}{R} \approx 2 \times 10^{-6} \approx 1000 \text{ kev}$$

This barrier is a thousand times higher than the average thermal energy. It has to be overcome if the nuclear forces are to enter into action.

Now a particle does not need to have a thermal energy exceeding the Coulomb potential to penetrate this barrier. Even at lower energies it has some chance of succeeding. However, the penetration probability, P_{p}, rapidly decreases with decreasing particle energy. On the other hand, as the energy increases above the thermal average the number of particles of a given energy rapidly decreases, according to Maxwell's Law. Clearly then, the main contribution to nuclear processes will arise from particles in an intermediate energy range, such as

$$E_{\text{Stellar}} \approx 20 \text{ kev}$$

where the energies are sufficiently large so that the penetration probability is still not too small and where at the same time the number of particles is still not vanishingly small. We will see that the battle between these two factors produces just the right rate of nuclear reactions in the stellar interior as long as we consider only particles of low charge. For particles of higher charge the Coulomb barrier becomes prohibitively high compared with the thermal energies at ten million degrees.

We have therefore to consider only reactions between particles of very low charge. Thus our ingredients are restricted to, say, the fourteen stable isotopes below oxygen, and the cookbook for these few and simple ingredients appears completely known even at this date. This abridged cookbook already contains the two most fundamental recipes, the proton-proton reaction and the carbon cycle. These two processes represent alternatives for the transmutation of hydrogen into helium, which is the most powerful source of stellar energy.

After a star has exhausted its hydrogen, it is forced to contract and its internal temperature will rise. When the central temperature reaches the order of one hundred million degrees, the thermal energies of the particles are sufficiently increased so that appreciably higher Coulomb barriers can be penetrated and a whole new set of nuclear reactions sets in. There may still exist some justified doubt whether our present list for this second course of nuclear processes is complete as yet. It already contains, however, one most important item, the transmutation of helium into heavier elements, which is the second—and possibly the last—of the main nuclear energy sources in the star.

When a star has also exhausted its helium fuel, it will again contract and again raise its internal temperature. When at the center the temperature reaches the order of one billion degrees, a wild multitude of new nuclear reactions breaks out. It appears not impossible that this event should be identified with the supernova phenomenon. Preliminary, speculative surveys of this event have indicated a fascinating set of possible phenomena. Neither the nuclear data nor the astronomical data regarding supernovae, however, appear sufficient at this time to permit a detailed analysis. We will restrict ourselves here, therefore, entirely to the nuclear reactions occurring at temperatures below a billion degrees.

We will now turn to the discussion of the three main energy-producing processes, the proton-proton reaction and the carbon cycle, both of which produce the transmutation of hydrogen into helium, and the triple-alpha process, which produces the transmutation of helium into heavier elements. Since an expert and up-to-date monograph on this subject is available we restrict ourselves here to a summary of the results.

The Proton-Proton Reaction

A proton-proton reaction consists of the following three steps

$$H^1 + H^1 \rightarrow D^2 + e^+ + \nu \qquad + 1.44 \text{ Mev } (14 \times 10^9 \text{ yrs})$$
$$D^2 + H^1 \rightarrow He^3 + \gamma \qquad + 5.49 \text{ Mev } (6 \text{ sec}) \qquad (10.1)$$
$$He^3 + He^3 \rightarrow He^4 + H^1 + H^1 \quad + 12.85 \text{ Mev } (10^6 \text{ yrs})$$

In the first step, two protons collide and produce a deuteron, a positron, and a neutrino. The positron immediately combines with an electron, and the pair disappears with the emission of two gamma rays. The neutrino, however, has a reaction cross-section that is virtually zero and thus will pass straight through the entire star, leaving it forever. The proton-proton interaction, because of a very unfavorable nuclear transmutation probability, is very slow in spite of the favorably low Coulomb barrier.

The build-up process is continued by a collision between the deuteron and another proton, which form He^3 with the emission of a gamma ray. This interaction has a very high probability and follows virtually immediately after the formation of the deuteron. The build-up of helium may be completed by a variety of reactions. The one given in the third line of Eq.(10.1) is by far the most frequent of the various alternatives under stellar conditions. It consists of the collision of two He^3 particles which form a He^4 nucleus with the re-emission of two protons. Since this third step uses two He^3 particles, the build-up of one He^4 nucleus necessitates that the first and second reactions of Eq.(10.1) each occur twice.

The total amounts of energy liberated in each of the three reactions are listed in Eq.(10.1). Not all of this energy, however, is available to the star since a fraction of the energy liberated in the first step (on the average 0.26 Mev) is carried away by the neutrino and thus is lost permanently to the star. If we subtract this little loss and remember that the first and the second reactions have to occur twice, we obtain for the total energy liberated per helium atom formed

$$E_{pp} = 26.2 \text{ Mev} = 4.2 \times 10^{-5} \text{ erg} \tag{10.2}$$

The Carbon Cycle

An alternative way of transmuting hydrogen into helium exists in the carbon cycle, which consists of the following six reactions

$$
\begin{array}{ll}
C^{12} + H^1 \longrightarrow N^{13} + \gamma & + 1.95 \text{ Mev } (1.3 \times 10^7 \text{ yrs}) \\
N^{13} \longrightarrow C^{13} + e^+ + \nu & + 2.22 \text{ Mev } (7 \text{ min}) \\
C^{13} + H^1 \longrightarrow N^{14} + \gamma & + 7.54 \text{ Mev } (2.7 \times 10^6 \text{ yrs}) \\
N^{14} + H^1 \longrightarrow O^{15} + \gamma & + 7.35 \text{ Mev } (3.2 \times 10^8 \text{ yrs}) \\
O^{15} \longrightarrow N^{15} + e^+ + \nu & + 2.71 \text{ Mev } (82 \text{ sec}) \\
N^{15} + H^1 \longrightarrow C^{12} + He^4 & + 4.96 \text{ Mev } (1.1 \times 10^5 \text{ yrs})
\end{array}
\tag{10.3}
$$

To start with, the collision of a proton with a common carbon nucleus produces a N^{13} particle with emission of a gamma ray. The N^{13} particle is not stable but decays—in seven minutes, on the average—into the heavy carbon isotope with the emission of a positron and a neutrino. Again, the positron disappears together with an electron and the neutrino leaves the star. The next build-up step is taken when a second proton collides with the heavy carbon isotope, forming a common nitrogen nucleus. The third step follows when another proton collides with the nitrogen and forms the unstable O^{15} which decays—in 82 seconds, on the average— into the heavy nitrogen isotope, N^{15}, with the emission of a positron and neutrino, whose fate we know. The final step in the build-up process is taken when a fourth proton collides with the heavy nitrogen nucleus, very rarely forming O^{16} but nearly always forming C^{12} plus a helium nucleus. It is this last step which makes the carbon cycle a true cycle in which the catalyst, C^{12}, is re-formed at the end so that the net effect is solely to produce one helium nucleus out of four protons—just the same as for the proton-proton reaction.

The energies liberated in the six reactions comprising the carbon cycle are listed in Eq.(10.3). In the decay of a N^{13} particle the neutrino takes away on the average 0.72 Mev and in the decay of a O^{15} particle 0.98 Mev. If we again add up the energies liberated minus the neutrino losses, we find per helium nucleus formed

$$E_{cc} = 25.0 \text{ Mev} = 4.0 \times 10^{-5} \text{ erg} \tag{10.4}$$

This value for the energy generation is slightly lower than the value in Eq.(10.2) for the proton-proton reaction because of the slightly larger losses by neutrinos in the carbon cycle. Both these values differ by small amounts from the value directly computed from the mass defect of one helium atom compared with four hydrogen atoms—a value which we have already used for our rough estimates at the end of §5. These small differences arise, of course, from the neutrino losses.

Nuclear Reaction Rates

It is not enough to have isolated the particular nuclear reactions which liberate the main store of stellar energy. It is the rates of these reactions which we need. The individual processes which make up the proton-proton reaction as well as the carbon cycle are all of one type—with the exception of the two beta-decays in the carbon cycle, for which the rates are well known. The type of process in question is the collision of two completely ionized atoms with subsequent nuclear interaction. The rate of such a process, that is the number of interactions per cubic centimeter per second, can be given by the equation

$$r = \int_0^\infty N_1 N_2 \, v q(v) \, P_P(v) \, P_N \, D(T, v) \, dv \tag{10.5}$$

The various factors in this equation arise as follows. The total number of reactions per cubic centimeter must be proportional to the number of atoms of the first kind, N_1; and the number of reactions per atom of the first kind must be proportional to the number of atoms of the second kind, N_2. These numbers of atoms follow the proportionalities

$$N_1 \propto \rho X_1 \quad \text{and} \quad N_2 \propto \rho X_2$$

where X_1 and X_2 are the abundances by weight of the two kinds of atoms.

The frequency of collisions between one atom of the first kind and one atom of the second kind is proportional to the product of the relative velocity v and the effective cross section q. The effective cross section for collisions which can lead to nuclear interactions is given by the square of the de Broglie wavelength so that

$$q \propto \frac{1}{v^2}$$

Not all collisions can lead to nuclear interactions, but only those in which the Coulomb barrier is penetrated. The penetration probability, P_P, depends on the relative velocity in the following form

$$P_P(v) \propto e^{-\frac{4\pi^2 \, Z_1' Z_2' e^2}{h} \cdot \frac{1}{v}}$$

This formula represents both the rapid drop of the penetration probability with decreasing velocity and the great disadvantage of reactions between atoms of high charges, Z_1' and Z_2', circumstances which we have already discussed above.

Even penetration does not assure nuclear interaction. The probability of nuclear interaction, P_N, does not depend appreciably on the relative velocity v in most cases, but does depend sensitively on the particular interaction and thus is the critical factor in Eq.(10.5). Our knowledge of the nuclear reaction probabilities comes from two sources, theoretical computations and laboratory experiments. Theoretical computations can now be made with good assurance for the simplest reactions. Thus a fairly accurate theoretical determination for the first step in the proton-proton reaction is available. That this particular theoretical determination is now possible is a lucky circumstance since the interaction between two protons is so rare an occurrence—however you arrange the laboratory conditions—that it is entirely outside the realm of experimental measurement. On the other hand, all the other reactions listed under Eqs.(10.1) and (10.3) are as yet hardly accessible to theoretical computations. For them, however, the available experimental data give the reaction probabilities with reasonable certainty.

Let us turn to the next factor in Eq.(10.5). Since the collision frequency as well as the penetration probability depends sensitively on the relative velocity v, we have to multiply by the frequency distribution of velocities and then integrate over all velocities. The Maxwellian velocity distribution is represented by

$$D\,(T,v) \propto \frac{v^2}{T^{\frac{3}{2}}} \cdot e^{-\frac{1}{2}HA_{1,2}\frac{v^2}{kT}}$$

where the reduced atomic mass is given by

$$A_{1,2} = \frac{A_1 A_2}{A_1 + A_2}$$

It is the dependence of the velocity distribution on temperature which causes the nuclear reactions to be temperature sensitive.

The evaluation of the integral of Eq.(10.5) is made easy by the fact that a rather narrow range in velocity contributes nearly the entire value —as we have already surmised above. This range is centered at the velocity for which the combined exponent of the two combatting factors P_P and D has its maximum. Using this simplifying circumstance, one finds in good approximation for the integral

$$r = C_{1,2} \cdot \rho^2 X_1 X_2 \frac{1}{T^{\frac{2}{3}}} e^{-3(2\pi^4 e^4 \frac{H}{h^2 k} \cdot \frac{Z_1'^2 Z_2'^2 A_{1,2}}{T})^{\frac{1}{3}}} \tag{10.6}$$

This formula for the reaction rate of nuclear processes of the type here considered shows explicitly the dependence on temperature, density, and composition while all the nuclear characteristics, including the critical factor P_N, are collected in the constant $C_{1,2}$, which varies from reaction to reaction.

Mean Reaction Times and Equilibrium Abundances

The main applications of Eq.(10.6) will be for the determination of the rate of energy generation ε. But before we enter upon this question let us use Eq.(10.6) to gain some insight into the relative speeds of the various reactions which make up the proton-proton reaction and the carbon cycle. We may define the "mean reaction time per particle of the first kind" by dividing the number of particles of the first kind per cubic centimeter by the number of reactions per cubic centimeter per second, i.e. by

$$t = \frac{N_1}{r} = \rho \frac{X_1}{H A_1} \cdot \frac{1}{r} \qquad (10.7)$$

Combining Eqs. (10.6) and (10.7) one sees that for each reaction the mean reaction time depends only on the temperature and the density of the particles of the second kind, ρX_2, besides of course the known nuclear constants occurring in Eq.(10.6). In the proton-proton reaction and the carbon cycle, as we have chosen to write them in Eqs.(10.1) and (10.3), the role of the second kind of particle is always played by the proton except in the last reaction of Eq.(10.1) where this role is taken by He^3. If we choose as representative sample conditions for the center of stars

$$T = 13 \times 10^6, \quad \rho X (H^1) = 100, \quad \rho X (He^3) = 0.01$$

(a justification for the last value we will find in a minute), we obtain for the mean reaction times the values listed in parentheses in Eqs.(10.1) and (10.3).

The mean reaction times differ from reaction to reaction enormously. In the carbon cycle, for example, N^{14} has a reaction time twenty times longer than C^{12}, one hundred times longer than C^{13} and three thousand times longer than N^{15}—not to mention the two unstable isotopes N^{13} and O^{15} for which we have already noted the extremely short decay times.

How can the reactions of the carbon cycle follow one another in a balanced manner if the reaction times are so greatly at variance? Imagine that we started with a mixture of C^{12} and N^{14} in the ratio 1 to 3 and with very little of the heavier isotopes, as would be fairly characteristic of the earth's crust. Then let the carbon cycle start. After some ten million years, much of the C^{12} will have undergone transmutations, but very little will as yet have happened to the N^{14}. In fact, most of the C^{12} will have changed into N^{14} through the first three reactions of Eq.(10.3) while very little of the N^{14} will have changed back into C^{12} through the last reactions of Eq.(10.3). Thus the abundance ratio of C^{12} to N^{14} will be steadily changing, and simultaneously, of course, a certain amount of the heavier isotopes will be built up. These changes will continue until the abundance of C^{12} has fallen so low that the reaction rate per cubic centimeter of the first step in Eq.(10.3), which transmutes C^{12}, has become just as low—in spite of its fast reaction time per particle—as the rate of the fourth step in Eq.(10.3), which transmutes N^{14}.

When equilibrium is finally reached, after a time of the order of one hundred million years, each isotope involved in the carbon cycle is formed just as fast by one reaction as it is destroyed by another one. This means that in equilibrium the reaction rates per cubic centimeter must be equal for all six steps in the carbon cycle. If we express these equilibrium conditions in terms of Eq.(10.6) for the four non-decay reactions of the carbon cycle, we may solve for the equilibrium abundance ratios for the four stable isotopes involved. Using the known nuclear constants for these four reactions we obtain

$$X\ (C^{12})/X\ (C^{13}) = 4.3 \pm 1.6$$
$$X\ (N^{14})/X\ (N^{15}) = 2800 \pm 1200 \qquad (10.8)$$
$$X\ (N^{14} + N^{15})/X\ (C^{12} + C^{13}) = 21 \pm 8$$

These values for the abundance ratios do not depend on the density since all four reactions depend on the same power of the density, which cancels out when the four rates are set equal to each other. The first two of these values do not depend perceptibly on the temperature either; however, the third one does, since the exponent in Eq.(10.6) differs slightly between the carbon isotopes and the nitrogen isotopes because of their difference in charge. Here we have used again a temperature of thirteen million degrees. The errors indicated in Eq.(10.8) represent rough estimates of the present uncertainties in the nuclear constant for the four relevant reactions.

The equilibrium value for the carbon isotope ratio is of interest in itself since it is very much smaller than the ratio found in the earth's crust. Thus, if we find a low ratio similar to the equilibrium value of Eq.(10.8), as we do in the majority of the carbon stars, rather than a high value characteristic for the earth and the atmospheres of most normal stars such as the sun and the normal red giants, we may confidently conclude that here we have direct evidence of the workings of the carbon cycle. Similar equilibrium considerations hold also for the proton-proton reaction. The only modification arises from the fact that two reactions of each of the first two steps of Eq.(10.1) are needed for one reaction of the last step. Hence, in equilibrium the reaction rates of the first two steps must be equal to each other and must be twice as large as the rate of the third step. From these conditions we obtain for the two relevant abundance ratios

$$X\ (D^2)/X\ (H^1) = 3 \times 10^{-17}$$
$$X\ (He^3)/X\ (H^1) \approx 10^{-4} \qquad (10.9)$$

These numerical values were again computed for a temperature of thirteen million degrees. The abundance ratio of He^3 to H depends sensitively on the temperature; for eight million degrees it is about one hundred times larger than for thirteen million degrees (its value for thirteen million degrees has already been used above in the computation of the mean reaction time).

Rate of Energy Generation by Proton-Proton Reaction and Carbon Cycle

Now we are ready to use Eq.(10.6) for determining the rate of energy generation by the proton-proton reaction and by the carbon cycle. Since the reaction rates are equal for the six steps of the carbon cycle, we may single out one of these six reactions, compute for it the rate per cubic centimeter from Eq.(10.6), multiply this rate by the total energy liberated per cycle according to Eq.(10.4), and thus obtain the rate of energy generation per cubic centimeter, i.e.

$$\varepsilon\rho = rE. \tag{10.10}$$

The same equation holds for the proton-proton reaction if we use for the rate half the reaction rate of the first or second step in Eq.(10.1) and if we use for the total liberated energy the value given by Eq.(10.2).

Which particular reaction shall we single out from the proton-proton process and from the carbon cycle for the determination of the rates? For the proton-proton reaction we shall obviously single out the first step since this permits us to express the energy generation in terms of no other abundance but that of hydrogen itself. In the carbon cycle we shall single out the fourth step—again the slowest—since its rate depends on the abundance of N^{14}, the most abundant among the isotopes involved. In fact, according to the equilibrium abundance ratios (10.8), we do not commit any appreciable error if we set the abundance of N^{14} equal to the abundance of all carbons and nitrogens together, X_{CN}. This substitution has the following great advantage. The onset of the carbon cycle does change the abundances of the individual carbon and nitrogen isotopes. It does not, however, change the total amount of carbon and nitrogen combined. Hence, we may use spectroscopically determined values for X_{CN} with some justification, a procedure which would not be safe for the individual isotopes.

With these selections we obtain from Eqs.(10.6) and (10.10) our final formulae for the energy production by the proton-proton reaction and by the carbon cycle

$$\varepsilon_{pp} = 2.5 \times 10^6 \times \rho X^2 \times \left(\frac{10^6}{T}\right)^{\frac{2}{3}} \times e^{-33.8\left(\frac{10^6}{T}\right)^{\frac{1}{3}}} \tag{10.11}$$

$$\varepsilon_{cc} = 9.5 \times 10^{28} \times \rho X X_{CN} \times \left(\frac{10^6}{T}\right)^{\frac{2}{3}} \times e^{-152.3\left(\frac{10^6}{T}\right)^{\frac{1}{3}}} \tag{10.12}$$

These equations give the rate of energy generation as a function of temperature, density and composition. The uncertainty in Eq.(10.11) caused by the uncertainty in the nuclear transmutation probability of a proton amounts to about 25 percent at present while that of Eq.(10.12), caused by the uncertainty in the nuclear transmutation probability of N^{14}, still amounts to about a factor 2.

The energy generation according to Eqs.(10.11) and (10.12) is represented as a function of temperature in Fig. 10.1. The graph shows that the proton-proton reaction dominates the carbon cycle at the lower temperatures while the reverse is true at the higher temperatures. That this should be so follows directly from the much higher Coulomb barriers acting in the carbon cycle compared with those of the proton-proton reaction. The two energy generation rates cross at a temperature somewhere between fifteen and twenty million degrees; the exact crossing temperature depends slightly on the carbon-nitrogen abundance. In this transition range the total energy generation is given by the sum of the two contributions. Thus, the proton-proton reaction and the carbon cycle together present the energy generation by hydrogen burning as one continuous function of temperature and density.

Fig. 10.1 permits us to estimate right now which stars have their main energy source provided by the proton-proton reaction, and which by the carbon cycle. The average rate of energy generation throughout a star is given by L/M. The rate in the central region in which the energy generation actually occurs must be higher than the average rate by an appreciable factor, say 30. We may thus compute an approximate value of the

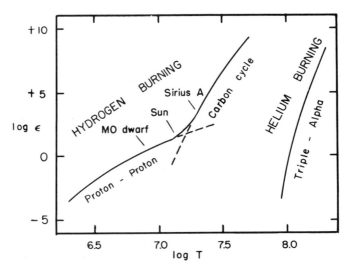

Fig. 10.1. Nuclear energy generation as a function of temperature (with $\rho X^2 = 100$ and $X_{CN} = 0.005X$ for the proton-proton reaction and the carbon cycle, but $\rho^2 Y^3 = 10^8$ for the triple-alpha process).

rate of energy generation in the central region of any star for which L and M are known. Such values are indicated in Fig. 10.1 for three representative stars. We may estimate from these indications that the carbon cycle is dominant in stars as bright or brighter than Sirius A, but that the proton-proton reaction is dominant in stars as faint or fainter than the sun. We shall see these estimates confirmed by the detailed models of §§15 to 25.

Following are two more minor modifications of the basic Eqs.(10.11) and (10.12). First, to relate the carbon-nitrogen abundance to the abundance of all the heavier elements, Z, we will frequently use the relation

$$X_{CN} = \frac{1}{3} Z .$$ (10.13)

This equation, based on earlier spectroscopic data, differs somewhat from Eq.(4.2), which we deduced from the most recent data; but this difference can hardly cause serious errors. Second, we may substitute a simple exponential form for the somewhat complicated temperature-dependent factors of Eqs.(10.11) and (10.12). This will be sufficiently accurate, in most cases, since in any one star the temperature range in which the bulk of the nuclear generation occurs is rather small. We may therefore replace the accurate equations by the interpolation formulae

$$\varepsilon_{pp} = \varepsilon_1 \times \rho X^2 \times \left(\frac{T}{10^6}\right)^\nu$$ (10.14)

$$\varepsilon_{cc} = \varepsilon_1 \times \rho X X_{CN} \times \left(\frac{T}{10^6}\right)^\nu$$ (10.15)

Appropriate values for the coefficients and the exponents in these interpolation formulae are given in Table 10.1 for various temperature ranges.

The Triple-Alpha Process

We leave now the temperature range of around ten million degrees, enter the next higher range of around one hundred million degrees, and

TABLE 10.1

Constants for interpolation equations (10.14) and (10.15) for various temperature ranges. (Bosman-Crespin, Fowler, Humblet, *Bull. Soc. Royale Sciences Liège*, No. 9–10, 327, 1954.)

	ε_{pp}			ε_{cc}	
$T/10^6$	$\log \varepsilon_1$	ν	$T/10^6$	$\log \varepsilon_1$	ν
4–6	−6.84	6	12–16	−22.2	20
6–10	−6.04	5	16–24	−19.8	18
9–13	−5.56	4.5	21–31	−17.1	16
11–17	−5.02	4	24–36	−15.6	15
16–24	−4.40	3.5	36–50	−12.5	13

turn to the transmutation of helium into carbon by the triple-alpha process. This process consists of the two reactions

$$He^4 + He^4 \rightarrow Be^8 + \gamma \quad - 95 \text{ kev}$$
$$Be^8 + He^4 \rightarrow C^{12} + \gamma \quad + 7.4 \text{ Mev}$$

(10.16)

In the first step the collision between two alpha particles leads to the formation of a Be^8 nucleus. This nucleus is not stable, however; the reaction is endothermic—in contrast to all other reactions discussed so far—and requires 95 kev. The unstable Be^8 nucleus will break down again into two alpha particles, which is the reverse process to the first reaction of Eq.(10.16). These continuous build-up and break-down processes will establish thermal equilibrium between the alpha particles and the Be^8 particles. Hence, we may use the usual Saha equation to determine the number of Be^8 particles per cubic centimeter, for which we find

$$N(Be^8) \propto N^2(He^4) \times \frac{1}{T^{\frac{3}{2}}} \times e^{-\chi_B/kT}$$

Here χ_B stands for the negative binding energy of Be^8, which amounts to 95 kev as we have already mentioned. Since this binding energy is about ten times larger than the average thermal energy of a particle at one hundred million degrees, the value of the exponent in the Saha equation is rather large and consequently the number of Be^8 particles will be quite small—about one in ten billion alpha particles. This small amount of the unstable beryllium isotope is nevertheless sufficient to permit the second step of the triple-alpha process to proceed at a sufficient rate.

This second step, according to Eq.(10.16), consists of a collision between a Be^8 particle and a third alpha particle, which leads to the formation of C^{12}. Again the rate of this process can be found with the help of Eq.(10.5)—with one important change, however. The carbon-forming reaction which we are just considering has a strong resonance at an energy of

$$\chi_R = 310 \text{ kev}$$

This means that the nuclear transmutation probability is not approximately constant, as it is for the other reactions we have discussed, but rather has a sharp and high maximum for particle velocities corresponding to the resonance energy. The main contribution to the integral of Eq.(10.5) arises therefore from a narrow range of velocities around the resonance velocity. In this narrow range we can in good approximation treat all factors of the integrand in Eq.(10.5) as constant except the factor P_N, which integrated over the critical velocity range gives a constant characteristic for the particular reaction. We note that the velocity distribution function D, though independent of the particle velocity over the critical range, depends on the temperature as follows

$$D\,(T, v) \propto \frac{\chi_R}{T^{\frac{3}{2}}} \times e^{-\chi_R/kT}$$

If we introduce into Eq.(10.5) this expression for D the above Saha equation for the number of Be^8 particles and also our usual designation for the number of alpha particles

$$N\,(He^4) \propto \rho\,Y$$

we obtain

$$r = C \times \rho^3\,Y^3 \times \frac{1}{T^3} \times e^{-\dfrac{\chi_B + \chi_R}{kT}} \tag{10.17}$$

This equation gives the reaction rate, that is the number of reactions per cubic centimeter per second, for the formation of carbon as a function of temperature, density, and helium abundance. The constant C for this reaction, which depends mainly on the width of the determining resonance, can be estimated but is still uncertain by easily a factor 10.

Before we apply Eq.(10.10) to determine the rate of energy generation from the triple-alpha process, we note that the total amount of energy liberated by the formation of a C^{12} particle according to Eq.(10.16) amounts to

$$E_{3\alpha} = 7.3 \text{ Mev} = 1.17 \times 10^{-5} \text{ erg} \tag{10.18}$$

With this value and with Eq.(10.17) we obtain from Eq.(10.10)

$$\varepsilon_{3\alpha} = 2 \times 10^{17} \times \rho^2\,Y^3 \times \left(\frac{10^6}{T}\right)^3 \times e^{-4670\frac{10^6}{T}} \tag{10.19}$$

This formula gives the rate of energy generation from the transmutation of helium into carbon, which is the second major source of stellar energy. This energy source is represented in Fig. 10.1 by the curve on the right.

For convenience in subsequent computations we may again represent the basic Eq.(10.19) by a simple exponential interpolation formula. Thus, we obtain for a temperature range around 140 million degrees.

$$\varepsilon_{3\alpha} = 10^{-8} \times \rho^2\,Y^3 \times \left(\frac{T}{10^8}\right)^{+30} \tag{10.20}$$

Rate of Change of Hydrogen and Helium Abundances

We have completed the consideration of nuclear reactions as far as energy generation is concerned. We still have to consider their equally important effects on the chemical composition. The principal effects of the nuclear reactions on the composition are, of course, the reduction of the hydrogen abundance X by the proton-proton reaction and the carbon cycle, the accompanying increase in the helium abundance Y, and finally the reduction of the helium abundance by the triple-alpha process. The corresponding rates of change of the hydrogen and helium abundances can be expressed in the simplest form if they are related to the rates of energy generation.

In Eqs.(10.2), (10.4), and (10.18) are given the amounts of energy released per particle formed in each of the three major processes. If we divide each of these values by the weight of the particle formed, we obtain the energy released per gram of transmuted matter. Thus, we have

$$E^*_{pp} = \frac{E_{pp}}{4H} = 6.3 \times 10^{18} \frac{erg}{g}$$

$$E^*_{cc} = \frac{E_{cc}}{4H} = 6.0 \times 10^{18} \frac{erg}{g} \qquad (10.21)$$

$$E^*_{3\alpha} = \frac{E_{3\alpha}}{12H} = 6 \times 10^{17} \frac{erg}{g}$$

Now the rate of change of a particular abundance by a particular process is obtained simply by dividing the rate of energy production of that process by the amount of energy released per gram. Hence, the rates of change of the hydrogen and helium abundances are given by

$$\frac{dX}{dt} = -\frac{\varepsilon_{pp}}{E^*_{pp}} - \frac{\varepsilon_{cc}}{E^*_{cc}} \qquad (10.22)$$

$$\frac{dY}{dt} = +\frac{\varepsilon_{pp}}{E^*_{pp}} + \frac{\varepsilon_{cc}}{E^*_{cc}} - \frac{\varepsilon_{3\alpha}}{E_{3\alpha}} \qquad (10.23)$$

Eqs.(10.22) and (11.23)—although as yet not often applied in their explicit form—contain the core of stellar evolution: the changes of the hydrogen and helium abundances produced by the nuclear reactions within a star, particularly within its core, are the main cause for the variations of the luminosity and radius of a star during its evolution.

Transmutations of the Heavier Elements

We conclude this section by a brief description of two groups of nuclear processes which neither contribute noticeably to the energy production nor affect seriously the hydrogen and helium abundances, but which do affect the abundances of the heavier elements and thus may provide us with critical clues. The first group involves only elements lighter than carbon and comprises the following processes

$$
\begin{aligned}
D^2 + H^1 &\rightarrow He^3 & (T_{cr} = 0.54 \times 10^6) \\
Li^6 + H^1 &\rightarrow He^3 + He^4 & (T_{cr} = 2.0 \times 10^6) \\
Li^7 + H^1 &\rightarrow 2\,He^4 & (T_{cr} = 2.4 \times 10^6) \\
Be^9 + 2\,H^1 &\rightarrow He^3 + 2\,He^4 & (T_{cr} = 3.2 \times 10^6) \\
B^{10} + 2\,H^1 &\rightarrow 3\,He^4 & (T_{cr} = 4.9 \times 10^6) \\
B^{11} + H^1 &\rightarrow 3\,He^4 & (T_{cr} = 4.7 \times 10^6)
\end{aligned}
\tag{10.24}
$$

Several of the processes given in Eq.(10.24) do not consist of a single nuclear reaction but rather of a group of them for which we have given only the net result. All these processes turn out to have very large nuclear transmutation probabilities so that they occur at appreciable rates in the low temperature range around one million degrees. In the parentheses of Eq.(10.24) we have given for each process the critical temperature at which the mean reaction time reaches the value of five billion years (the critical temperatures depend but slightly on the density; the values here given are based on the relevant densities in the solar envelope).

We conclude from the numerical values listed in Eq.(10.24) that deuterium, lithium, beryllium, and boron have long been burned up in the interior of stars like the sun and can exist now only in shallow surface layers where the temperature is lower than the respective critical temperatures. But even in the surface layers these light elements may have disappeared if the surface layers are mixed by convection with deeper layers hotter than the critical temperatures. Now, spectroscopic observations show that in the solar atmosphere lithium—though of an extremely low abundance even compared with the earth's crust—as well as beryllium do exist. We have to conclude that the solar atmosphere cannot be mixed by convection with those deeper layers in which the temperature exceeds 2.5 million degrees. We shall make use of this essential clue when we construct a model for the sun in §23.

The second group of processes still to be mentioned occur in the high temperature range around one hundred million degrees. They are processes which are capable of building up really heavy elements. We have already seen how the triple-alpha process (10.16) builds up carbon. It may be followed by subsequent captures of additional alpha particles leading to the formation of O^{16}, then to Ne^{20}, and so on. From the point of view of energy production these processes following the triple-alpha process are not very important since they would amount at most to a 50 percent addition to Eq.(10.19). But they may build up heavier elements, possibly up to Ca^{40}. Because of the higher Coulomb barrier of the heavier

elements, their build-up will occur, of course, only at temperatures somewhat higher than that required by the triple-alpha process.

A quite different chain of reactions may, however, lead to the build-up
of really heavy elements at temperatures even a little lower than the
critical temperature for the triple-alpha process. The starting reaction is

$$C^{13} + He^4 \rightarrow O^{16} + n \quad (T_{cr} \approx 10^8) \tag{10.25}$$

The importance of this alpha particle captured by a heavy carbon isotope
lies not so much in the formation of oxygen but rather in the production
of free neutrons. These neutrons are not handicapped by any Coulomb
barriers. They may be captured by any atom, with a special preference
for the heavier atoms because of their larger cross sections. Thus a
continuous build-up of the heavy elements appears possible, limited only
by the supply of C^{13} which, as we have seen, occurs in an appreciable
abundance in the equilibrium established by the carbon cycle. It does
not appear impossible that this will build up elements as heavy as those
occurring with exceptionally high abundance in the S-stars. If this is
true, we may take the S-star phenomenon as a clue for stars which have
reached central temperatures as high as one hundred million degrees or
higher.

Summary

Our nuclear cookbook is turning out somewhat complex. In the future
the basic principles may become more unified and simpler, but the particular data needed for the theory of stellar evolution and of the formation of the elements is bound to become even more manifold and complex.
For the time being let us summarize our various recipes as follows.

In the low temperature range around one million degrees no major
energy sources are available. Deuterium, lithium, beryllium, and boron
all burn up.

In the medium temperature range around ten million degrees hydrogen
burns into helium either by the proton-proton reaction or by the carbon
cycle. This provides the greatest energy source for stars.

In the high temperature range around one hundred million degrees helium
burns into carbon by the triple-alpha process, which provides the second
major stellar energy source. Heavier elements can be produced by a
succession of further alpha particle captures. The alpha particle capture
by the heavy carbon isotope produces free neutrons, and successive
captures of these neutrons may lead to the formation of very heavy
elements.

The extreme temperature range around one billion degrees we exclude
from this book.

11. Surface Layers

In the preceding six sections we have assembled the broad physical conditions which must hold throughout the stellar interior. We still have to clear up the one untidy section of a star, its surface layers.

The Zero Boundary Conditions

The temperature at the surface of most stars is in the order of a few thousand degrees. This is about one thousandth of the internal temperatures. An even smaller ratio is found for the pressures in the atmospheres of stars compared with the internal pressures. It is tempting therefore to neglect the surface temperatures and surface pressures entirely and to accept the following boundary conditions for the theory of the stellar interior.

$$\text{At } r = R: \quad T = 0, \, P = 0. \tag{11.1}$$

Obviously these "zero boundary conditions" will not lead to a good representation of the surface layers. Will they lead to a good representation of the interior from the center to where the temperature has dropped to, say, three hundred thousand degrees? To answer this question we have first to formulate the accurate physical boundary conditions and next to investigate how sensitive the solution in the interior is to changes in the boundary values.

The Physical Boundary Conditions

The first surface condition, that for the temperature, is easily formulated precisely. Let us define the "surface" of the star as that layer in which the actual temperature T_R is equal to the effective temperature T_e. According to the definition of the effective temperature in terms of the luminosity and the radius of the star, we have then

$$L = 4\pi R^2 \times \frac{ac}{4} \, T_R{}^4. \tag{11.2}$$

The second surface condition, that for the pressure, follows from the fact that the atmospheric layers above the "surface" must have a certain optical thickness; an excessive transparency would cause too large a radiation from the surface while too high an opacity would cause too small a radiation. This condition on the optical depth of the atmosphere may be written as

$$\tau_R = \int_R^\infty \varkappa \rho \, dr = \frac{2}{3}$$

where the numerical value is taken from the theory of radiation transfer in stellar atmospheres, in its roughest approximation.

Generally, the absorption coefficient will be a function of height in the atmosphere. If we here permit ourselves the approximation of setting the absorption coefficient equal to its value at the surface, we may take $\varkappa$ outside the above integral and thus express the optical depth at the surface by the product of the absorption coefficient times the mass of the atmosphere above the "surface" per square centimeter. On the other hand, the mass of the atmosphere is related to the pressure by the condition that the pressure at the surface must be equal to the gravitational acceleration multiplied by the mass of the atmosphere per square centimeter, i.e.

$$P_R = \frac{GM}{R^2} \int_R^\infty \rho \, dr.$$

If we now eliminate the mass of the atmosphere between these two equations, we obtain

$$\varkappa_R \, P_R = \frac{2}{3} \frac{GM}{R^2} \tag{11.3}$$

This equation is not exact because of the approximate value used for the optical depth at the surface and because of the neglect of the variation of absorption coefficient with height. However, by comparing this relation with the results of accurate computations for stellar atmospheres one finds that Eq. (11.3) appears to hold rather generally with the error not exceeding a factor 1.5—an accuracy sufficient for our present purpose. Since the atmospheric absorption coefficient can be considered a known function of pressure, temperature, and composition, we can take Eq. (11.3) as the looked-for condition for the surface pressure.

The Pressure-Temperature Relation in Stellar Envelopes

Eqs. (11.2) and (11.3) are the physical boundary conditions which determine the surface temperature and the surface pressure for a star of given mass, luminosity, radius, and atmospheric composition. Now we have to investigate how seriously the solution for the interior is affected if these proper boundary conditions are replaced by the zero conditions (11.1).

One might fear that for this purpose we will get involved in the full set of equilibrium equations assembled in the preceding sections. Lucky circumstances, however, permit us to study the relationship between the pressure and the temperature throughout the envelope of a star without simultaneously considering all the other factors which determine the complete structure of the envelope. If we divide the hydrostatic equilibrium condition (5.2) for the pressure gradient by the radiative equilibrium condition (6.12) for the temperature gradient and if we represent the absorption coefficient for bound-free and free-free transitions, according to Eqs. (9.16) and (9.18), by

$$\varkappa = \varkappa_0 \rho \, T^{-3.5}$$

We obtain the differential equation

$$\frac{dP}{dT} = \frac{4ac}{3} \times \frac{4\pi GM}{\varkappa_0 L} \times \frac{T^{6.5}}{\rho}$$

The distance from the center, r, does not appear in this equation. Furthermore, we have replaced M_r by its surface value M since in the envelope where the density is rather low M_r is substantially constant, according to Eq. (5.1). Similarly, we have replaced L_r by its surface value L since in the envelope the temperatures are too low for nuclear energy generation to occur so that according to Eq. (5.8), L_r is constant. Thus, the only variables occuring in this differential equation—after the density has been eliminated with the help of the equation of state (8.2) for an ideal gas—are the pressure and the temperature. The solution of the differential equation is

$$P^2 = \frac{2}{8.5} \frac{4ac}{3} \frac{k}{H} \frac{4\pi GM}{\mu \varkappa_0 L} \times T^{8.5} + C \tag{11.4}$$

where C is the integration constant.

Eq. (11.4) represents the general relation between pressure and temperature in stellar envelopes in which radiative equilibrium holds and in which the absorption coefficient is given by Kramers' law. For the particular solution which fulfills the zero conditions (11.1) at the surface, C must be zero and therefore

$$P = \left(\frac{2}{8.5} \frac{4ac}{3} \frac{k}{H} \frac{4\pi GM}{\mu \varkappa_0 L} \right)^{\frac{1}{2}} \times T^{4.25} \tag{11.5}$$

This radiative zero solution is shown in the temperature-density diagram of Fig. 11.1, with numerical values chosen roughly appropriate for the sun.

Let us consider the general solution (11.4) first for negative values of C and then for positive values. If C has a negative value, the temperature approaches a finite limiting value as the pressure goes to zero. These solutions have the approximately isothermal character shown by the four samples at the left in Fig. 11.1. They all converge extremely strongly toward the radiative zero solution. This is caused by the high exponent of T occurring in Eq. (11.4); following a solution inwards, we find that a small rise in the temperature causes a big rise in the temperature-dependent term on the righthand side of (11.4), which makes the integration constant C negligible. Hence the solution, after a small rise in the temperature, is indistinguishable from the radiative zero solution for which C is strictly zero. This strong convergence has the happy consequence that two solutions fulfilling quite different surface conditions approach rapidly the same solution for the interior.

A similar convergence holds to some extent for the solutions with positive values of C. In this case the pressure approaches a limiting finite value as the temperature goes to zero. However, long before this limiting status is reached these radiative solutions have to be terminated because they become convectively unstable. For any particular solution (11.4) with a definite value of C it is easy to determine the termination point according to the stability condition (7.2). Fig. 11.1 shows four such radiative solutions with their termination points; they fall very close to the radiative zero solution.

To follow these solutions further out we must use the adiabatic relation (7.9) which is appropriate for convective equilibrium. If we consider for the moment only temperatures above, say, fifty thousand degrees where

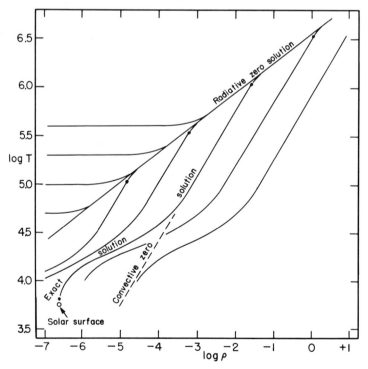

Fig. 11.1. Temperature-density diagram for envelope of sun (left lower quadrant of Fig. 8.1).

hydrogen and helium are practically completely ionized, we may use for γ the constant value 5/3. This permits the integration of the adiabatic equation (7.9) with the result

$$P \propto T^{\frac{\gamma}{\gamma-1}} = T^{\frac{5}{2}}$$

or with the help of the equation of state (8.2)

$$\rho \propto T^{\frac{1}{\gamma-1}} = T^{\frac{3}{2}} \tag{11.6}$$

With these relations we may continue the solutions beyond the termination points as shown in Fig. 11.1 by the set of five steep straight lines. Altogether therefore we find for temperatures above fifty thousand degrees a strong convergence of all possible solutions (11.4) to the one particular radiative zero solution (11.5)

The Ionization Zone of Hydrogen and Helium

Unhappily, this convenient convergence does not hold at lower temperatures—and obviously we have to concern ourselves with these lower temperatures since most stars have surface temperatures far below fifty thousand degrees. At temperatures around twenty thousand degrees hydrogen is partially ionized, and the same is true for helium at somewhat higher temperatures, as shown by Fig. 8.1. Under these conditions the ionization energy must be taken into account in the derivation of the adiabatic equation. If this is done, one finds for the adiabatic temperature-density relation the form shown by the lower portions of the five adiabats of Fig. 11.1. The shallow slopes of these curves reflect the fact that in an adiabatic compression of a partially ionized gas the temperature rises relatively slowly since most of the compression energy goes into the increase of the ionization. Fig. 11.1 clearly shows that the peculiar character of the adiabatic relation in the ionization zones of hydrogen and helium badly delays the otherwise rapid convergence of the entire family of solutions to the one radiative zero solution.

Consider the case of the sun. The physical boundary conditions (11.2) and (11.3) give for the sun the surface temperature and the surface density represented by the circle in the lower left corner of Fig. 11.1. Proceeding inwards in the solar atmosphere, one finds that after a short rise of the temperature—marked by the dot in the figure—radiative equilibrium becomes unstable and convection commences. Following from this point inwards along the appropriate adiabatic relation we find that the solution becomes radiatively stable again and converges to the radiative zero solution, but only at a temperature of about one million degrees.

It is clear that this exact solution for the structure of the solar envelope is complicated. One additional serious complication has still to be mentioned. At the point just below the surface where radiative equilibrium becomes unstable, the temperature and density are still so low that our order-of-magnitude estimates at the end of §7 do not apply at all. In fact, convection there is not a very efficient mechanism for the energy transport and in consequence the excess of the actual temperature gradient over the adiabatic value will not be negligible. Therefore, one has to use in this layer the explicit Eq. (7.7) for the convective heat transport. This involves us directly with the poorly known average mixing length l. It appears at present fairly certain that the mixing length in the outer convection zone of a star is determined by the scale height, that is by the distance over which the pressure falls by a factor e. But it has not yet been possible to determine the exact numerical factor which relates the mixing length and the scale height. This uncertainty in one of our basic physical relations produces an uncomfortable uncertainty in the present theoretical models of the envelopes of many stars.

The Practical Boundary Conditions

Under these circumstances, what recipes shall we apply for determining the appropriate—and practicable—boundary conditions? Our first recipe refers to stars of spectral type earlier than, say, A5. From the theory of stellar atmospheres it appears fairly likely that stars of early spectral type have either no outer convection zone or only such a shallow one that it is of practically no consequence to the structure of the interior. We may therefore bank for these spectral types on the fine convergence of all the radiative solutions (11.4) to the one radiative zero solution (11.5). Hence, our recipe for the early spectral types is: use the zero boundary conditions (11.1) and assume that the stellar envelope is in radiative equilibrium all the way to the edge.

For the later spectral types, from say G0 on, the theory of stellar atmospheres makes it appear likely that deep convection zones are the rule. In these cases we can clearly not use the radiative zero solution. This does not mean, however, that we have to take into account the full complexity of the exact solution for the outermost layers; an intermediate way appears most effective at present. This intermediate way works as follows. Compare the exact solution—for example for the sun as given in Fig. 11.1—with the "convective zero solution," which is the straight extrapolation of the simple adiabatic relation (11.6) all the way out to the surface. You find that the exact solution and the convective zero solution differ from each other appreciably only at temperatures lower than fifty thousand degrees. Since errors restricted to such very low temperatures will have virtually no consequences for our results for the stellar interior, we may use the simple adiabatic relation (11.6) throughout the convection zone out to the very edge. This gives us

$$P = KT^{\frac{5}{2}} \tag{11.7}$$

The constant K of this relation must in principle always be determined by the theory of the stellar atmosphere. In practice, however, we will consider K a free parameter for which the theory of stellar atmospheres provides only approximate values.

Our second recipe, applicable to the later spectral types, is then: apply again the zero condition (11.1) at the surface, but assume that the stellar envelope is in simple convective equilibrium so that Eq. (11.7) holds. For the parameter K use values of the order indicated by the theory of stellar atmospheres—that is, if you cannot determine more accurate values directly by considerations of the stellar interior alone.

We have tidied up the last complication we had left, the star's surface layers. We are now ready to look at the problem presented by all the physical conditions as a whole and to develop the mathematical techniques necessary for its solution.

CHAPTER III
MATHEMATICAL TECHNIQUES

12. The Over-all Problem

The physical relations and processes which we have assembled in the preceding chapter pose the following over-all problem for the theory of the stellar interior. The four basic equilibrium conditions—the hydrostatic equilibrium conditions (5.1) and (5.2), the thermal equilibrium condition (5.8), and the energy transport condition (6.12) or (7.9)—give the four differential equations

$$\frac{dP}{dr} = - \rho \frac{GM_r}{r^2} \tag{12.1}$$

$$\frac{dM_r}{dr} = 4\pi r^2 \rho \tag{12.2}$$

$$\frac{dL_r}{dr} = 4\pi r^2 \rho \varepsilon \tag{12.3}$$

$$\left.\begin{array}{l} \dfrac{dT}{dr} = -\dfrac{3}{4ac}\dfrac{\varkappa\rho}{T^3}\dfrac{L_r}{4\pi r^2} \text{ (radiative)} \\[3mm] \dfrac{dT}{dr} = \left(1-\dfrac{1}{\gamma}\right)\dfrac{T}{P}\dfrac{dP}{dr} \text{ (convective)} \end{array}\right\} \tag{12.4}$$

The last of these differential equations has two alternative forms. Which of these forms is applicable under any given circumstances is uniquely determined by the stability condition (7.2). (The case of energy transport by electron conduction in a degenerate gas is here not listed separately since it can be expressed in the form of the first of Eqs.[12.4] if the absorption coefficient is given the equivalent, very low values.)

In addition to these differential equations which characterize general conditions we have three explicit relations which characterize more specifically the behavior of the interior gases. They are the equation of state, the equation for the absorption coefficient, and the equation for energy generation by nuclear processes. We may represent them formally by

$$\begin{array}{l} P = P\,(\rho,\,T,\,X,\,Y) \\[2mm] \varkappa = \varkappa\,(\rho,\,T,\,X,\,Y) \\[2mm] \varepsilon = \varepsilon\,(\rho,\,T,\,X,\,Y) \end{array} \tag{12.5}$$

96

The specific forms which these relations for the gas characteristics take under various circumstances have been discussed in §§8, 9, and 10 respectively. All three of these "gas characteristics relations" depend directly on the composition, which we represent here by the hydrogen abundance X and the helium abundance Y. We do this with the understanding that more composition parameters have to be introduced in those special cases in which serious deviations from the normal proportions among the heavier elements need be considered.

Eqs.(12.1) to (12.5) must all be fulfilled in every layer of the star. We have still to add the particular conditions for the boundaries. For the center must hold

$$\text{at } r = 0: \quad M_r = 0, \quad L_r = 0, \tag{12.6}$$

which follows directly from the definitions of M_r and L_r. At the surface holds, according to §11,

$$\text{at } r = R: \quad \text{either } T = 0, \quad P = 0$$
$$\text{or } T = 0, \quad P = K T^{\frac{5}{2}} \tag{12.7}$$
$$\text{with } K = K\ (M, L, R, X, Y).$$

The first of these two alternative surface conditions applies for stars with radiative envelopes and the second for stars with convective envelopes. Which of the two alternatives applies to a given star is uniquely determinable—at least in principle—from mass, luminosity, radius, and composition from the theory of stellar atmospheres. The same is true of the parameter K, which occurs in the second alternative.

Uniqueness of the Solution

Eqs.(12.1) to (12.7) comprise the basic problem of the stellar interior. What is its mathematical character? If we use the gas characteristics relations (12.5) to eliminate ρ, $\varkappa$, and ε from the differential equations (12.1) to (12.4), these four equations contain, in addition to well-known constants, nothing but the five variables P, M_r, L_r, T, and r. Thus our basic problem consists of four simultaneous, total, non-linear first order differential equations for four variables, all functions of the fifth variable. These four differential equations, together with the four boundary conditions (12.6) and (12.7), two at the center and two at the surface, represent a typical, well-defined boundary value problem. If we remember that all the five variables, according to their physical definitions, must be everywhere positive, and if we do not worry about unlikely cases of mathematical degeneracy, we conclude that our basic problem has a unique solution.

The physical meaning of this conclusion is most easily formulated if we first perform a simple transformation of the equations. Let us change the independent variable from r to M_r. The corresponding changes in the four differential equations are achieved by replacing Eq.(12.2) by its reciprocal and by dividing Eqs.(12.1), (12.3), and (12.4) each by Eq.(12.2). The three relations (12.5) are not at all affected by the change in the independent variable. In the boundary condition at the center, (12.6), the two statements $r = 0$ and $M_r = 0$ exchange their logical roles; the latter now defines the location of this boundary while the former becomes the first of the two boundary conditions. The boundary condition at the surface (12.7) remains unchanged except that the location of this boundary is now defined by $M_r = M$ instead of by $r = R$. Thus the character of the boundary value problem is not changed by this switch in the independent variable.

The uniqueness of its solution now implies that, given the total mass M (needed to fix the location of the surface condition [12.7]) and given the composition parameters X and Y (needed for the relations [12.5]), then the pressure P, the distance from the center r, the flux L_r, and the temperature T are completely determined as functions of M_r throughout the star. In other words, a star of a given mass and composition has only one equilibrium configuration in which its physical state and structure are completely fixed.

From the point of view of the star the "given quantities," which it cannot change at will, are obviously the total mass and the composition. From the point of view of the astronomer, however, the "given quantities" are those obtainable from observations. For many stars, for example, the most accurately observed quantities are the luminosity and the radius. Since, according to the above discussion, L and R—just as all other features of the stellar structure—are uniquely determined by M, X, and Y, there must be two relations connecting L and R with M, X, and Y. In the subsequent chapters we will frequently derive explicit forms for these two key relations and use them for a variety of purposes. In whatever form these two key relations appear they are always a specific consequence of the fact that the equilibrium configuration of a star is unique.

Inhomogeneities in Composition and
Evolutionary Model Sequences

Thus far we have discussed the over-all problem represented by Eqs.(12.1) to (12.7) under the implicit assumption that the composition of the star does not vary from layer to layer throughout the star, that is that the star is homogeneous in composition. We know of no physical reason, however, to believe that this is necessarily so. On the contrary, we have good reasons to believe that inhomogeneities in composition play a decisive role in the evolution of a star. We should therefore consider X and Y not as constants for a star but as continuous functions of M_r from the center to the surface. If we reconsider the over-all problem under this more general aspect for the composition, we find that the character of the problem is not at all changed. The equilibrium configuration of a star is still completely determined if its total mass and composition are given, where we now understand under "given composition" not just the over-all composition but the complete, detailed run of composition from the center to the surface.

The appearance of the two new functions $X(M_r)$ and $Y(M_r)$ does, however, introduce a new feature. These two functions are clearly not determined for a particular evolutionary state of a star by the equilibrium conditions applied to that state, but rather by the action of the nuclear transmutations throughout the preceding evolutionary phases. Thus it is clear that it is not possible—except for the initial state of a star—to consider a particular evolutionary state by itself. Instead we are led to the necessity of deriving step by step the sequence of stellar models which describes the internal structure of the star as a function of time during its evolution.

How do we then proceed to develop such a model sequence? To start with we have to construct a model for the initial state, when the star has just finished its contraction out of interstellar matter and is beginning to burn its main nuclear fuel, hydrogen. For this state we shall assume that the star is homogeneous in composition; it seems certain that the turbulence in the interstellar medium will mix well the matter from which the star originates, and it seems reasonable to assume—as long as no evidence to the contrary has appeared—that the various elements are not separated from one another during the contraction to the initial state. Thus the over-all problem for the initial state takes the simple form in which we have first discussed it, with X and Y both constant throughout the star.

This happy situation cannot last long. The nuclear processes will transmute elements and thus change the composition. Since the nuclear reactions are much stronger in the hot inner core than in the cooler outer portions an inhomogeneous composition will result. This phenomenon complicates matters but it does not introduce anything indeterminate. With the help of the transmutation equations (10.22) and (10.23), which give the rates of change in the composition for every point in the star, we may compute the composition in a new state from the composition in a previously determined state (subscript p). If the time interval elapsed between the previous state and the new state is Δ, we get in first approximation

$$X = X_p + \Delta \times \left.\frac{dX}{dt}\right|_p , \ Y = Y_p + \Delta \times \left.\frac{dY}{dt}\right|_p . \tag{12.8}$$

We do not want to discuss here the mathematical inaccuracy of Eqs. (12.8); we may use all the usual methods for the numerical solution of first order total differential equations to improve these equations to any arbitrarily high accuracy. The essential physical point of Eqs.(12.8) is that they determine the new composition everywhere in the star on the basis of the solution of the over-all problem for a previous state in the evolution.

Two modifying conditions have to be observed in the application of Eqs.(12.8). First, we note that the transmutation equations (10.22) and (10.23) give the rates of change in composition—which are used in Eqs.(12.8)—in a fixed element of matter, i.e. for a fixed value of M_r, irrespective of whether or not this element changes its position r during the time interval considered. If we want to apply Eqs.(12.8) in their simplest form we must clearly use not r but M_r as the space-like independent variable.

Second, Eqs.(12.8) give the local change in composition only if there is no mixing with the neighboring layers. We have seen in §7, however, that in a convection zone mixing is very active, indeed practically instantaneous. This mixing can be taken into account, in the same first order approximation used for Eqs.(12.8), by first applying Eqs.(12.8) to every layer throughout the convection zone and then taking the straight averages of the X values and of the Y values over all these layers. These averages represent the new composition throughout the convection zone.

With these two modifications in mind, we can use Eqs.(10.22), (10.23), and (12.8) to determine the composition uniquely layer by layer throughout the star for one evolutionary state after we have solved the over-all problem for the previous state. Then with X and Y known as functions of M_r the solution of the over-all problem for the new state is again uniquely determined. Thus we see that the occurrence of inhomogeneities in composition during the evolution of a star, which would lead to serious arbitrariness were we to study individual evolutionary states in a disconnected manner, does not cause any arbitrariness at all if we investigate the evolutionary states of a star in the same sequence in which the star evolves through them.

The Over-all Problem for Fast Evolutionary Phases

There is still one question left: how can we handle the rare but important phases in which the evolution proceeds so fast that the thermal equilibrium condition (12.3) has to be replaced by the more general equation (5.10)?

The complicating feature of Eq.(5.10) is that it contains a term with a time derivative which represents the release of gravitational energy in a contraction. This term can be taken care of easily if we permit ourselves to represent the time derivative by the difference in the relevant quantity between the model under consideration and the previous model in the evolutionary sequence, i.e. if we use

$$\frac{d}{dt}\left(\frac{P}{\rho^{\frac{5}{3}}}\right) = \frac{1}{\Delta}\left[\left(\frac{P}{\rho^{\frac{5}{3}}}\right) - \left(\frac{P}{\rho^{\frac{5}{3}}}\right)_p\right]. \tag{12.9}$$

With this representation for the time derivative the basic equation (5.10) can be written in the form

$$\frac{dL_r}{dM_r} = \varepsilon - \frac{3}{2}\frac{\rho^{\frac{2}{3}}}{\Delta}\left[\left(\frac{P}{\rho^{\frac{5}{3}}}\right) - \left(\frac{P}{\rho^{\frac{5}{3}}}\right)_p\right]. \tag{12.10}$$

Eq.(12.10) has the same character as the thermal equilibrium condition (12.3) which it replaces; it is a first-order differential equation for the flux L_r.

On the right-hand side of Eq.(12.10) all the quantities depend on the new state with the exception of the term $(P/\rho^{5/3})_p$, which depends entirely on quantities of the previous state. Hence after the over-all problem for the previous state has been solved $(P/\rho^{5/3})_p$ can be computed as a function of M_r for all layers of the star. In this way $(P/\rho^{5/3})_p$ represents a known quantity in Eq.(12.10) and this equation can be used for the new state in exactly the manner in which the thermal equilibrium condition (12.3) had been used before.

We conclude that we can handle the unusually fast evolutionary phases with the help of Eq.(12.10) in exactly the same manner in which we have handled the normal evolutionary phases. In all cases a fourth-order boundary-value problem has to be solved at each state. Between states the change of composition has to be determined according to Eqs.(12.8) and, during the fast phases, $(P/\rho^{5/3})_p$ has to be computed from the previous state in preparation for Eq.(12.10).

We have now finished the review of the over-all problem and shall turn to the mathematical details in the following sections.

13. Transformations and Invariants

We have seen in the preceding section that the over-all problem of the stellar interior has the mathematical character of a boundary-value problem. How do we obtain the actual solution of this problem for any given evolutionary state of a given star? No analytical solutions are known for the basic differential Eqs.(12.1) to (12.4) for any physically significant case. Hence, we have to rely completely on numerical integrations.

Integrations from Center and from Surface;
Fitting Conditions

We may start an integration at one of the two boundaries (center or surface) and proceed step by step towards the other boundary. If we start at the center and integrate outwards, we encounter a bad divergence of the solutions as we approach the surface. We have already discussed in §11 this divergence in the outwards direction, corresponding to a convergence in the inward direction. It is caused essentially by the occurrence of the temperature, T—which becomes smaller and smaller as we approach the surface—in the denominator of the radiative Eq.(12.4). The divergence means that a very small change in the solution in the deep interior causes a very large change in the solution in the envelope. It is therefore not practicable to integrate from the center outwards to the very surface—at least not if the envelope is mostly in radiative equilibrium.

On the other hand, if we start at the surface and integrate inwards, we first encounter the good convergence of the solutions in a radiative envelope in this direction. But when we approach the center a divergence appears. It is caused here by the occurrence of r—which vanishes at the center—in the denominator of Eq.(12.1). Thus it is also impossible to integrate from the surface inwards all the way to the center.

In consequence, the numerical determination of a complete stellar model will have to consist of three parts: first, an integration from the center outwards to an appropriately chosen point; second, an integration from the surface inwards to the same point; and third, the fitting together of the two solutions.

Now let us consider the three parts of this procedure for a definite star of given mass and composition. For the outward integration starting at the center, we have the boundary conditions (12.6) which provide us with starting values for two of the four dependent variables. However, the starting values for the two other variables, P_c and T_c, are not determined by the boundary conditions at the center; they are therefore trial parameters. Hence, the integration work from the center outwards does not consist of just one definite solution but rather of a whole family of solutions depending on the values of the two trial parameters at the center.

A similar situation holds for the inward integrations from the surface. Here the boundary conditions (12.7) give us starting values for P and T. But the starting values L and R are not given by the boundary conditions at the surface; they therefore play the role of trial parameters, and the integration work from the surface inwards again consists not of one definite solution but of a whole family of solutions depending on the values of the two trial parameters at the surface.

To make it possible to fit the outward and inward integrations together, we will choose a convenient value of r and we will carry forward all integrations from both sides to this point. At this fitting point we must insure that all physical quantities go over continuously from the interior solution to the exterior solution. Thus we have the five fitting conditions

$$r_{if} = r_{ef}, \; P_{if} = P_{ef}, \; T_{if} = T_{ef}, \; M_{r\,if} = M_{r\,ef}, \; L_{r\,if} = L_{r\,ef} \quad (13.1)$$

Here the subscript "if" designates the fitting point approached from the interior while the subscript "ef" designates the fitting point approached from the exterior. The first of these fitting conditions has been automatically fulfilled by our choice of the same value of r for the termination of both the interior and the exterior integrations. But the four remaining fitting conditions (13.1) have to be explicitly fulfilled, by trial integrations for a variety of values of the four trial parameters P_c, T_c, L, and R, until the unique set of values for these four parameters is found which leads to the fulfillment of the fitting conditions (13.1).

It is clear from this general formulation of our problem that a rather substantial number of numerical trial integrations is necessary if one wants to obtain the equilibrium solution for a particular star in a particular evolution stage. If one wants to consider a variety of stars with different masses and compositions, the number of necessary integrations is multiplied by still another substantial factor. When all the numerical work is to be performed by a large electronic computing machine, the most efficient way to attack our problem may be by following the general formulation discussed above, in spite of the large number of integrations necessary. When, however, the numerical work has to be performed by more modest means—as has been the case for most of the investigations which we will review in the following chapters—it appears necessary for efficiency's sake to apply all methods which reduce the number of necessary integrations, even though the miscellany of these labor-saving methods spoils the simplicity of the general formulation. We shall show in this section how in many circumstances the number of necessary integrations can be greatly reduced by simple transformations of the five principle variables.

Standard Transformation

Let us replace the physical variables M_r, L_r, r, P, and T by the non-dimensional variables q, f, x, p, and t with the help of the following transformation

$$M_r = qM, \quad L_r = fL, \quad r = xR,$$

$$P = p \, \frac{GM^2}{4\pi R^4}, \quad T = t\mu_s \frac{H}{k} \frac{GM}{R}. \tag{13.2}$$

The symbol μ_s in the last of these equations represents the molecular weight of the standard composition; for a homogeneous star we will always use for this standard composition the actual composition of the star and for an inhomogeneous star we will usually use as the standard the composition of the outermost layers. The factors with which the non-dimensional variables are multiplied in Eqs.(13.2) are all constants for a particular star in a particular state of evolution.

The specific choice for the values of the five constant factors as given in Eqs.(13.2) has its advantages, as we will see. But certain

other choices have similar advantages. Thus, if we make the specific
choice of transformation (13.2) and if we use the five new variables
defined by these equations throughout this book as our basic non-dimensional variables, we run the risk of being accused of local prejudice and
narrow-mindedness. We hope, however, that the reader will agree, first,
that it is to his advantage if we use as much as possible a uniform set of
variables, and second, that one may earn the right of making one's own
choice of variables by presenting a substantial set of numerical integrations and stellar models in terms of the chosen variables.

Let us now introduce the transformation (13.2) into the over-all problem
formulated by Eqs.(12.1) to (12.7). Let us choose for the gas characteristics relations (12.5) the following specific forms: the equation of
state (8.2) for an ideal gas, Kramers' law of opacity (9.16) or (9.18) for
bound-free and free-free transitions, and the interpolation formulae
(10.14) or (10.15) for the proton-proton reaction and the carbon cycle.
Thus, we have

$$P = \frac{1}{\mu} \frac{k}{H} \rho \, T \quad \text{with} \quad \mu = l\mu_s$$

$$\varkappa = \varkappa_o \rho \, T^{-3.5} \quad \text{with} \quad \varkappa_o = \frac{j}{l} \varkappa_{os} \tag{13.3}$$

$$\varepsilon = \varepsilon_o \rho \, T^{\nu} \quad \text{with} \quad \varepsilon_o = \frac{i}{l} \varepsilon_{os}$$

The three constants μ_s, $\varkappa_{os}$, and ε_{os} refer again to the standard composition while the three functions l, j and i represent the variations
caused by inhomogeneities in the composition for the molecular weight,
the absorption coefficient, and the energy generation respectively.
According to these definitions we have the particular simple case

$$l = j = i = 1 \quad \text{for a homogeneous star.} \tag{13.4}$$

In any case, we consider l, j, and i as known for every layer in the star
considered.

If we now introduce the transformation (13.2) and the relations (13.3)
into the four differential Eqs.(12.1) to (12.4) we obtain

$$\frac{dp}{dx} = -l \times \frac{p}{t} \frac{q}{x^2} \tag{13.5}$$

$$\frac{dq}{dx} = +l \times \frac{p}{t} x^2 \tag{13.6}$$

$$\frac{dt}{dx} = -C \times jl \times \frac{p^2}{t^{8.5}} \frac{f}{x^2} \quad \text{or} \quad \frac{dt}{dx} = \frac{2}{5} \frac{t}{p} \frac{dp}{dx} \tag{13.7}$$

$$\frac{df}{dx} = +D \times il \times p^2 \, t^{\nu-2} \, x^2 \tag{13.8}$$

$$\text{with } C = \left[\frac{3}{4ac}\left(\frac{k}{HG}\right)^{7.5}\left(\frac{1}{4\pi}\right)^3\right]\left[\frac{\chi_{0\,s}}{\mu_s{}^{7.5}}\right]\left[\frac{LR^{0.5}}{M^{5.5}}\right] \qquad (13.9)$$

$$\text{and } D = \left[\left(\frac{HG}{k}\right)\nu\,\frac{1}{4\pi}\right]\left[\varepsilon_{0s}\mu{}^{\nu}_{\,s}\right]\left[\frac{M^{\nu+2}}{LR^{\nu+3}}\right] , \qquad (13.10)$$

and the boundary conditions (12.6) and (12.7) become

$$\text{At } x = 0: \quad q = 0, \, f = 0, \qquad\qquad\qquad (13.11)$$

$$\text{At } x = 1: \quad q = 1, \, f = 1, \, t = 0, \, p = 0$$

$$\text{or} \quad q = 1, \, f = 1, \, t = 0, \, p = E \, t^{\frac{5}{2}} \qquad (13.12)$$

$$\text{with } E = \left[4\pi\left(\frac{H}{k}\right)^{2.5}G^{1.5}\right]\left[\mu{}_{s}{}^{2.5}\right]\left[M^{0.5}\,R^{1.5}\,K\right].$$

The over-all problem appears in a purified form in Eqs.(13.5) to (13.12) in the sense that all references to the particular star under consideration occur only in the three constants C, D, and E—at least for homogeneous stars for which (13.4) holds. In the defining equation of each of these three parameters the first bracket contains well-known physical constants; the second, the composition; and the third, the mass, luminosity, and the radius of the star considered.

Reduction in the Numerical Work by the Standard Transformation

The advantage of the transformed version for the over-all problem becomes clearer if we consider the inward integrations starting at the surface. Let us take as a first case a star with a radiative envelope, so that the first alternative of the boundary condition (13.12) applies, and let us pursue the inward integrations only through the envelope layers in which the nuclear energy is negligible, so that the right hand side of Eq.(13.8) vanishes, and let us restrict ourselves to homogeneous envelopes so that condition (13.4) holds. In this case the boundary conditions (13.12) uniquely specify starting values for all variables, and the relevant differential Eqs.(13.5) to (13.7) contain only one parameter, C, which depends on a certain combination of the mass, luminosity, radius, and composition of the star considered. Hence, we conclude that any homogeneous radiative envelope in which the equation of state and the absorption law are represented by Eqs.(13.3) can be expressed in terms of the non-dimensional variables defined by Eq.(13.2) by a family of integrations depending on the single parameter C. Thus, a very large reduction in the number of necessary integrations is found for this case.

As a second case let us take stars with convective envelopes, so that only the second alternative of Eq.(13.7) has to be applied. Let us again restrict ourselves to the envelope layers in which nuclear energy generation is negligible, so that the righthand side of Eq.(13.8) vanishes, and let us again consider only homogeneous envelopes so that condition (13.4) holds. Now the surface conditions(13.12), which we have to apply in their second form, contain the parameter E while the relevant differential equations in this case do not contain any extra parameter. Hence, we conclude that all homogeneous convective envelopes can be expressed in terms of the non-dimensional variables defined by Eqs.(13.2) by a family of integrations depending on the single parameter E. Again a large reduction in the number of necessary numerical integrations is obtained.

This advantageous situation becomes somewhat less happy in cases in which physical circumstances introduce complications. For example, if a stellar envelope is convective in its outer portion and radiative in its inner portion, both parameters, C and E, occur in the over-all problem, and the general solution is represented by a family of integrations depending on these two parameters. Similarly, if electron scattering has to be added to Kramers' opacity, an additional parameter occurs in the over-all problem. The same thing happens if radiation pressure has to be added to the gas pressure. In cases in which only one of these various complications occurs so that the general solution depends on not more than two parameters, the advantage of the transformation (13.2) is still large. Only if two or more of such complications occur simultaneously so that three or more parameters occur in the over-all problem, and if all these parameters have to be varied over wide ranges to cover the stars under consideration, the advantage of a transformation like that given by Eq.(13.2) becomes questionable.

Supplementary Transformation for Integrations from the Center

We turn next to the outward integrations starting at the center. The center conditions in the form of Eq.(13.11) is much less advantageous than the surface condition (13.12) because the former leaves two starting values, p_c and t_c, indeterminate. This disadvantage of the center compared with the surface arises from the fact that the first three equations of the standard transformation (13.2) are obviously tailored for the surface.

If we are to reduce the number of integrations from the center outwards to its minimum, we are forced to follow up the transformation (13.2) by a supplementary transformation tailored for the center. This additional transformation we may write in the general form

$$q = q^* q_0, \quad f = f^* f_0, \quad x = x^* x_0,$$
$$p = p^* p_0, \quad t = t^* t_0 \tag{13.13}$$

where the quantities with asterisks are the new variables and the quantities with zero subscript are five arbitrary constants. We may choose these constants partly so as to eliminate extra constants from the differential equations and partly so as to obtain convenient starting conditions. Specifically, we may use four of the five free choices to eliminate all constants from the four differential Eqs.(13.5) to (13.8), and use the last free choice to obtain a definite starting value for t. Thus, we obtain the five conditions

$$\frac{q_0}{t_0 x_0} = 1, \quad \frac{p_0 x_0^3}{t_0 q_0} = 1, \quad C\frac{p_0^2 f_0}{t_0^{9.5} x_0} = 1, \quad D\frac{p_0^2 t_0^{\nu-2} x_0^3}{f_0} = 1, \quad t_0 = t_c \qquad (13.14)$$

where t_c designates the value of t at the center. The conditions (13.14) completely define the constants occurring in the transformation (13.13).

If we apply this transformation to the differential Eqs.(13.5) to (13.8), we obtain

$$\frac{dp^*}{dx^*} = -l \times \frac{p^*}{t^*} \frac{q^*}{x^{*2}} \qquad (13.15)$$

$$\frac{dq^*}{dx^*} = +l \times \frac{p^*}{t^*} x^{*2} \qquad (13.16)$$

$$\frac{dt^*}{dx^*} = -jl \times \frac{p^{*2}}{t^{*8.5}} \frac{f^*}{x^{*2}} \quad \text{or} \quad \frac{dt^*}{dx^*} = \frac{2}{5} \frac{t^*}{p^*} \frac{dp^*}{dx^*}, \qquad (13.17)$$

$$\frac{df^*}{dx^*} = +il \times p^{*2} t^{*\nu-2} x^{*2} \qquad (13.18)$$

and the center conditions (13.11) become

$$\text{at } x^* = 0: \quad q^* = 0, \quad f^* = 0, \quad t^* = 1. \qquad (13.19)$$

The result is a set of differential equations in which no extra parameters occur and a set of starting conditions which leaves indeterminate only one starting value, namely p_c^*.

We conclude that any stellar core—whether radiative or convective or mixed—can be represented by a family of integrations depending on the single parameter p_c^*, as long as the gas characteristics relations are given by Eqs.(13.3) and as long as the variation of the composition functions l, j, and i is fixed and known (as is the case for homogeneous cores, for example). Again we have found a large reduction in the number of necessary integrations by the use of an appropriate transformation.

Let us consider as a final example the frequently occurring case of an exhausted, isothermal, homogeneous stellar core in which the densities are so high that the gases are partially degenerate. In this case we may write the equation of state according to Eqs.(8.8), (8.9), (8.12), and (8.13), in the form

$$P = \frac{4\pi}{h^3} (2m)^{\frac{3}{2}} (k T)^{\frac{5}{2}} \times p^*$$

$$\rho = \frac{4\pi}{h^3} (2m)^{\frac{3}{2}} (k T)^{\frac{3}{2}} \mu_E H \times s^* \qquad (13.20)$$

with $p^* = \frac{2}{3} F_{\frac{3}{2}} (\psi) + \frac{\mu_E}{\mu_A} F_{\frac{1}{2}} (\psi)$

and $s^* = F_{\frac{1}{2}} (\psi)$.

Since the quantities $F_{\frac{1}{2}}$ and $F_{\frac{3}{2}}$ are known functions of the quantity ψ, we may consider s^* as a known function of p^* for any given composition, i.e. for any given value of μ_E/μ_A. Next, we may re-define the variables q^* and x^* by the transformation

$$M_r = (4\pi)^{-1} h^{\frac{3}{2}} G^{-\frac{3}{2}} (2m)^{-\frac{3}{4}} (k T)^{\frac{3}{4}} (\mu_E H)^{-2} \times q^*$$

$$r = (4\pi)^{-1} h^{\frac{3}{2}} G^{-\frac{1}{2}} (2m)^{-\frac{3}{4}} (k T)^{-\frac{1}{4}} (\mu_E H)^{-1} \times x^* . \qquad (13.21)$$

If we introduce the Eqs.(13.20) and (13.21) into the hydrostatic equilibrium conditions (12.1) and (12.2)—the conditions for thermal equilibrium and for the energy transport are of no consequence for isothermal cores—we obtain

$$\frac{dp^*}{dx^*} = - s^* \frac{q^*}{x^{*2}}, \quad \frac{dq^*}{dx^*} = + s^* x^{*2}, \qquad (13.22)$$

and the center condition becomes

$$\text{at } x^* = 0 : \quad q^* = 0. \qquad (13.23)$$

Again we have found a set of differential equations in which no extra parameters occur and a starting condition which leaves only the starting value p_c^* free, and we conclude that all isothermal partially degenerate homogeneous cores with a given μ_E/μ_A value are represented by a family of integrations depending on the single parameter p_c^*.

The Invariants and the UV Plane

The final step in the construction of a complete stellar model consists of the fitting together of the inward and outward integrations in accordance with the fitting conditions (13.1). This step is simple if all the integrations have been performed in terms of the basic physical variables. In this case we will terminate all the integrations at one pre-chosen value of r and hence know exactly where in the integrations we have to read off the various quantities needed for the fitting conditions

(13.1). The fitting, however, is not this simple if we use different variables for the inward and the outward integrations; for example, the non-dimensional variables defined by Eqs.(13.2) for the inward integrations and the variables with asterisks defined by Eqs.(13.13) for the outward integrations.   How are we to know which value of x^* corresponds to a pre-chosen termination value of x?

This difficulty can be overcome by introducing the following three quantities

$$U = + \frac{r}{M_r} \frac{dM_r}{dr} = \frac{4\pi r^3 \rho}{M_r}$$

$$V = - \frac{r}{P} \frac{dP}{dr} = \frac{\rho}{P} \frac{G M_r}{r} \tag{13.24}$$

$$n + 1 = + \frac{T}{P} \frac{dP}{dT}.$$

The physical meaning of the first two of these quantities becomes apparent if we write them in the form

$$U = 3 \frac{\rho}{M_r / \frac{4}{3}\pi r^3} = 3 \times \frac{\text{local density}}{\text{mean interior density}}$$

$$V = \frac{3}{2} \frac{G M_r / r}{\frac{3}{2} P / \rho} = \frac{3}{2} \times \frac{\text{``gravitational energy''}}{\text{internal energy}}. \tag{13.25}$$

The third quantity, n, is called the polytropic index. It characterizes the relation between the pressure and the temperature within a star. Stellar models in which the polytropic index is constant from center to surface have played an important role in the development of the theory of the stellar interior.   Here, however, we shall consider n as varying from layer to layer just as U and V do. We may note that with the help of the polytropic index the stability condition for radiative equilibrium (7.2) may be written in the simple form

$$n + 1 > \frac{\gamma}{\gamma - 1} = 2.5 \tag{13.26}$$

and that the adiabatic relation (7.9) which holds in a convective zone becomes

$$n + 1 = 2.5. \tag{13.27}$$

For our purposes, however, the most essential characteristic of these three quantities lies in the fact that they are invariant against transformations of the type of Eqs.(13.2) or (13.13). That this is so follows directly from Eqs.(13.24) since they define the three quantities in terms of

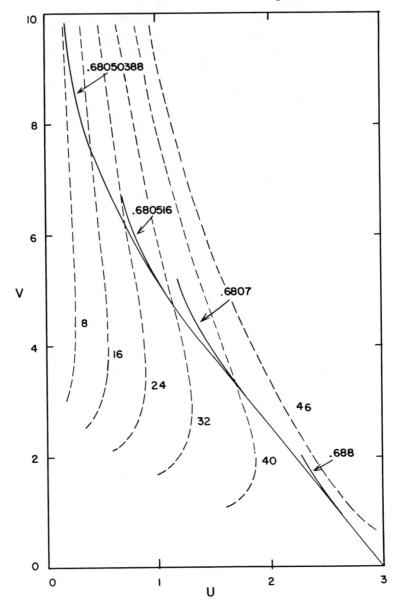

Fig. 13.1. Sample solutions in UV plane. The dashed curves represent convective envelopes, designated by their E values. The solid curves represent radiative cores ($\nu = 4.5$, $\alpha = 0.25$), designated by their p_c^* values.

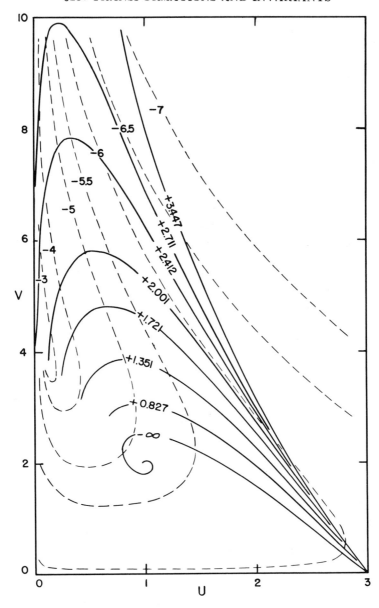

Fig. 13.2. Sample solutions in UV plane. The dashed curves represent radiative envelopes with Kramers' opacity, designated by their $\log C$ values. The solid curves represent partially degenerate, isothermal helium cores, designated by their $\log p_c^*$ values.

logarithmic derivatives which are automatically invariant against multiplication of any variable by a constant factor. Indeed, if we introduce the transformations (13.2) and (13.13) into the definitions (13.24) we obtain respectively for U and V

$$U = + \frac{x}{q} \frac{dq}{dx} = \frac{px^3}{tq}, \qquad V = -\frac{x}{p} \frac{dp}{dx} = \frac{q}{tx}$$

$$U = + \frac{x^*}{q^*} \frac{dq^*}{dx^*} = \frac{p^* x^{*3}}{t^* q^*}, \qquad V = -\frac{x^*}{p^*} \frac{dp^*}{dx^*} = \frac{q^*}{t^* x^*}. \tag{13.28}$$

We see from these relations that we can compute U and V at every point of a particular integration from the very variables used in the integration without any knowledge of the constants occurring in the transformation.

Before we employ the invariants for their main purpose, that is for the fitting together of the inward and the outward integrations, let us have a first look at the UV plane, which we will find to be a convenient tool for our present purposes. Samples of the UV plane are shown in Figs. 13.1 and 13.2. Any particular integration is represented in this plane by a curve. The curves representing inward integrations from the surface start at a point high above the left upper corner of these figures while the curves representing outward integrations from the center start at the right lower corner of the figures. This follows from the behavior of U and V at the two boundaries:

$$\text{for } r \to 0: \quad U \to 3, \quad V \to 0,$$
$$\text{for } r \to R: \quad U \to 0, \quad V \to \infty. \tag{13.29}$$

As a sample of inward integrations, Fig. 13.1 shows the one-parameter family of convective envelopes and, as a sample of outward integrations, the one-parameter family of radiative cores (with a specific value of the temperature exponent for the energy generation, $\nu = 4.5$, and a specific value of the density exponent for the guillotine factor, $\alpha = 0.25$). Similarly, Fig. 13.2 shows the one-parameter family of radiative envelopes with Kramers' opacity and the one-parameter family of partially degenerate isothermal helium cores.

The Fitting Procedure in the UV *Plane*

With the help of the UV plane the fitting procedure takes a simple form. Since the invariants are defined as direct functions of the basic physical variables and since these physical variables must be continuous at the fitting point according to conditions (13.1), the invariants must also be continuous at the fitting point. This circumstance we may formally represent by

$$U_{if} = U_{ef}, \quad V_{if} = V_{ef}, \quad (n+1)_{if} = (n+1)_{ef}. \tag{13.30}$$

Let us now pick as an example one particular envelope solution of Fig. 13.1 and one particular core solution and try to combine these two integrations into one complete stellar model. Since U and V must be continuous, we can switch from the envelope to the core only at the one point in which the two integrations cross each other in the UV plane. This determines uniquely the termination point in the envelope integration as well as the termination point in the core integration. Thus we have overcome the difficulty introduced by the use of different transformations for the envelope and the core.

But this does not complete the fitting procedure, since we have fulfilled thus far only two fitting conditions, those for U and V, instead of the five fitting conditions (13.1) for all the physical variables. The best form for the remainder of the fitting procedure depends very much on the particular situation. Hence we will have to postpone the discussion of it to later chapters in which we describe the derivation of specific stellar models for particular stars and particular evolution states.

Altogether we have seen in this section that the number of numerical integrations necessary to solve the over-all problem for a particular star in a particular state of evolution can often be much reduced by the application of appropriate transformations, and that the use of these transformations does not seriously complicate the fitting together of the interior and exterior integrations if we use for this purpose invariant variables, such as U and V. Clearly, therefore, such transformations are quite generally used, as long as the integrations are performed by hand computations. On the other hand, if the entire computation of stellar models is to be performed fully automatically by large electronic machines, future experience may well indicate that the introduction of a variety of transformations applicable for a variety of physical circumstances introduces a serious complication and that it is therefore best for completely automatized calculations to solve the over-all problem in its original form with no transformations.

14. Numerical Integrations

One point follows inevitably from the discussions of the preceding two sections: however we turn the over-all problem of stellar structure and whatever transformations we apply, eventually we have to resort to numerical integrations if we are to obtain definite stellar models. On occasions in the past appreciable efforts have been expended in the attempt to avoid numerical integrations. In practice, however, the effort necessary for solving by numerical integrations the essential problems occurring in the theory of the stellar interior has never been large in comparison with many other astrophysical research programs. Recently the necessary computing work has been growing because of the need for extensive evolutionary sequences of stellar models. By a most fortunate coincidence the available computing equipment has also been growing by leaps and bounds. It appears therefore that we may confidently regard numerical integrations as remaining an adequate and efficient tool for our purposes.

The general character of the boundary-value problem for the stellar interior differs in no way from many other well-known boundary-value problems, and hence all the usual methods for the numerical solution of such problems apply. Here we shall restrict ourselves to a review of the major necessary steps and to a description, for each step, of one or two particular procedures which have proven useful in the past.

Developments at the Center

Neither for the integrations from the center outward nor for the integrations from the surface inward can one start the numerical step-by-step integration procedure right at the boundary. This complication arises from the fact that at both boundaries vanishing denominators occur in the basic differential Eqs.(12.1) to (12.4). Therefore one has to develop the solution at the boundary, use this development to compute the solution at a point a little distance from the boundary, and start at this point the step-by-step integration procedure.

If one develops the solution at the center in powers of r, if one ignores terms of fourth or higher order, and if one designates all quantities at the center with the subscript c, one obtains from Eqs.(12.1) to (12.4)

$$M_r = + \frac{4}{3} \pi \rho_c \times r^3$$

$$P = P_c - \frac{2}{3} \pi G \rho_c^2 \times r^2$$

$$L_r = + \frac{4}{3} \pi \rho_c \varepsilon_c \times r^3 \tag{14.1}$$

$$T = T_c - \frac{\varkappa_c \varepsilon_c \rho_c^2}{8 \, ac \, T_c^3} \times r^2 \text{ (radiative)}$$

$$T = T_c - \frac{4}{15} \pi G \frac{T_c \rho_c^2}{P_c} \times r^2 \text{ (convective)}$$

This development holds quite generally since it does not depend on the specific form of the laws for the equation of state, the absorption coefficient, and the energy generation. But if we extend these developments to higher order terms, the particular forms of the three gas characteristics relations would enter.

The developments (14.1) can be expressed in a more convenient way with the help of non-dimensional variables. For example, in cases in which the variables with asterisks defined by Eqs.(13.13) and (13.14) are applicable and in which $l = j = i = 1$, we obtain from the differential Eqs.(13.15) to (13.18) and the boundary conditions (13.19)

$$q^* = + \frac{1}{3} p_c^* \times x^{*3}$$

$$p^* = p_c^* - \frac{1}{6} p_c^{*2} \times x^{*2}$$

$$f^* = + \frac{1}{3} p_c^{*2} \times x^{*3} \tag{14.2}$$

$$t^* = 1 - \frac{1}{6} p_c^{*4} \times x^{*2} \quad \text{(radiative)}$$

$$t^* = 1 - \frac{1}{15} p_c^* \times x^{*2} \quad \text{(convective)}.$$

We note that the developments (14.2) contain only the one parameter p_c^*, as we should have expected from the results of the preceding section. We will see that for most cases of interest p_c^* is of the order of unity. We may therefore apply the developments (14.2) with good accuracy at values of x^* of a few times 0.01. The procedure then is to choose an appropriate value of x^* of this order, to compute the solution at this point from Eqs.(14.2), and to proceed from this point outward by a step-by-step integration method.

Developments at the Surface

For the inward integrations from the surface we may obtain the necessary developments as follows. The density and, even more, the energy generation are extremely low in the outermost layers. Hence, according to Eqs.(12.2) and (12.3), the quantities M_r and L_r will be practically constant and we may use for the outermost layers

$$M_r = M; \quad L_r = L. \tag{14.3}$$

With these simplifications we can obtain explicitly the pressure-temperature relation in a radiative envelope by dividing the hydrostatic equilibrium Eq.(12.1) by the radiative equilibrium Eq.(12.4)—just as we have already done in §11. The resulting differential equation can be solved and, with due regard to the first alternative of the boundary conditions (12.7), gives for the case of a homogeneous envelope with Kramers' opacity

$$P = \left(\frac{2}{8.5} \frac{4ac}{3} \frac{k}{H} \frac{4\pi GM}{\mu \varkappa_0 L} \right)^{\frac{1}{2}} \times T^{4.25} \quad \text{(radiative)}$$

$$P = K \times T^{2.5} \quad \text{(convective)}. \tag{14.4}$$

Here we have added to the pressure-temperature relation for the radiative case the corresponding relation for the convective case, which follows directly from the second of Eqs.(12.4) or from the second alternative of the boundary conditions (12.7).

Eqs.(14.4) give P as a function of T. It still remains to determine T as a function of r. We can use the hydrostatic equilibrium condition (12.1) and eliminate from it the pressure with the help of Eq.(14.4) and the density with the help of the equation of state of an ideal gas. We obtain thus a differential equation for the temperature which can be solved explicitly with due regard to the boundary condition (12.7) for the temperature. The solution gives

$$T = \frac{1}{4.25} \frac{\mu H}{k} GM \times \left(\frac{1}{r} - \frac{1}{R}\right) \text{ (radiative)}$$

$$T = \frac{1}{2.5} \frac{\mu H}{k} GM \times \left(\frac{1}{r} - \frac{1}{R}\right) \text{ (convective).}$$

$$(14.5)$$

Eqs.(14.3) to (14.5) give all the physical variables as functions of r with high accuracy as long as we apply these equations only to the outermost, essentially massless layers of a star.

Again we may rewrite these equations in terms of non-dimensional variables, for example those defined by Eqs.(13.2), for which we get

$$q = 1, \quad f = 1$$

$$p = \left(\frac{8.5}{2} C\right)^{-\frac{1}{2}} \times t^{4.25}, \quad t = \frac{1}{4.25}\left(\frac{1}{x} - 1\right) \text{ (radiative)}$$

$$p = E \times t^{2.5} \qquad\qquad t = \frac{1}{2.5}\left(\frac{1}{x} - 1\right) \text{ (convective).}$$

$$(14.6)$$

In general it is found that these equations hold with good accuracy from the surface inwards to $x = 0.95$ or even $x = 0.90$. At some such point q, f, p, and t are taken from Eqs.(14.6) and the step-by-step inward integration is commenced.

Logarithmic Variables

Before we start the actual numerical integrations we have to decide on the particular variables to use. The physical variables P and T, as well as the corresponding non-dimensional variables, have the disadvantage of varying in many cases over a large number of powers of ten. Such large variations are not convenient in numerical integrations. Therefore it is often better to use logarithmic variables. For example, we may replace the non-dimensional variables defined by the standard transformation (13.2) by their logarithms

$$g_1 = \log p, \quad g_2 = \log q, \quad g_3 = \log t$$

$$g_4 = \log f, \quad y = \log x.$$

$$(14.7)$$

If we introduce these logarithmic variables into the differential Eqs.(13.5) to (13.8) and consider only the more complicated radiative case, we obtain

$$\log \left(-\delta \frac{dg_1}{dy}\right) = +g_2 - g_3 - y + \log (l\,\delta)$$

$$\log \left(+\delta \frac{dg_2}{dy}\right) = +g_1 - g_2 - g_3 + 3y + \log (l\,\delta)$$

$$\log \left(-\delta \frac{dg_3}{dy}\right) = +2g_1 - 9.5g_3 + g_4 - y + \log (C\,jl\,\delta)$$

$$\log \left(+\delta \frac{dg_4}{dy}\right) = +2g_1 + (\nu - 2)\,g_3 - g_4 + 3y + \log (D\,il\,\delta).$$

(14.8)

Here we have multiplied each of the four derivatives by the step value δ since it is this product which we will need in the numerical integrations. The right-hand sides of Eqs.(14.8) contain simple linear combinations of the four dependent and the one independent variables. The one additional term which occurs on the right-hand side of each of these equations is completely fixed at the start of the integration and therefore does not introduce any complications in the numerical procedure. Thus, the only non-linear operation needed for solving Eqs.(14.8) for the four derivatives is to look up four antilogarithms corresponding to the logarithms occurring on the left-hand sides of these equations.

Altogether, therefore, logarithmic variables, such as those defined by Eqs.(14.7) and the corresponding form of the basic differential Eqs.(14.8), are highly advantageous for numerical integrations. This advantage is, however, somewhat reduced in those physical situations in which it is necessary to take account of complicating circumstances, such as the addition of radiation pressure to the gas pressure or the addition of electron scattering to Kramers' opacity.

Two more remarks may be added with regard to the logarithmic variables. First, the invariants defined by Eqs.(13.24) can be expressed in terms of the logarithmic variables by

$$U = +\frac{dg_2}{dy}, \quad V = -\frac{dg_1}{dy}, \quad n + 1 = \frac{dg_1}{dy} \Big/ \frac{dg_3}{dy}. \tag{14.9}$$

These relations make the computation of the invariants from integrations using the logarithmic variables extremely simple. Second, the developments (14.6) near the surface suggest that it would be better for envelope integrations to replace the independent variable y by the variable z defined by

$$z = \log \left(\frac{1}{x} - 1\right). \tag{14.10}$$

If we replace y by z on the left-hand side of Eqs.(14.8), we have to add on the right-hand side

$$\log\left(-\frac{dy}{dz}\right) = y + z.$$

This eliminates y in the first and the third of Eqs.(14.8), but produces a term $4y$ in the second and fourth of these equations. This term is a known function of z, can be tabulated once and for all, and hence introduces no complication in the numerical integrations. The introduction of z for the inward integrations from the surface has often been found to produce an appreciable speedup in the numerical integrations.

Procedures for Step-by-Step Integrations

The number of different procedures all of which can be used for the numerical integration of first order differential equations is large, and the choice often appears to depend on the personal preference of the integrator. Here we shall describe only two such procedures, one convenient for integrations by hand and the other for integrations by electronic machines, both of which have been extensively used in integrations for the stellar interior.

The first procedure is most easily explained with the help of the integration scheme shown in Table 14.1. The first column of this tabulation lists the independent variable; the second column one of the dependent variables; the third column the logarithm of the derivative (or whatever quantities need to be written down during the computation of the derivative from the differential equation); the fourth column the derivative multiplied by the step value; and the remaining columns the successive differences of the values in the fourth column.

Assume that we have progressed in the numerical integration to the point where all the quantities marked by X in Table 14.1 are known. We

TABLE 14.1

Scheme for numerical integration with the help of Eq.(14.11).

y	g	$\log\ \delta\dfrac{dg}{dy}$	$\delta\dfrac{dg}{dy}$	^1D	^2D	^3D	^4D
XX	XXXX	XXX	XXX		XX		X
				XX		X	
XX	XXXX	XXX	XXX		XX		X
				XX		X	
XX	XXXX	XXX	XXX		XX		$\boxed{P}$
				XX		$\boxed{P}$	
XX	$\boxed{XXXX}$	XXX	$\boxed{XXX}$		$\boxed{PP}$		
				$\boxed{PP}$			
XX	$\boxed{????}$	AAA	AAA				
XX							

start the next integration step by predicting the values for the differences marked by P in Table 14.1. This may be done by extrapolating smoothly, for example, the fourth difference ^4D, from which the lower differences can be computed successively down the diagonal. Now we can compute the next value of the dependent variable, marked with question marks in Table 14.1, with the help of the extrapolation formula

$$g_{n+1} = g_n + \left(\delta\frac{dg}{dy}\right)_n + \frac{1}{2}\,{}^1D_{n+\frac{1}{2}} - \frac{1}{12}\,{}^2D_n - $$

$$\frac{1}{24}\,{}^3D_{n-\frac{1}{2}} - \frac{19}{720}\,{}^4D_{n-1}. \quad (14.11)$$

All the terms occurring in this formula are shown in blocks in Table 14.1. Finally, we compute the new derivative, marked by A, from the differential equation. Comparing this new derivative with the preceding one we can check our prediction of the differences. If we find them inaccurate, we have to improve them and have to repeat the integration step. When the check turns out satisfactory, the integration step is completed.

Since our basic problem contains four dependent variables and four differential equations, we have to carry simultaneously side by side four integration schemes like that shown by Table 14.1.

The high accuracy of the extrapolation formula (14.11) permits the use of rather large step values. In the envelope step values as large as 0.05 in z can often be used, though in the core somewhat smaller step values, say 0.01 in y, are frequently needed.

Another integration procedure, useful for electronic machines, is given by the following two formulae

$$g_{n+1}^{\circ} = g_n + \left(\delta\frac{dg}{dy}\right)_n$$

$$g_{n+1} = g_n + \frac{1}{2}\left(\delta\frac{dg}{dy}\right)_n + \frac{1}{2}\left(\delta\frac{dg}{dy}\right)_{n+1}^{\circ}. \quad (14.12)$$

The first formula determines from the value of the dependent variable at a given point, g_n, and from the derivative at this same point, a preliminary value for the same variable at the next point, g_{n+1}°. With the help of this preliminary value we then derive from the differential equation a preliminary value for the derivative at the new point $(dg/dy)_{n+1}^{\circ}$. Finally, we use this value to obtain the definitive value of the dependent variable at the new point, g_{n+1}, with the help of the second of Eq.(14.12).

The formulae (14.12) are only of second order, and hence if they are employed one has to use appreciably smaller step values, say five times smaller, than those required by formulae (14.11).

For many problems in the theory of the stellar interior the speed of numerical integrations by hand is entirely sufficient. A person can usually

accomplish more than twenty integration steps per day for a set of differential equations like Eq.(14.8). Thus for a typical single integration
consisting of, say, forty steps less than two days are needed. Correspondingly, if, for example, a set of models is to be determined and if
these models are to be constructed of a one-parameter family starting
from the surface and a one-parameter family starting from the core, and if
each of these two families can be represented with sufficient accuracy
by, say, six individual integrations, then the entire numerical work for
this fairly typical case can be accomplished by one person in one month.
However, if extensive evolutionary model sequences including a variety
of physical complications are to be derived, then numerical integrations
by hand may become prohibitive and the advantage of large electronic
machines will be incontestable.

CHAPTER IV
INITIAL STELLAR STRUCTURE

15. The Upper Main Sequence

In the three preceding chapters we have assembled all the astronomical data, the physical conditions, and the mathematical tools which we need to construct theoretical stars. In the four chapters which we are now beginning we shall discuss the actual construction of model stars of various masses and compositions and we shall follow the evolution of the stars—to the best of our present knowledge—from their initial state to their final state.

Much of the four coming chapters will be devoted to technical details. The main results of these investigations will be summarized and reviewed in the last chapter of this book.

According to our discussion of the over-all problem in §12, we have to start our program of model construction with the initial state, that is with homogeneous stellar models. We shall see that stars in the initial state belong to the main sequence or are subdwarfs. Consequently, the subject of this chapter will be the construction of homogeneous models for main-sequence stars and subdwarfs.

We shall divide the main sequence into two sections, the upper and lower main sequence. This division is a natural one for two reasons. First, the main energy-liberating process will be the carbon cycle in stars of the upper main sequence while on the lower main sequence this role will be played by the proton-proton reaction. Second, it follows from the discussion of the surface layers in §11 that we may use purely radiative envelopes for stars on the upper main sequence while we have to count with deep convection zones in the envelopes of lower main-sequence stars. We shall concentrate in this section on models for the upper main sequence.

Choice of Gas Characteristics Relations

Whenever one wants to compute detailed models for specific stars, one has first to decide what particular forms of the three gas characteristics relations apply. These decisions will depend on the temperatures and densities to be expected in the stars in question. Thus, one has to start by making a rough guess at the temperature and density run in the star considered; next one has to choose the appropriate gas characteristics relations; then one has to construct the model; and finally one has to compute from the model the actual run of temperature and density. If the original guess and the final result disagree, one has to repeat the procedure. If they agree, one has obtained a correct model. In this and the subsequent sections we shall skip the description of any unsuccessful guesses and start in each case immediately with the listing of those particular forms for the gas characteristics relations which were proven to be appropriate by the checks after the construction of the models.

For upper main-sequence stars the following choices for the three gas characteristics relations appear satisfactory. For the equation of state the ideal gas law augmented by the radiation pressure, Eq. (8.3), is appropriate. If we introduce the ratio β of the gas pressure to the total pressure we may write this equation in the form

$$P = \frac{k}{\mu H} \rho \, T \, \frac{1}{\beta} \tag{15.1}$$

$$\text{with } 1 - \beta = \frac{a}{3} \, T^4 / P. \tag{15.2}$$

For the opacity the main contribution comes from the bound-free processes of the heavier elements as given by Eqs. (9.16) and (9.17) and from electron scattering as given by Eq. (9.19). If we use the straight sum of these two contributions and if we introduce the ratio δ of the total absorption coefficient to that of the bound-free transitions then we can represent the absorption coefficient by the formula

$$\varkappa = \frac{4.34 \times 10^{25}}{(t/\overline{g})_0} \times Z(1 + X)^{1-\alpha} \times \frac{\rho^{1-\alpha}}{T^{3.5}} \times \delta \tag{15.3}$$

$$\text{with } \delta - 1 = 0.19(1 + X) \left/ \left[\frac{4.34 \times 10^{25}}{(t/\overline{g})_0} \times Z(1 + X)^{1-\alpha} \times \frac{\rho^{1-\alpha}}{T^{3.5}} \right] \right. \tag{15.4}$$

Finally, the energy generation is provided by the carbon cycle, which can be represented by the interpolation formula (10.15). With the carbon-nitrogen abundance given by Eq. (10.13) and the constants listed in Table 10.1 for temperatures around twenty-five million degrees, we obtain

$$\varepsilon = 8 \times 10^{-114} \times \frac{1}{3} \, XZ \times \rho T^{16}. \tag{15.5}$$

We may now proceed to the construction of the models. Our present task is greatly simplified by the following feature. The rate of the carbon cycle depends sensitively on the temperature, as is shown by the high exponent of T in Eq. (15.5). This has the effect that the energy generation is highly concentrated toward the center of the star. Hence, even at a small distance from the center the energy flux is great, which in turn produces a steep radiative temperature gradient—in fact so steep that it is unstable and convection occurs. The star will have a convective core, and it turns out that this convective core reaches far enough out so that practically all the nuclear energy sources of the star are contained in the core and nuclear processes can be ignored outside the core. The star consists therefore of a convective core and a radiative envelope in which L_r is constant, i.e. equal to L. We will cover the core by integrations from the center outwards and the envelope by integrations from the surface inwards, and we will choose the interface between the core and the envelope as the fitting point at which we will join the inward and the outward integrations together.

Equations for the Envelope

Let us first consider the envelope. If we use the gas characteristics relations (15.1), (15.3), and (15.5) to eliminate ρ, $\varkappa$, and ε from the basic differential Eqs. (12.1) to (12.4); if we remember that the condition (12.3) for thermal equilibrium is automatically fulfilled in the envelope by setting $L_r = L$; and if we apply the transformation (13.2), we obtain

$$\frac{dp}{dx} = -\frac{p}{t}\frac{q}{x^2}\beta \tag{15.6}$$

$$\frac{dq}{dx} = +\frac{p}{t}x^2\beta \tag{15.7}$$

$$\frac{dt}{dx} = -C\frac{p^{2-\alpha}}{x^2 t^{8.5-\alpha}}\delta\beta^{2-\alpha} \tag{15.8}$$

with $C = \dfrac{3}{4ac}\left(\dfrac{k}{HG}\right)^{7.5}\dfrac{1}{(4\pi)^{3-\alpha}}\dfrac{4.34\times10^{25}}{(t/g)_0}\dfrac{Z(1+X)^{1-\alpha}}{\mu^{7.5}}\times$

$$\frac{LR^{0.5+3\alpha}}{M^{5.5+\alpha}} \tag{15.9}$$

If we furthermore apply the same transformation to the two auxiliary relations (15.2) and (15.4), we obtain

$$\beta = 1 - B\frac{t^4}{p} \tag{15.10}$$

$$\text{with } B = \frac{4}{3}\pi a\left(\frac{H}{k}\right)^4 G^3\mu^4 M^2 \tag{15.11}$$

$$\delta = 1 + AB\frac{t^{4.5-\alpha}}{(\beta p)^{1-\alpha}} \tag{15.12}$$

with $A = \dfrac{3}{a(4\pi)^\alpha}\left(\dfrac{kG}{H}\right)^{0.5}\dfrac{0.19(t/g)_0}{4.34\times10^{25}}\dfrac{(1+X)^\alpha}{Z\mu^{0.5}}\dfrac{M^{0.5+\alpha}}{R^{0.5+3\alpha}}$ (15.13)

The three differential Eqs. (15.6) to (15.8) and the two auxiliary relations (15.10) and (15.12) contain the three parameters A, B, and C. The parameter B is completely determined when the mass and the composition of a star are given. The parameter A depends not only on the mass and the composition but also on the radius. This latter dependence is, however, not very strong, so that a reasonable estimate of the radius in general suffices to fix A with sufficient accuracy. Hence we may consider A and B as given and C as the only free parameter. Furthermore, the surface conditions, which here take the form of the first alternative of Eqs. (13.12), uniquely determine all the starting values. We conclude that the only unknown quantity which enters the envelope solution is C.

Equations for the Core

Next, let us consider the core. The hydrostatic equilibrium conditions (15.6) and (15.7) apply in the core as they stand, but the radiative equilibrium condition (15.8) has to be replaced by the convective condition, that is by the second alternative of Eq. (12.4), and the thermal equilibrium condition (12.3) has to be taken into account explicitly. If we apply the transformation (13.2) to the two equations just mentioned we obtain

$$\frac{dt}{dx} = \left(1 - \frac{1}{\gamma}\right) \frac{t}{p} \frac{dp}{dx} \tag{15.14}$$

$$\frac{df}{dx} = + D \, p^2 \, t^{14} \, x^2 \, \beta^2 \tag{15.15}$$

$$\text{with } D = \left(\frac{HG}{k}\right)^{16} \frac{1}{4\pi} \frac{8 \times 10^{-114}}{3} XZ \, \mu^{16} \frac{M^{18}}{LR^{19}}. \tag{15.16}$$

The radiation pressure appreciably complicates the equations for the core. First, it introduces the additional variable factor β into the differential Eqs. (15.6), (15.7), and (15.15). Second, it affects the ratio of the specific heats γ in the adiabatic relation (15.14) since γ is found to be somewhat smaller than its usual value of 5/3 under those circumstances in which the pressure and energy of radiation provide important contributions to the total pressure and energy. The radiation pressure does not play an important role, however, in the specific cases which we will consider here, and therefore we may permit ourselves the simplifying assumption that γ has its normal value of 5/3 and that β is constant throughout the core.

With these approximations for the core, we may apply the supplementary transformation

$$p = p^* \, p_c, \quad t = t^* \, t_c, \quad q = q^* \frac{t_c^2 (2.5)^{\frac{3}{2}}}{\beta^2 \sqrt{p_c}}, \quad x = x^* \frac{t_c \sqrt{2.5}}{\beta \sqrt{p_c}}. \tag{15.17}$$

The center conditions now become

$$\text{at } x^* = 0: \quad p^* = 1, \quad t^* = 1, \quad q^* = 0. \tag{15.18}$$

The adiabatic relation (15.14) can be integrated and, in view of the boundary conditions (15.18), gives

$$p^* = t^{*2.5} \qquad (15.19)$$

If we now introduce the transformation (15.17) also into the hydrostatic equilibrium conditions (15.6) and (15.7), if we eliminate p^* with the help of Eq. (15.19) and if we eliminate q^* between the two hydrostatic conditions, we obtain

$$\frac{d^2 t^*}{dx^{*2}} + \frac{2}{x^*} \frac{dt^*}{dx^*} + t^{*1.5} = 0. \qquad (15.20)$$

This is the well-known second-order differential equation for the "polytrope" of index 1.5. Its solution is uniquely determined by the center conditions (15.18). This unique solution was derived long ago by numerical integration and is available in tabular form.

Thus far we have entirely ignored the thermal equilibrium condition (15.15). We were able to derive the structure of the core without considering this equation since it is a condition on the energy flux and the energy flux has no direct effect on the structure of a convective region. We may now introduce the solution for the core in the right-hand side of Eq. (15.15) and by a simple quadrature obtain the run of f. This run of f is not of particular interest in itself. But we have to ensure that f, which starts at the center with the value 0, reaches its boundary value 1 at the surface or in physical terms to ascertain that the nuclear processes produce just as much energy flux as passes through the envelope. We may put this condition into a simple form by introducing the transformation (15.17) into Eq. (15.15) and by integrating this equation from the center to the surface. Thus we obtain

$$1 = D \times \frac{(2.5)^{\frac{3}{2}}}{\beta} p_c^{\frac{1}{2}} t_c^{17} \times \int_0^R t^{*19} x^{*2} \, dx^*. \qquad (15.21)$$

The integral on the right-hand side gains practically its entire value from a range of rather small x^* values. The integral can be computed by numerical quadrature from the solution for the convective core. Its value is 0.0776. Eq. (15.21) is the form which the thermal equilibrium condition takes for the particular models which we are here investigating.

Choice of Mass and Composition

Now that we have finished the transformation of all the basic equations into a form appropriate for upper main-sequence stars, we can go on to describe the actual construction of models which have been carried out with the help of these equations. Three cases have been considered, with stellar masses equal to 10, 5, and 2.5 solar masses respectively. For all three cases the composition was taken to be

$$X = 0.90, \quad Y = 0.09, \quad Z = 0.01(\mu = 0.53), \tag{15.22}$$

where the molecular weight was computed according to Eq.(8.2). For each case the constants $(t/\overline{g})_0$ and α which occur in Eq.(15.3) for the opacity were chosen so that this equation fitted the accurate opacity tables as well as possible for the temperatures and densities in question, and the parameters A and B were computed with the help of Eqs. (15.11) and (15.13) form the mass, composition, and estimated radius. The numerical values for these four quantities are listed in Table 15.1.

Fitting Procedure and Computation of Model Characteristics

The fitting procedure is best described in terms of the UV plane as shown in Fig. 15.1. To start with, the unique solution for the convective core is plotted in this plane. Next, a trial value for C is chosen, a numerical integration of the differential Eqs.(15.6) to (15.8) is performed for the envelope, and this envelope solution is plotted in the UV plane. If the envelope solution does not cross the core solution at all (the case for $\log C = -5.700$ in Fig. 15.1), a smaller C value is chosen and the procedure repeated. If the envelope solution does cross the core solution (the case for $\log C = -6.600$ in Fig. 15.1) the crossing point is accurately determined by interpolation in the two solutions, and the value for $n + 1$ at that point is read from the envelope solution. According to the last of the three fitting conditions (13.30)—of which the first two are automatically fulfilled at the crossing point—the quantity $n + 1$ must be continuous at the crossing point and hence must have there the value 2.5 which it has throughout the convective core. If the value of $n + 1$ found from the envelope solution at the crossing point does not agree with this core value, another C value has to be chosen and the procedure repeated until finally that value of C is found which gives an envelope solution for which $n + 1$ reaches 2.5 just at that point where the envelope solution reaches the core solution (the case for $\log C = -6.140$ in Fig. 15.1). This final envelope solution does not actually cross the core solution but only touches it, since the fulfillment of the fitting conditions (13.30) means that all physical variables have continuous gradients and

TABLE 15.1

Parameters used for upper main-sequence models.

$M =$	$10M_\odot$	$5M_\odot$	$2.5M_\odot$
α	0.0	0.1	0.2
$\log (t/\overline{g})_0$	+0.07	+0.22	+0.30
$\log B$	+0.8	+0.2	−0.4
$\log A$	+2.25	+2.25	+2.25

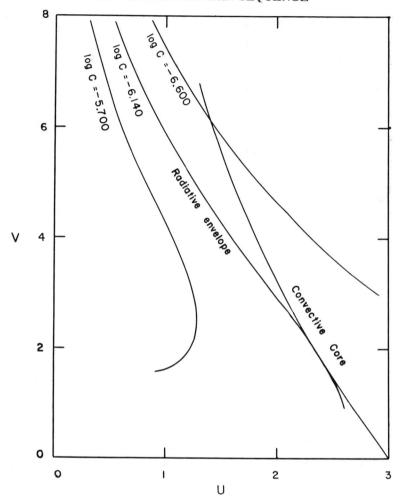

Fig. 15.1. *UV* plane for upper main-sequence ($M = 5$). (R. S. Kushwaha, *Ap.J. 125*, 242, 1957)

that therefore also U and V according to their definitions (13.24) must have continuous gradients.

This procedure for obtaining the correct envelope solution by successive trial integrations may appear very cumbersome on first sight. In practice, however, it converges quite rapidly as long as only one unknown parameter is involved, as in the present problem. One usually finds that after the first two trial integrations one reduces the error in C by a full decimal place with every additional integration.

After the correct value of C and the correct envelope solution had been obtained the model construction was completed by the following sequence of computations. The values at the fitting point of all the various non-dimensional variables were read off from the envelope integrations and the core integrations. In particular p_f, t_f, $p_f{}^*$, and $t_f{}^*$ were thus determined. By introducing these quantities into the first two of Eqs.(15.17), p_c and t_c could be computed. This in effect insured the continuity of P and T at the fitting point. Thus together with the conditions (13.30) five continuity conditions in all were observed at the fitting point, just as is prescribed by the basic conditions (13.1). By introducing p_c, t_c and β_f into Eq. (15.21), D was obtained.

The switch from the non-dimensional variables to the physical variables was performed with the help of the key relations (15.9) and (15.16). With C and D known, they represent two relations between M and the composition on the one hand and L and R on the other hand. From them L and R were obtained. Next, T_e followed from its definition (1.3), and the spectral type from its relation with the effective temperature. Finally, the transformations (13.2), in which all the constants were now known, could be used to determine any of the physical quantities at any point in the star from the corresponding non-dimensional quantity. In particular, the central temperature and central density were thus obtained.

Theoretical Results and Comparison with Observations

The numerical results for the three chosen masses are given in Table 15.2. From these data we note the following. The three cases cover the spectral type range from B1 to A2. The central temperatures run from 28 to 20 million degrees. The importance of radiation pressure, measured by $1 - \beta_f$, increases with increasing mass but does not reach decisive values even for a main sequence star of ten solar masses. The high value of δ_f for the heavy star shows that in the deep interior of heavy stars electron scattering provides the main opacity, though in the outer parts where δ has much smaller values Kramers' opacity dominates. For the lightest of the three sample stars Kramers' opacity dominates throughout. Lastly, according to the q_f values, the convective core of the heaviest of the three stars contains about $\frac{1}{4}$ of the total mass while for the lightest of the three stars it contains only $\frac{1}{6}$ of the mass.

Now we come to the critical question: How do these theoretical stars compare with real ones? Let us make this comparison first in terms of the mass-luminosity relation and second in terms of the Hertzsprung-Russell diagram. Fig. 15.2 gives the mass-luminosity relation for the three theoretical stars, indicated by crosses, as compared with the main sequence observations from Fig. 2.1, represented by dots. The agreement is as good as one could hope for. Fig. 15.3 shows the three theoretical stars, again indicated by crosses, in the Hertzsprung-Russell diagram. The crosses fall very close to the line, which represents the observed initial main sequence from Table 1.2. We may therefore conclude that the three stellar models in fact represent satisfactorily the observed upper main sequence stars.

TABLE 15.2

Results for upper main-sequence models. (Kushwaha, *Ap.J.* *125*, 242, 1957.)

$M =$	$10M_\odot$	$5M_\odot$	$2.5M_\odot$
$\log C$	-6.579	-6.140	-5.793
U_f	2.404	2.478	2.547
V_f	1.791	1.551	1.333
x_f	0.232	0.192	0.155
q_f	0.244	0.201	0.162
$\log p_f$	$+1.439$	$+1.677$	$+1.942$
$\log t_f$	-0.242	-0.174	-0.106
$1 - \beta_f$	0.025	0.007	0.002
δ_f	4.41	2.51	1.69
x^*_f	1.432	1.338	1.244
t^*_f	0.705	0.738	0.770
$\log p_c$	$+1.818$	$+2.007$	$+2.227$
$\log t_c$	-0.090	-0.042	$+0.008$
$\log D$	$+1.124$	$+0.198$	-0.780
$\log L/L_\odot$	$+3.477$	$+2.463$	$+1.327$
$\log R/R_\odot$	$+0.559$	$+0.376$	$+0.202$
$\log T_e$	$+4.350$	$+4.188$	$+3.991$
Sp. T.	B1	B5	A2
T_f	1.95×10^7	1.74×10^7	1.52×10^7
ρ_f	4.62	12.3	32.4
T_c	2.76×10^7	2.36×10^7	1.98×10^7
ρ_c	7.80	19.5	48.3

Dependence of Results on Composition

There is one major question left: How far do the theoretical results depend on the composition chosen for the models? Let us investigate this question only for the smallest of the three chosen masses, because in this case two simplifying circumstances occur. The radiation pressure is negligible and hence the parameter B has practically no influence on the model. Similarly, since electron scattering is only a minor contributor to the opacity the parameter A has little influence on the model. With A and B eliminated, the construction of the model could have proceeded as described, without any reference to a pre-chosen composition, up to the point where the values of C and D (which are unaffected by the composition) are introduced into the key relations (15.9) and (15.16). But the solution for L and R from these relations will be affected by the composition chosen. If in Eq.(15.16) we neglect the variation of all factors except those with high exponents, we find for the dependence of the radius on the composition approximately

$$R \propto \mu. \tag{15.23}$$

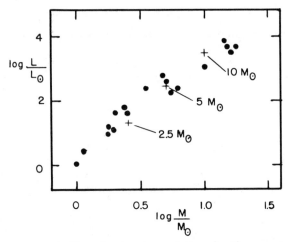

Fig. 15.2. Mass-luminosity relation for the upper
main-sequence models (crosses) compared with
observations (dots, see Fig. 2.1)

In Eq.(15.9) the variation of R just about compensates the variation of
$1 + X$, and thus we obtain from this equation for the dependence of L
on the composition

$$L \propto \frac{\mu^{7.5}}{Z}. \tag{15.24}$$

As an example, let us replace the composition (15.22) previously
chosen by

$$X = 0.70, \quad Y = 0.27, \quad Z = 0.03(\mu = 0.62),$$

which is well within the range of possible compositions suggested by
the spectroscopic data. According to Eqs.(15.23) and (15.24) these
changes in composition produce the following changes in the observables

$$\Delta \log L \approx 0, \quad \Delta \log R \approx + 0.06, \quad \Delta \log T_e \approx - 0.03.$$

Thus, the luminosity has not been changed—because of a compensation
of the increase in μ and the increase in Z in Eq.(15.24)—and the mass-
luminosity plot of Fig. 15.2 is unaltered. On the other hand, the effective
temperature is somewhat reduced, producing a small, hardly significant
discrepancy in the Hertzsprung-Russell diagram of Fig. 15.3.

We might be tempted to build up the last discussion into an actual
determination of the composition of upper main-sequence stars. In principle it is true that relations (15.9) and (15.16) permit a unique determination of the composition if C and D are known from theoretical models and
if L, M, and R are known from observations.

To obtain reliable compositions for upper main-sequence stars by this
method requires, however, still further improvements: improvements in

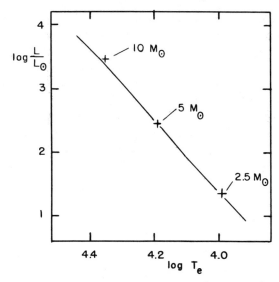

Fig. 15.3. Hertzsprung-Russell   diagram   for
upper main-sequence models (crosses) compared
with observations (line, see Table 1.2).

the models, particularly regarding the representation of the opacity; improvements in the observations of L, M, and R for stars which we know
to be young enough so that we do not have to worry about possible inhomogeneities in composition; and improvements in the relations between
the effective temperature, the color index, and the bolometric correction
which always enter the comparison between theoretical and observational
results.

For the time being we conclude that with reasonable composition the
present models for upper main-sequence stars give results which are in
entirely satisfactory agreement with observations. At the same time,
however, it is clear that a repetition of these models with a better opacity
representation and an extension of these models to higher masses would
be much worthwhile.

16. *The Lower Main Sequence*

We now leave the heavy stars and turn to the lighter ones. In this section we shall discuss the construction of homogeneous models for the initial state of stars with masses about one solar mass or less. More precisely, we shall consider here main sequence stars of Population I with
spectral types between G0 and M0. For specific examples we will use
the sun (G2) and Castor C (= YY Gem, M0) for which the observed data
are listed in Table 2.1.

Choice of Gas Characteristics Relations

What are the appropriate forms of the gas characteristics relations for these lighter stars? We may estimate the temperatures with which we will have to deal by extrapolating the central temperatures given in Table 15.2. We find that the relevant temperatures will lie below fifteen million degrees. This has three consequences. First, we can completely ignore radiation pressure. Second, we may in fair approximation ignore electron scattering. Third, we have to count with the proton-proton reaction rather than with the carbon cycle as the main energy source. We thus have, according to Eqs.(8.2), (9.16), and (10.14)

$$P = \frac{k}{\mu H} \rho T \tag{16.1}$$

$$\varkappa = \frac{4.34 \times 10^{25}}{(t/\bar{g})_0} \times Z (1 + X)^{1-\alpha} \times \frac{\rho^{1-\alpha}}{T^{3.5}} \tag{16.2}$$

$$\varepsilon = 2.8 \times 10^{-33} \times X^2 \times \rho \, T^{4.5} \tag{16.3}$$

For the two parameters $(t/\bar{g})_0$ and α which occur in the opacity equation (16.2) we use the following numerical values

$$
\begin{aligned}
&\text{for the sun:} \quad \alpha = 0.25, \;\; \log (t/\bar{g})_0 = 0.82 \\
&\text{for Castor C:} \quad \alpha = 0.50, \;\; \log (t/\bar{g})_0 = 0.38.
\end{aligned}
\tag{16.4}
$$

These values were adjusted so that Eq.(16.2) represents the accurate opacity tables as well as possible in the expected ranges of temperature and density. Eq.(16.2) for the opacity as it stands takes into account only the bound-free transitions of the heavier elements. The addition of the free-free transitions of hydrogen and helium can be accomplished with fair accuracy by replacing Z in Eq.(16.2) by $Z + (X + Y)/60$. This modification has been employed in the computations for the sun, but not yet in the older computations for Castor C.

The most consequential difference between these gas characteristic relations and those for the upper main-sequence stars is the relatively low temperature sensitivity of the proton-proton reaction. Accordingly, the energy sources will not be highly concentrated towards the center, the energy flux will not have very great values in the core, the radiative temperature gradient will not be dangerously steep, and hence the core in the lower main-sequence stars will be in stable radiative equilibrium. On the other hand, we had seen in §11 that the outermost parts of lower main-sequence stars must be in convective equilibrium. Hence, the initial models for lower main-sequence stars are just the inverse of those for upper main-sequence stars; they consist of radiative cores and convective envelopes.

We will construct these models again by covering the core with integrations from the center outward, by covering the envelope with integrations from the surface inward, and by choosing for the fitting point the interface between core and envelope where the equilibrium switches from radiative to convective.

Equations for the Envelope

Let us start with the envelope. If we apply the standard transformation (13.2), the basic differential equations (13.5) to (13.7) take the form

$$\frac{dp}{dx} = -\frac{pq}{tx^2}, \quad \frac{dq}{dx} = +\frac{px^2}{t}, \quad p = E\,t^{\frac{5}{2}} \qquad (16.5)$$

$$\text{with } E = 4\pi \left(\frac{H}{k}\right)^{2.5} G^{1.5} \times \mu^{2.5}\, M^{0.5}\, R^{1.5} \times K. \qquad (16.6)$$

The last of Eqs.(16.5) is the convective equilibrium condition (13.7) after integration with due regard to the boundary conditions (13.12). The parameter E depends on the parameter K, which is determined by the photospheric boundary conditions and can be obtained by computations of detailed model atmospheres, as we have seen in §11. Such model atmospheres have been computed for our two sample stars with the result

$$\begin{aligned}
&\text{for the sun, if } l = H\text{:} \quad K = 0.0009, \quad E = 1.0, \\
&\text{for the sun, if } l = 2H\text{:} \quad K = 0.008, \quad E = 10, \qquad (16.7) \\
&\text{for Castor C:} \quad\quad\quad\quad K = 0.07, \quad\; E = 14.
\end{aligned}$$

For the sun two values are given here for K and E corresponding to two values for the mixing length l expressed in terms of the scale height H. The value of l enters into Eq.(7.7) for the convective energy flux which has to be explicitly employed in the subphotospheric layers. The influence of the particular value of l used is quite large in G dwarfs like the sun, but fairly small in fainter dwarfs like Castor C.

Clearly, we must consider the numerical values of (16.7) for E with great caution since model atmospheres are still quite uncertain, particularly in the sub-photospheric turbulent layers, and since therefore the E values are uncertain by at least a factor 2. Accordingly, for the time being we shall not use the E values of the tabulation (16.7) but instead consider E a free, unknown parameter. Thus the convective envelopes which we have to consider are represented by the one-parameter family of solutions of Eqs.(16.5). These solutions are available in tabular form for a large number of E values.

Equations for the Core

Turning next to the radiative cores, we may apply again the supplementary transformation (13.13). Now, however, the third and the fourth of the defining conditions (13.14) become

$$C\,\frac{p_0^{2-a}\,f_0}{t_0^{9.5-a}\,x_0} = 1, \quad D\,\frac{p_0^2\,t_0^{2.5}\,x_0^3}{f_0} = 1 \qquad (16.8)$$

with $C = \dfrac{3}{4ac}\left(\dfrac{k}{HG}\right)^{7.5}\dfrac{1}{(4\pi)^{3-\alpha}}\dfrac{4.34\times 10^{25}}{(t/\bar{g})_0}\dfrac{Z\,(1+X)^{1-\alpha}}{\mu^{7.5}}\dfrac{LR^{0.5+3\alpha}}{M^{5.5+\alpha}}$ (16.9)

and $D = \left(\dfrac{HG}{k}\right)^{4.5}\dfrac{2.8\times 10^{-33}}{4\pi}X^2\,\mu^{4.5}\dfrac{M^{6.5}}{LR^{7.5}}.$ (16.10)

The four basic differential equations, in terms of the asterisked variables, are

$$\frac{dp^*}{dx^*} = -\frac{p^*q^*}{t^*x^{*2}},\quad \frac{dq^*}{dx^*} = +\frac{p^*x^{*2}}{t^*}$$

$$\frac{dt^*}{dx^*} = -\frac{p^{*2-\alpha}f^*}{t^{*8.5-\alpha}x^{*2}},\quad \frac{df^*}{dx^*} = +p^{*2}\,t^{*2.5}\,x^{*2}.$$ (16.11)

These differential equations together with the center conditions (13.19) define the solution for the radiative core—for a given value of α—as a function of the single parameter p_c^*, as we have already seen in § 13.

Fitting Procedure and Computation of Model Characteristics

The fitting together of the envelope and core solutions can again be accomplished most easily with the help of the UV plane. Fig. 13.1 shows in this plane the envelope solutions (dashed curves) as a function of E and the core solutions for $\alpha = 0.25$ (solid curves) as a function of p_c^*. The fitting condition (13.30) for $n+1$ requires that this quantity reach the value 2.5 in the radiative core just at the fitting point where the core joins the convective envelope. This condition fixes uniquely the termination point for each core solution. If we now consider one specific value of p_c^* and hence one specific core solution and termination point, then the corresponding envelope solution is uniquely determined by the fitting conditions (13.30) for U and V since these conditions prescribe that in the UV plane the envelope solution must go through the termination point of the core solution. The value of E for this envelope solution may be computed by interpolation between the available neighboring envelope integrations.

After the fitting in the UV plane has thus been completed the various non-dimensional quantities can be computed, for each chosen value of p_c^*, as follows. The values of the non-dimensional variables without asterisks can be obtained for the fitting point by interpolation between neighboring envelope solutions. The values of the non-dimensional variables with asterisks can be read off at the fitting point from the core solution. The introduction of all these quantities into the transformation (13.13) gives the values for the five constants with zero subscripts. With the help of these constants all the asterisked quantities can be translated into non-asterisked quantities. In particular one thus obtains p_c and t_c. Finally, with the help of the constants with zero subscripts, one can determine C and D from Eqs.(16.8). The results of such computations are listed in Table 16.1 for a series of values of $p_c{}^*$, separately for the two opacity formulae corresponding to the two values of α chosen in Eq.(16.4). The last column in the lower half of Table 16.1 gives the limiting model for $E = 0$, which corresponds to completely radiative stars.

Relation of Model Characteristics to Observed L, M, and R Values

How do we now translate the non-dimensional quantities of Table 16.1 into physical quantities? Let us for the moment assume that we know the exact value of E from atmospheric considerations, for example for the sun. We would then know exactly which particular model of those given in the upper half of Table 16.1 would apply to the sun. In particular we could read from this table the values of C and D which correspond to the given value of E. We might then introduce these values of C and D into the key equations (16.9) and (16.10) and attempt to determine from these equations the composition parameters X and Y, using, of course, the known values of L, M, and R for the sun. This procedure does not work well, however, for lower main-sequence stars which live on the proton-proton reaction. This failure is caused not only by the uncertainty in E but also by the following circumstance. In consequence of the moderate dependence of the proton-proton reaction rate on the temperature, D varies only with a moderate power of μ according to Eq.(16.10)—quite in contrast with Eq.(15.16) which for upper main-sequence stars living on the carbon cycle involves a very high power of μ. To make things worse, the moderate power of μ in Eq.(16.10) is largely compensated by the additional factor depending directly on X. Altogether D changes only by about 30 percent if we change the composition all the way from pure hydrogen to a mixture of half hydrogen and half helium (the abundance of the heavy elements has very little influence on D as long as it does not exceed 10 percent). This near-independence of D on the composition makes it obviously a poor quantity by which to determine the composition.

The very characteristic which makes D of poor use for determining the composition makes it of excellent use for determining E. For a star with known L, M, and R, and with even the roughest guess for the composition, we can compute D from Eq.(16.10) with good accuracy. With this value of D we can then enter Table 16.1 and determine which particular model is appropriate to the star in question. This model fixes—among other quantities—the value of E, which can then be compared with the approximate value derived from the atmosphere. It also fixes the value of C, which we then can introduce into Eq.(16.9) and thus gain at least one condition for the composition. In physical terms we may express this procedure as follows: for a given star we determine the depth of the convection zone (governed by the value of E) so that the corresponding model gives a central temperature as needed by the proton-proton reaction. Then we use the mass-luminosity relation, here represented by Eq.(16.9), for the resulting model to determine one composition parameter.

TABLE 16.1

Mathematical properties of lower main-sequence models. (Osterbrock, *Ap.J.* *118*, 529, 1953, and Schwarzschild, Howard and Harm, *Ap.J.* *125*, 233, 1957.)

$\alpha = 0.25$

p_c^*	0.688	0.6807	0.680516	0.6805077	0.68050388	0.680503833	0.680503825	0.680503822
E	44.83	34.73	23.87	19.66	8.20	5.70	4.14	1.68
U_1	2.274	1.186	0.660	0.502	0.152	0.092	0.059	0.016
V_1	2.040	5.173	7.059	7.817	10.99	12.47	13.94	19.33
x_f	0.413	0.605	0.675	0.698	0.775	0.801	0.821	0.871
q_f	0.310	0.730	0.877	0.914	0.982	0.991	0.995	0.999
$\log p_f$	+0.564	−0.039	−0.460	−0.647	−1.433	−1.754	−2.038	−2.845
$\log t_f$	−0.435	−0.632	−0.735	−0.776	−0.939	−1.003	−1.061	−1.227
$\log p_c$	+1.010	+1.238	+1.466	+1.562	+1.890	+1.988	+2.061	+2.214
$\log t_c$	−0.277	−0.254	−0.207	−0.186	−0.107	−0.083	−0.064	−0.026
$\log C$	−5.430	−5.678	−5.714	−5.708	−5.641	−5.611	−5.590	−5.539
$\log D$	+0.887	+0.610	+0.239	+0.072	−0.524	−0.705	−0.843	−1.129

$\alpha = 0.50$

p_c^*	0.6338	0.63294	0.63274	0.632720	0.632715	0.6327111	0.6327099	0.6327095
E	37.67	30.65	21.43	16.82	14.69	11.05	8.02	0
U_1	1.432	1.017	0.620	0.455	0.385	0.272	0.184	—
V_1	4.247	5.403	6.650	7.280	7.592	8.205	8.860	—
x_f	0.558	0.610	0.655	0.676	0.686	0.705	0.723	—
q_f	0.636	0.766	0.873	0.913	0.930	0.953	0.971	—
$\log p_f$	+0.148	−0.097	−0.415	−0.603	−0.705	−0.913	−1.144	—
$\log t_f$	−0.571	−0.633	−0.698	−0.732	−0.749	−0.783	−0.819	—
$\log p_c$	+1.195	+1.358	+1.595	+1.737	+1.810	+1.954	+2.100	+3.342
$\log t_c$	−0.269	−0.242	−0.193	−0.161	−0.144	−0.110	−0.075	+0.234
$\log C$	−5.385	−5.435	−5.420	−5.385	−5.361	−5.309	−5.248	−4.645
$\log D$	+0.691	+0.456	+0.066	−0.180	−0.310	−0.570	−0.836	−3.156

This procedure may be carried through as shown in Fig. 16.1. In this graph D is plotted as a function of C. Two curves are shown, corresponding to the two sequences of models listed in Table 16.1. The two curves therefore correspond to slightly different opacity laws, a difference which luckily turns out to be of little consequence. For a particular star with given L, M, and R and with an assumed value of X we can compute C and D from Eqs.(16.9) and (16.10) as functions of Y. Thus, we can draw for each star and each chosen value of X a curve in Fig. 16.1. The intersection of this curve with the curve of the model series with the appropriate α represents the solution for the given star and for the chosen X value. For this intersection we can read the values of C and D from Fig. 16.1. With their help we can select the correct model from Table 16.1 and obtain all the non-dimensional quantities for this model by interpolation. At the same time we can enter C in Eq.(16.9) and compute from this equation Z—or Y—since L, M, R, and X are all given. Now the composition, and hence μ, is completely determined. Thus, we know all the factors in the basic transformation (13.2) and we can finally translate all the non-dimensional data into physical quantities.

Checks and Results

Results of such computations are shown in Table 16.2 for our two sample stars and for three values of X. As a first check we note that the temperature and density data given in the last four columns of this table, when compared with Figs. 8.1, 9.1, and 10.1, show that we have properly chosen the forms of the gas characteristic relations (16.1) to (16.3). As a second check we compare the E values of Table 16.2 with those of Eq.(16.7) which were derived from model atmospheres. In view of the great uncertainty in the latter values the agreement is entirely satisfactory. Additional checks for our models cannot be gained by comparing them with the observed mass-luminosity relation and the Hertzsprung-Russell diagram since we have consistently used the observed values of L, M, and R for the sample stars so that it is a priori certain that our models will fulfill the observed relations.

After our models have satisfactorily withstood the possible checks we may note some of their essential physical properties. We shall here concentrate on the results for Castor C since the results for the sun are somewhat altered if the effects of evolution are taken into account, as we shall do in §23. The evolution of Castor C is too slow to have had any noticeable effects.

According to the values of x_f shown in Table 16.2, the convection zone in Castor C covers fully one third of the radius, but contains only 11 percent of the mass. As was to be expected, the central temperatures for our two lower main-sequence sample stars show a continuation of the temperature decrease with decreasing mass which our upper main-sequence models had already indicated; for Castor C, with a spectral type of M0,

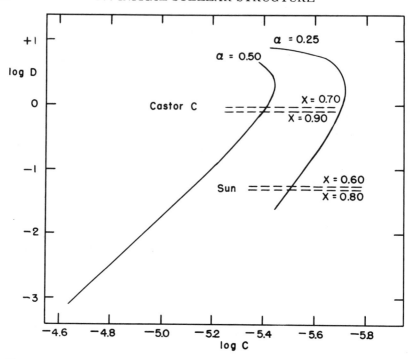

Fig. 16.1. Interrelation of the parameters C and D for lower main-sequence models.

the central temperature falls below ten million degrees. Finally, the central densities, though higher than those of the upper main sequence, remain sufficiently low on the lower main sequence for degeneracy to be unimportant.

There remains the uncomfortable indeterminateness in the value of X for any given star. Luckily this indeterminateness does not introduce large uncertainties in most of the physical characteristics we have just discussed, as shown by Table 16.2. Nevertheless, it would obviously be desirable if this indeterminateness could be eliminated. This can be achieved if we accept for Z the value deduced from spectroscopic observations. For Castor C, for example, we may assume according to Table 4.2 an abundance of the heavier elements of about 3 percent. With this value for Z we read from Table 16.2

$$X \approx 0.7 \quad \text{and} \quad Y \approx 0.3.$$

We shall discuss the problem of the internal composition of the sun in §23 where we will investigate the changes in the solar model caused by the nuclear processes during the past five billion years.

We have completed our discussion of homogeneous models for the main sequence of Population I. The investigations which we have described have produced theoretical stars which are in satisfactory agreement with

TABLE 16.2

Physical properties of the sun and Castor C as deduced from lower main-sequence models for various assumed hydrogen contents. (See Table 23.2 for improved solar data.)

	Sun (G2)			Castor C (M0)		
X	0.6	0.7	0.8	0.7	0.8	0.9
Y	0.344	0.276	0.197	0.271	0.184	0.091
Z	0.056	0.024	0.003	0.029	0.016	0.009
E	1.02	0.86	0.68	19.9	19.4	18.7
x_f	0.887	0.891	0.896	0.663	0.666	0.669
q_f	0.9997	0.9998	1.0000	0.888	0.894	0.900
T_f	0.8×10^6	0.7×10^6	0.6×10^6	2.6×10^6	2.4×10^6	2.2×10^6
ρ_f	0.0068	0.0058	0.0051	1.94	1.88	1.77
T_c	15.0×10^6	13.8×10^6	12.9×10^6	8.9×10^6	8.3×10^6	7.8×10^6
ρ_c	87	88	90	76	79	81

the observed main-sequence stars over a wide range of stellar masses. It is easy to lose sight of this over-all result in the forest of computational details. We may not even be surprised by this favorable over-all result since with the usual—and necessary—optimism of scientists we may not have expected any worse results. In review, however, we may justly consider the favorable over-all result a good cause for increased confidence in the correctness of the present theory for the stellar interior.

On the other hand, it is clear that the theoretical investigation even of just the main sequence of Population I is far from complete. It would certainly be desirable to develop a more accurate and detailed model for the sun, including the critical turbulent phenomena in the subphotospheric layers. Similarly, we should fill the gap between the upper main sequence and the lower main sequence by deriving models for the late A-stars and the F-stars in which the transition from the carbon cycle to the proton-proton reaction and from the convective core to the convective envelope occurs. Last, and perhaps most important, we should extend the present investigations to stars of lower masses and luminosities for which up till now no satisfactory models have been devised.

17. The Subdwarfs

In the two preceding sections we have covered the homogeneous models appropriate for the initial state of Population I stars. These models cannot be applied to the initial state of Population II stars because of the great difference in the abundance of the heavier elements between the two populations. Therefore, we have to consider anew the construction of homogeneous models, this time for very low values of Z, say $Z = 0$. We may concentrate entirely on the lower main sequence of Population II since in this old population the brighter stars have long ago left the initial state, run through the evolution and finished with their nuclear fuel.

Effects of Low Abundance of the Heavier Elements

Before we start with a detailed construction of appropriate models let us investigate for a moment what effects we should expect when we lower the heavy element abundance, Z. Let us use for this purpose the key equations (16.9) and (16.10) for lower main-sequence stars of Population I. Let us consider stars of a fixed mass M, a fixed hydrogen content X, and a sufficiently small heavy element content ($Z \leq 0.05$) so that the molecular weight μ is essentially fixed by X and independent of Z. Let us further assume that a change in Z will not change the model—a rather dangerous assumption—so that C and D remain fixed. If we now solve the two key equations for the luminosity and the radius we obtain in rough approximation

$$L \propto Z^{-1}, \quad R \approx \text{Const.}$$

Thus a reduction in Z leaves the radius fairly unchanged but increases the luminosity.

The physical reason for this increase in luminosity is clearly the fact that a decrease in the heavy element abundance means a decrease in opacity. This opacity decrease can not be pushed indefinitely far, however, since for very small values of Z the bound-free absorption processes of the heavier elements, represented by Eq.(9.16), will fall below the free-free absorption processes of hydrogen and helium, represented by Eq.(9.18), which set a lower limit to the opacity. If we accept as typical figures for Population I $Z = 0.03$, an average guillotine factor of 10, and Gaunt factors of the order of 1, then we find by comparing Eq.(9.16) with (9.18) that the opacity can be lowered by at most a factor 3. Accordingly the luminosity can be increased at most by this factor 3.

The results of such a change are shown in Fig. 17.1. This graph gives the main sequence passing through a K1 dwarf (the cross). If we change this dwarf—a thought experiment, not an evolution—by keeping its mass, hydrogen content, and model fixed but by reducing its heavy element abundance, we move the star along a line of constant R towards increasing luminosity. When its luminosity has increased by the maximum factor of 3, this dwarf has become an F2 star (the circle in Fig. 17.1) and lies about two thirds of a magnitude below the main sequence, as shown by the length of the dotted line in our graph. Thus, according to Eq.(1.5), the star has become a typical subdwarf. This very approximate result encourages us to expect that we will find models fitting the subdwarf sequences of Population II if we construct theoretical stars with negligible content of the heavier elements.

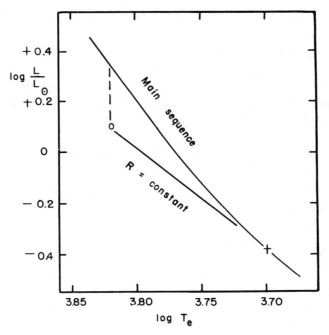

Fig. 17.1. Relation of subdwarfs to the main-sequence in
the Hertzsprung-Russell diagram.

Gas Characteristics Relations and Boundary Conditions

We start the model building as always with the selection of appropriate
forms for the three gas characteristics relations. For the equation of
state we shall use that of an ideal gas. For the opacity we shall use the
interpolation formula (9.20), which takes account of the free-free absorption processes of hydrogen and helium as well as of electron scattering,
and is thus exactly appropriate for cases of negligible abundance of the
heavier elements. For the energy generation we shall use

$$\varepsilon = 10^{-29} X^2 \rho T^4 \qquad (17.1)$$

which, according to Eq.(10.14) and Table 10.1, represents the proton-
proton reaction in the temperature range of about fifteen million deegrees.

Next we have to decide upon the proper boundary conditions at the surface. Little appears to be known as yet about the possible existence or
depth of a convection zone below the atmosphere in a subdwarf, that is
the value of the parameter E has not yet been estimated for subdwarfs
from considerations of model atmospheres. Nor can we determine this
parameter for the subdwarfs by considerations of the interior models, as
we did for the ordinary dwarfs, since this procedure requires observed
masses which are not available for subdwarfs. We shall therefore proceed by first constructing a purely radiative model and by afterwards in-

vestigating the changes which would be caused by the introduction of convective envelopes.

Construction of Radiative Model

For the entirely radiative model we may use the standard transformation (13.2). The two hydrostatic equilibrium conditions are then represented by Eqs.(13.5) and (13.6), while the conditions of radiative and thermal equilibrium become

$$\frac{dt}{dx} = -C\,\frac{p^{1.5}f}{t^{6.25}x^2}, \quad \frac{df}{dx} = +D\,p^2\,t^2\,x^2 \tag{17.2}$$

$$\text{with } C = \frac{3}{4ac}\left(\frac{k}{HG}\right)^{5.75}\left(\frac{1}{4\pi}\right)^{+2.5} \times 1.6 \times 10^{11} \times \frac{1+X}{\mu^{5.75}} \times \frac{LR^{0.25}}{M^{4.25}} \tag{17.3}$$

$$\text{and } D = \left(\frac{HG}{k}\right)^4 \frac{10^{-29}}{4\pi} \times X^2\,\mu^4 \times \frac{M^6}{LR^7}. \tag{17.4}$$

We shall have to solve the four differential equations again by numerical integrations from the surface inwards and from the center outwards. For the fitting let us choose a point so that all of the energy generation occurs inward from the fitting point. Accordingly, the right-hand side of the second of Eqs.(17.2) is negligible in the outer parts from the surface to the fitting point, and the parameter D therefore does not occur in the inward integrations from the surface. Since these integrations have completely defined starting values by the first alternative of the boundary conditions (13.12), the inward integrations depend only on the parameter C.

For the outward integrations from the center, just as for the ordinary dwarfs we may use the supplementary transformation (13.13) with the exponents in the third and the fourth of the defining conditions (13.14) appropriately modified for the present case. We will find again that the resulting differential equations and the boundary conditions (13.19) completely define the integrations except for the one parameter $p_c{}^*$.

The fitting conditions (13.30) require that the integration from the surface and the integration from the center, when plotted in the UV plane, have the same value of $n + 1$ at the intersection. This will occur only for one specific value of C and one specific value of $p_c{}^*$. These values can be found by successive trial integrations both from the surface and from the center. After the correct values have been found and the two corresponding integrations performed, we can read from the integrations all the values of the asterisked and non-asterisked variables at the fitting point. By introducing these into the transformation (13.13) we obtain the five constants with zero subscript which in turn permit us the translation of all the asterisked quantities into those without asterisks— giving us in particular p_c and t_c. Finally, we obtain the value of D from the fourth of the defining Eqs.(13.14).

The results of these calculations are

$$\log p_c = +1.91, \ \log t_c = -0.07, \ \log C = -4.63, \ \log D = -0.75. \quad (17.5)$$

These values are completely unique and depend on nothing but the chosen form for the three gas characteristics relations. This situation is in contrast with the general situation for the ordinary dwarfs discussed in the preceding section where we had introduced the depth of the convection zone, i.e. E, as an arbitrary parameter. The present model for the subdwarfs, however, corresponds exactly to the limiting case of $E = 0$ presented in the last column of Table 16.1 in the preceding section.

Let us apply our subdwarf model to six sample stars, the first three stars of one solar mass and the last three of 0.6 solar masses. For the three samples of each mass we shall use hydrogen contents of 1.0, 0.75, and 0.5 respectively. The hydrogen content completely defines the composition since we are assuming $Z \approx 0$. For each of these six sample stars we can compute L and R from the key equations (17.3) and (17.4) with the help of the numerical values for C and D of Eq.(17.5). From L and R follows the effective temperature. The central temperature and density can be obtained from the transformation Eqs.(13.2) with the help of the numerical values for p_c and t_c of Eqs.(17.5). The results of these computations are listed in Table 17.1.

Fig. 17.2 represents our six sample stars in the Hertzsprung-Russell diagram. It shows that the subdwarf model has in fact given us subdwarfs.

Effects of Possible Convective Envelopes

Before we consider this result in detail we still have to investigate how these results will have to be changed if the subdwarfs should turn out to have convective envelopes of appreciable depth. No subdwarf models with convective envelopes are as yet available. This circumstance has the disadvantage that we cannot give well-defined numerical results. It has the advantage, however, of forcing us to study the effects

TABLE 17.1

Properties of model subdwarfs without heavy element content and without convective envelopes. (A. Reiz, *Ap.J. 120*, 342, 1954.)

	$1 M_\odot$			$0.6 M_\odot$		
X	1.0	0.75	0.5	1.0	0.75	0.5
$\log \dfrac{L}{L_\odot}$	-0.172	$+0.326$	$+0.926$	-1.101	-0.602	-0.003
$\log \dfrac{R}{R_\odot}$	-0.062	-0.126	-0.212	-0.119	-0.184	-0.269
$\log T_e$	$+3.748$	$+3.905$	$+4.098$	$+3.545$	$+3.702$	$+3.894$
$T_c \times 10^{-6}$	11.33	15.57	23.29	7.76	10.67	15.93
ρ_c	68.9	107.3	194.3	61.3	96.1	172.8

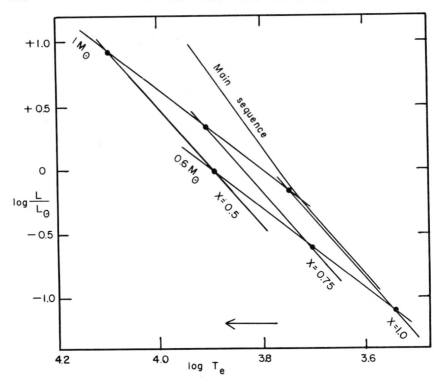

Fig. 17.2. Hertzsprung-Russell diagram for model subdwarfs (data from Table 17.1).

of convective envelopes in more general terms. From this point of view let us look once more at the models with convective envelopes discussed in the preceding section.

For a star of given mass and composition the luminosity and the radius are determined by the key equations (16.9) and (16.10) in which the parameters C and D are functions of E, the depth of the convection zone. If we use the purely radiative model ($E = 0$) as a standard of comparison and denote it by a subscript zero, we obtain from the two key equations, for $\alpha = 0.5$,

$$\frac{L}{L_o} = \left(\frac{C}{C_o}\right)^{\frac{15}{11}} \times \left(\frac{D}{D_o}\right)^{\frac{4}{11}}, \quad \frac{R}{R_o} = \left(\frac{C}{C_o}\frac{D}{D_o}\right)^{-\frac{2}{11}} \qquad (17.6)$$

Thus we can compute the variation of the luminosity and the radius with depth of the convection zone from the numerical data listed in Table 16.1. The results are given in Table 17.2. We see that with increasing depth of the convection zone (from right to left in Table 17.2) the luminosity increases but very slightly while the radius decreases strongly. This result is not surprising. On the one hand, the luminosity is determined by the radiative core, which comprises the bulk of the mass of the star. On the other hand, the radius is determined not only by the size of the core but also by the extent of the envelope, which in turn critically depends on the character of the equilibrium in the envelope.

Therefore the introduction of a convective envelope will move a stellar model in the Hertzsprung-Russell diagram horizontally to the left. For the sun ($E \approx 3$) as an example, the amount of this shift to the left is about 0.1 in $\log T_e$, as indicated by the arrow at the bottom of Fig. 17.2. If we estimate that similar shifts to the left will apply to the subdwarfs, we find that our six sample stars lie even further to the left of, or below, the main sequence than shown in Fig. 17.2.

We conclude that the preliminary models for stars with negligible content of the heavier elements represent well the main features of the subdwarf sequences of Population II in the Hertzsprung-Russell diagram.

Problem of Helium Abundance in Subdwarfs

Before we leave the subdwarfs we should still discuss one question which at present remains unanswered, but which if solved would provide an essential cue for cosmological problems: what is the helium abundance of the subdwarfs? If we were to find that the original composition of these old stars contained virtually no helium, we would probably have to conclude that our galaxy in its initial state consisted essentially purely of hydrogen. If on the other hand we were to find that the oldest stars in the initial state contained an appreciable amount of helium, we probably would have to conclude that the origin of helium preceded the first star formation in our galaxy.

It does not seem likely that we will be able to answer this question by spectroscopic observations of bright Population II stars with early spectral types. These stars are nearly certainly in advanced stages of their evolution so that their present helium content—however large we may find it—may have been produced by nuclear transmutations during the stellar lifetime and hence could tell us nothing about their initial helium content. This complication is avoided if we use the subdwarfs. Their spectral types are too late to permit direct spectroscopic observation of their helium content. On the other hand, Fig. 17.2 indicates that the position of a subdwarf in the Hertzsprung-Russell diagram depends on its helium content. Since, however, the position in the Hertzsprung-

TABLE 17.2

Variation of luminosity, radius, and effective temperature with depth of convection zone. (Data from Table 16.1, lower half.)

$E =$	37.67	30.65	21.43	16.82	14.69	11.05	8.02	0
$\log L/L_0$	$+0.390$	$+0.236$	$+0.115$	$+ .073$	$+0.058$	$+0.035$	$+0.021$	0
$\log R/R_0$	-0.565	-0.513	-0.445	-0.407	-0.387	-0.349	-0.312	0
$\log T_e/T_{eo}$	$+0.380$	$+0.316$	$+0.251$	$+0.222$	$+0.208$	$+0.183$	$+0.161$	0

Russell diagram depends on the depth of the convective envelopes as well, we can use the observed positions for our present purpose only after accurate model atmosphere have been constructed and the depth of the convective zone thus determined with a reasonable certainty.

The difficulties and uncertainties introduced by the convective envelope could be circumvented if we had an accurately observed mass for a subdwarf. We could then eliminate R between the two key equations (17.3) and (17.4) and thus obtain a relation for L which is little affected by the convective envelope, as we have seen above. We could then introduce into this relation the observed values for L and M and thus obtain one condition on the composition. This one condition would be sufficient in this case—in contrast to the ordinary dwarfs—to determine the composition completely since with $Z \approx 0$ the composition contains only one free parameter.

At present we do not have the answer to our question about the helium content of the subdwarfs with all its cosmological implications since neither theoretical model atmospheres nor observed masses of sufficient precision appear to be available as yet. But the answer does not seem far out of our grasp.

18. The Apsidal Motion Test

Even though homogeneous models appear to represent the observed stars in their initial state very well, we must admit that the points of comparison between theory and observation are still rather few. Any additional tests for the theory are therefore of great value.

For the purpose of testing a theory the physicist will often subject the object of his studies to perturbations and then watch its reactions. Well-known examples of this method are the application of electric and magnetic fields to atoms and the observations of the resulting Stark and Zeeman effects. Astronomers do not have the freedom of the physicists to apply perturbations to the objects of their study. Nature herself, however, has subjected a fair sample of stars to various perturbations, and all that astronomers need do is to observe carefully these natural experiments.

We find stars in four specific types of perturbations: rotation, large-scale magnetic fields, close companions in binaries, and pulsations. The first two types until now have not lead to actual tests of the theory of the stellar interior, partly because the effects of these perturbations are difficult to observe and partly because it does not seem possible at present to make definite theoretical predictions. This latter difficulty is due to the serious indefiniteness of the distribution of the angular momentum in the case of the rotation perturbation and of the geometry of the magnetic fields in the case of the magnetic perturbation. In contrast, the third type of perturbation, that by a close companion, has led to a significant test of the theory for main-sequence stars. This test is the subject of this section. The fourth perturbation, that of stellar pulsations, we shall take up in §27, in connection with the stability problem of white dwarfs.

The two companions in a close binary will distort each other. The amount of this distortion depends on the internal structure of the stars, and thus provides a possibility of testing stellar models. The distortion is difficult to observe directly, however, even in the favorable cases of eclipsing variables. Luckily there is a secondary effect of the distortion which is more readily observable. The distortion alters slightly the gravitational attraction between the two components and hence perturbs the binary orbit. In particular, it causes the major axis, or line of apsides, to rotate slowly.

Observation of Apsidal Motion

Apsidal motion can be observed in eclipsing binaries, as illustrated by Fig. 18.1. On the right-hand side of this figure is drawn the relative orbit of the secondary component around the primary one. The four diagrams show four successive phases in the rotation of the major axis. The two circles in each diagram indicate the positions in which eclipses occur, the lower circle (towards the observer) for the primary eclipse and the upper circle for the secondary eclipse.

The corresponding light curves are shown in the left-hand side of Fig. 18.1. The only feature of these light curves which is of interest here is the position of the secondary minimum relative to the primary minimum. In the first diagram of Fig. 18.1 clearly the radius vector in the relative orbit covers a much smaller area from primary eclipse to secondary

Fig. 18.1. Eclipsing variable with apsidal motion in four phases of the apsidal period. At left: light curve. At right: orbit.

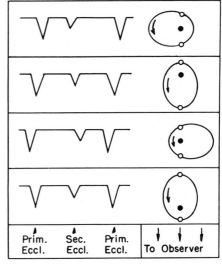

eclipse than from secondary eclipse to primary eclipse. Accordingly, the time interval from primary to secondary minimum is shorter than that from secondary to primary minimum. In the second diagram the orbit is oriented symmetrically with respect to the observer's line of sight and hence the minima follow each other after equal time intervals. In the third diagram again an asymmetric situation is encountered. This time, however, the interval from primary to secondary minimum is longer than that from secondary to primary minimum. Finally, in the fourth phase another symmetrical orientation is obtained. Thus one can determine the period of the apsidal motion of an eclipsing variable just by observing the slow variation of the time of the secondary minima relative to the primary minima.

The relevant observational data are collected in Table 18.1. The binaries in which apsidal motion is observed with satisfactory accuracy are still few in number and are entirely limited to the upper main sequence. In spite of these limitations, the available sample constitutes a powerful test for the theory of the stellar interior.

To carry through this test three steps have to be executed. First, the distortion of one component by the other has to be derived. Second, the change in the mutual gravitational attraction caused by the distortion has to be determined. Finally, the perturbation of the orbit has to be computed. We shall here carry out the first two steps in full detail, but for the last step, which belongs to the field of celestial mechanics, we shall only report the result.

Simplifying Approximations

To simplify our task let us permit ourselves the following approximations. If we are to deal with apsidal motion that is observable we have to consider binary orbits with appreciable eccentricity. In these orbits the distance between the two components is far from being constant and in consequence the distortion of the components will vary with time. Therefore our problem is, strictly speaking, a hydrodynamic rather than a hydrostatic one. It turns out, however, that the reaction time of a star, that is the periods of its free oscillations, is rather short compared with the orbital period so that the star is capable of adjusting itself at any time fairly closely to the instantaneous distorting force. We may therefore assume with fair accuracy quasi-stationary equilibrium, that is we may compute the distortion at any time from the equilibrium condition using the instantaneous distance between the two components.

As a second simplification, let us consider only those cases in which the distortions are sufficiently small so that a first-order perturbation theory suffices. This has, besides the usual advantage of greatly simplified equations, the additional advantage of permitting us to consider the distortion of the first component by the second separately from the distortion of the second component by the first.

As a final simplification let us ignore the rotation of the stars. The effects of the distortion caused by rotation can, in first-order theory, be computed by much the same method which we will use here for the main effects. To correct for the effects of rotation we have simply to add to our final Eq.(18.18) additional terms which are always of moderate size.

Hydrostatic Equilibrium Conditions

Let us now consider the hydrostatic equilibrium conditions for the distortion of the first component by the second component. The geometry of the situation is represented by Fig. 18.2. According to our simplifying approximations, we may forget the orbital motion as well as the rotation, and hence our problem has axial symmetry around the line connecting the centers of the two components. We may then characterize the position of the arbitrary point P by the two coordinates r and θ as defined in Fig. 18.2.

Since our present problem does not reduce to just one coordinate—as it always does in an unperturbed star—we have to use here the hydrostatic equilibrium condition in its vectorial form

$$\nabla P = -\rho \nabla V \tag{18.1}$$

where V is the gravitational potential. This potential consists of two parts,

$$V = V_s + V_d, \tag{18.2}$$

where V_s is the potential arising from the mass of the first component itself while V_d is the distorting potential caused by the second component. The Poisson equation gives us for V_s

$$\nabla^2 V_s = 4\pi G \rho \tag{18.3}$$

where ρ represents the density as a function of position throughout the first component. The second component may in first-order theory be considered a point mass and hence its potential is given by $-GM_2/d$, where d is defined by Fig. 18.2. We may express this potential in terms of our two coordinates by the following development

$$-\frac{GM_2}{d} = -\frac{GM_2}{D}\left[1 + \frac{r}{D}P_1(\theta) + \frac{r^2}{D^2}P_2(\theta) + \ldots\right],$$

where the first two Legendre polynomials are given by

$$P_1(\theta) = \cos\theta \quad \text{and} \quad P_2(\theta) = \frac{3}{4}\cos 2\theta + \frac{1}{4}$$

and where D represents the distance between the centers of the two components. The first term in the above development is constant and hence does not represent a force and may be ignored. The second term

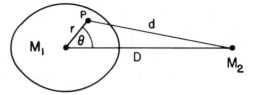

Fig. 18.2. Distortion of one component (M_1) by the other component (M_2).

is proportional to $r \cos \theta$, the gradient of which is a constant vector. Therefore this term does not produce a distorting force, but gives exactly the orbital acceleration GM_2/D^2. Thus for the distorting potential we are left with

$$V_d = - \frac{GM_2}{D^3} r^2 P_2(\theta). \qquad (18.4)$$

Eqs.(18.1) to (18.4) represent the hydrostatic equilibrium conditions for our present problem. We shall see later that for the present purposes we do not need to consider the perturbations in thermal equilibrium.

From Eq.(18.1) we can derive one very general result. According to this equation, the pressure gradient is everywhere parallel to the gradient of the potential. Since the pressure gradient is always perpendicular to the surfaces of constant pressure and the potential gradient perpendicular to the surfaces of constant potential, it follows that these two types of surfaces must be everywhere tangential to each other. This, however, can only be so if in fact the two types of surfaces are identical. Thus we find that the pressure must be constant on an equipotential surface and that instead of considering the pressure a function of the two coordinates we may consider it a function of the one variable V, i.e.

$$P = P(V) \qquad (18.5)$$

We can go further than this. From Eq.(18.5) we obtain the relation between the pressure gradient and the potential gradient:

$$\nabla P = \frac{dP}{dV} \nabla V.$$

If we introduce this relation into Eq.(18.1) and then cancel the potential gradient on both sides of this equation, we find

$$\frac{dP}{dV} = -\rho.$$

Now the left-hand side of this equation is a pure function of V and thus we get for the density on the right-hand side

$$\rho = \rho(V). \qquad (18.6)$$

We see that the density, just as the pressure, is a function of the potential only and hence is constant on the equipotential surfaces.

Transformation of the Poisson Equation

The fact that the pressure and the density are constant on the equipotential surfaces suggests that it would be useful for our present problem if we replaced the two geometrical coordinates r and θ by two other coordinates chosen so that one of them is constant on the equipotential surfaces. We can achieve this replacement as follows.

Let us describe the form of the equipotential surfaces by

$$r = \bar{r}[1 + c_1(\bar{r})P_1(\theta) + c_2(\bar{r})P_2(\theta) + \dots] \qquad (18.7)$$

where $\bar{r}$ is the mean distance of a particular equipotential surface from the center. Eq.(18.7) gives for any direction θ the distance from the center, r, of any particular equipotential surface characterized by $\bar{r}$. The functions $c_n(\bar{r})$ describe the deviations of the equipotential surfaces from spherical symmetry.

To simplify our equations let us introduce right here

$$c_1(\bar{r}) = 0. \qquad (18.8)$$

If we did not, but carried the $P_1(\theta)$ terms through all our equations, we would find at the end that Eq.(18.8) is in fact true as a consequence of the lack of a $P_1(\theta)$ term in the distorting potential (18.4). Let us furthermore ignore all the Legendre terms higher than $P_2(\theta)$, as we have done already in the distorting potential (18.4).

We now may use Eq.(18.7) to replace the coordinate r by $\bar{r}$ while continuing to use θ as the second coordinate. Since by definition $\bar{r}$ is constant on the equipotential surfaces and since by Eqs.(18.5) and (18.6) the same is true for P and ρ, we may now write our main deductions from the hydrostatic equilibrium condition in the form

$$V = V(\bar{r}), \quad P = P(\bar{r}), \quad \rho = \rho(\bar{r}). \qquad (18.9)$$

It still remains to transform the Poisson Eq.(18.3) into the new coordinates. In this equation we may replace the partial potential V_s by the total potential V, since the Laplacian of the difference, i.e. of V_d given by Eq.(18.4), is zero. To carry through the transformation it is useful to note that in first order

$$\left.\frac{\partial \bar{r}}{\partial r}\right|_{\theta = \text{const}} = 1 - \frac{d(\bar{r}c_2)}{d\bar{r}}P_2(\theta), \quad \left.\frac{\partial \bar{r}}{\partial \theta}\right|_{r = \text{const}} = -\bar{r}c_2\frac{dP_2(\theta)}{d\theta}$$

and that the Legendre polynomials fulfill the relation

$$\frac{\partial^2 P_n}{\partial \theta^2} + \frac{\cos\theta}{\sin\theta}\frac{\partial P_n}{\partial \theta} = -n(n+1)P_n.$$

If we furthermore remember that V does not depend on θ but only on $\bar{r}$, we obtain for the Poisson equation in the new coordinates

$$\frac{d^2V}{d\bar{r}^2} + \frac{2}{\bar{r}}\frac{dV}{d\bar{r}} - \left[\frac{d^2(\bar{r}c_2)}{d\bar{r}^2} + \left(\frac{2}{\bar{r}} + 2\frac{d^2V}{d\bar{r}^2} \Big/ \frac{dV}{d\bar{r}}\right)\frac{d(\bar{r}c_2)}{d\bar{r}} - \frac{4}{\bar{r}^2}\bar{r}c_2\right]\frac{dV}{d\bar{r}}P_2(\theta)$$

$$= 4\pi G\rho(\bar{r}). \quad (18.10)$$

Differential Equation for the Distortion Function c_2

The Poisson equation in the form (18.10) permits us to draw one conclusion immediately, as follows. The first two terms of the left-hand side of Eq.(18.10) do not depend on θ, nor does the right-hand side. Hence the only remaining term, the one with the long bracket, must not depend on θ either. It does, however, contain the factor $P_2(\theta)$. To resolve this discrepancy the term must vanish, that is the bracket must be zero.

All the terms in the bracket contain c_2 which is already of first order. We may therefore use zero-order approximations for the other quantities occurring in the bracket. Thus we may use for the quantities the values given by the unperturbed model—without having to worry about the perturbations in the thermal equilibrium of the star. In particular, we may use the values of the invariant U of the unperturbed model, which according to its definition (13.24) is related to the potential V by

$$\frac{d^2V}{d\bar{r}^2} = \frac{dV}{d\bar{r}}\frac{U-2}{\bar{r}}.$$

With the help of this relation we find for the condition that the bracket in Eq.(18.10) must be zero

$$\frac{d^2c_2}{d\bar{r}^2} + \frac{2U}{\bar{r}}\frac{dc_2}{d\bar{r}} - 2\frac{3-U}{\bar{r}^2}c_2 = 0. \quad (18.11)$$

Thus we have a condition for the distortion function $c_2(\bar{r})$.

Eq.(18.11) is a second-order, homogeneous differential equation for c_2. Since the coefficient of the first derivative is singular at the center, we have the center condition

$$\frac{dc_2}{d\bar{r}} = 0 \quad \text{at} \quad \bar{r} = 0. \quad (18.12)$$

The differential equation (18.11) and the center condition (18.12) uniquely define the run of c_2 for all values of $\bar{r}$ from the center to the surface, except for a constant factor which has to be determined by the following surface conditions.

The Surface Conditions

As always in potential problems, the surface conditions are obtained by fitting the potential for the inside of the body considered to the appropriate potential for the outside. The potential V_s arising from the mass of the first component can be represented outside this component by the development

$$V_s = -\frac{a_0}{r} - \frac{a_1}{r^2} P_1(\theta) - \frac{a_2}{r^3} P_2(\theta) + \cdots .$$

Let us again introduce $a_1 = 0$ and ignore the terms higher than a_2. Since furthermore the first term must give the unperturbed potential, we get

$$V_s = -\frac{GM_1}{r} - \frac{a_2}{r^3} P_2(\theta). \tag{18.13}$$

The distorting potential V_d may be represented outside the first component by Eq.(18.4) as before. The total potential V is therefore given outside the first component by the sum of Eqs.(18.4) and (18.13).

This total potential for the outside must be identical at the surface with the total potential $V(\bar{r})$ for the inside, and the same must be true for the first derivative with respect to r. Thus we have at the surface $(\bar{r} = R_1)$

$$-\frac{GM_1}{r} - \frac{a_2}{r^3} P_2(\theta) - \frac{GM_2}{D^3} r^2 P_2(\theta) = V(\bar{r}),$$

$$+\frac{GM_1}{r^2} + 3\frac{a_2}{r^4} P_2(\theta) - 2\frac{GM_2}{D^3} r P_2(\theta) = \frac{dV}{d\bar{r}} \left(1 - \frac{d(\bar{r}c_2)}{d\bar{r}} P_2(\theta)\right)$$

On the left-hand side of these equations we have to replace r by $\bar{r}$ according to Eq.(18.7). This has to be done explicitly, however, only in the first term of each of the two equations since the remaining terms are already of first order so that in them r can be set equal to $\bar{r}$. After this replacement the zero-order terms cancel and the first-order terms, divided by $P_2(\theta)$, give for $\bar{r} = R_1$

$$+\frac{GM_1}{\bar{r}} c_2 - \frac{a_2}{\bar{r}^3} - \frac{GM_2}{D^3} \bar{r}^2 = 0,$$
$$-2\frac{GM_1}{\bar{r}^2} c_2 + 3\frac{a_2}{\bar{r}^4} - 2\frac{GM_2}{D^3} \bar{r} = -\frac{GM_1}{\bar{r}^2} \frac{d(\bar{r}c_2)}{d\bar{r}} . \tag{18.14}$$

If we eliminate a_2 from these two equations we obtain

$$\frac{GM_1}{R_1^3} \left(2c_2 + \bar{r}\frac{dc_2}{d\bar{r}}\right)_{R_1} = 5\frac{GM_2}{D^3} . \tag{18.15}$$

This condition fixes the one constant factor in the distortion function c_2 which Eqs.(18.11) and (18.12) had left undetermined.

We have completed the determination of the distortion of the first component by the second component and now have to derive the effect of this distortion on the binary orbit.

The Perturbing Potential and the Rate of Apsidal Motion

The potential arising from the first component is given by Eq.(18.13). The first term in this equation is the usual potential which causes the Keplerian motion of the two components. The second term is the perturbing potential which causes the motion of the line of apsides. The coefficient a_2 of the perturbing potential can be derived from the first of Eqs.(18.14). If in this equation M_1 is eliminated with the help of Eq.(18.15), one finds

$$a_2 = 2k \times R_1{}^5 \, \frac{GM_2}{D^3} \tag{18.16}$$

$$\text{with } k = \left[\frac{3c_2 - \overline{r} \, dc_2/d\overline{r}}{4c_2 + 2\overline{r} \, dc_2/d\overline{r}} \right]_{R_1} \tag{18.17}$$

It is solely through this coefficient k that the internal structure of the components affects their binary motion.

As the last step of this lengthy derivation we have to introduce the potential (18.13), with the coefficient a_2 given by Eqs.(18.16) and (18.17), into the equations of motion for the binary orbit and determine the rate of apsidal motion. We shall not enter here into this problem in orbit perturbations, but just report the result. The rate of apsidal motion, as measured by the reciprocal of the apsidal motion period, is found to be

$$\frac{\text{Orbital period}}{\text{Apsidal period}} = 15 \left(k_1 \frac{R_1{}^5}{D^5} \frac{M_2}{M_1} + k_2 \frac{R_2{}^5}{D^5} \frac{M_1}{M_2} \right) \left(\frac{1 + \dfrac{3}{2}e^2 + \dfrac{1}{8}e^4}{(1 - e^2)^5} \right), \tag{18.18}$$

where e is the orbital eccentricity. The first term in the first parentheses of this equation arises from the distortion of the first component by the second component, which we have derived above, while the second term arises from the distortion of the second component by the first component. The two terms have, of course, identical forms; only the subscripts are interchanged. The values of k_1 and k_2 are in general not identical since they refer to the two components respectively. We shall, however, apply Eq.(18.18) only to those binaries in which the two components are rather similar so that we may assume one average value of k for both components.

Theoretical Results and Comparison with Observations

We are finally ready to put our theoretical stellar models to the apsidal motion test. It is easy to compute the value of k for each of the three models for upper main-sequence stars which we have discussed in §15. The model gives U as a function of x, which we introduce into the differential Eq.(18.11). A single numerical integration of this equation for each model from the center to the surface gives c_2 and its derivative at the surface, which we introduce into Eq.(18.17). Thus we obtain k. The results of these computations are

$$\begin{aligned}
\log k &= -1.70 &\text{for } M &= 10 \; M_\odot, \\
\log k &= -1.87 &\text{for } M &= 5 \; M_\odot, \\
\log k &= -2.07 &\text{for } M &= 2.5 \; M_\odot,
\end{aligned} \qquad (18.19)$$

On the other hand we can determine k for each of the well-observed apsidal motion binaries listed in Table 18.1. By introducing the various observed quantities into Eq.(18.18) and by solving this equation for k, we obtain the values given in the last column of Table 18.1.

The comparison of theory and observations is given in Fig. 18.3, which shows k as a function of spectral type or effective temperature. The dots represent the observational values of Table 18.1 while the circles represent the theoretical values (18.19). The agreement between theory and observations, though not yet perfect, appears entirely satisfactory for the present status of the stellar models. The theoretical k values are larger than the average observed values, but only by a factor of about 3, and at least part of this factor may be explained by evolutionary effects, as we shall see in §22. The observed variation of k with spectral type is well reproduced by the theoretical values.

The satisfactory outcome of the apsidal motion test is encouraging but at the same time proves that much would be gained for the certainty of this basic test if the theoretical models of the upper main sequence could be improved, if the accuracy of the apsidal motion observation of upper main sequence binaries could be increased, and—most important of all— if it were possible to extend the apsidal motion observations to other

TABLE 18.1

Observations of apsidal motion in eclipsing variables. (Russell, *Ap.J.* 90, 641, 1939. Rediscussion by Keller, *Ap.J.* 108, 347, 1948. Spectra by Wood, *Publ. Univ. Pennsylv., Astr. Series*, Vol. 8, 1953. New data for GL Car: van Wijk, et al., *A.J.* 60, 95, 1955, and for YY Sgr: Keller and Limber, *Ap.J.* 113, 637, 1951.)

Star	Spec. Type	$\log T_e$	Orb. Per. (days)	Ecc. e	$\dfrac{M_{Pr.}}{M_{Sec.}}$	R/D Pr.	R/D Sec.	Aps. Per. (yrs.)	$\log k$
Y Cyg	O9	4.44	3.00	0.14	1.01	0.206	0.206	46	−1.9
GL Car	B3	4.25	2.42	0.16	1.	0.216	0.216	25	−1.9
AG Per	B3	4.25	2.03	0.07	1.13	0.23	0.21	72	−2.4
RU Mon	B9	4.07	3.58	0.44	1.	0.13	0.11	600.	−2.3
HV 7498	A0	4.04	3.47	0.55	1.08	0.123	0.147	600	−2.9
YY Sgr	A0	4.04	2.63	0.16	1.	0.17	0.16	330	−2.4
CO Lac	A0	4.04	1.54	0.03	1.	0.24	0.24	40	−2.4
V523 Sgr	A5	3.94	2.32	0.17	1.	0.22	0.22	200	−2.8

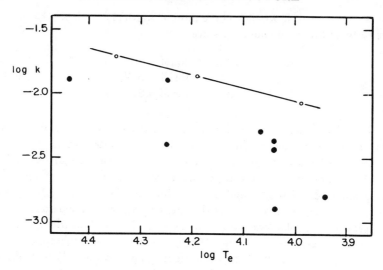

Fig. 18.3. The apsidal motion coefficient k as function of effective temperature. The dots represent the observed binaries of Table 18.1 and the circles the theoretical models of § 15.

parts of the Hertzsprung-Russell diagram, such as the lower main sequence and the subgiant region.

19. Pre-main-sequence Contraction

Now that we have constructed and checked the initial models for main-sequence stars of various masses and compositions, logically the next step is to follow the evolution of these stars from the main sequence on forward in time. This we shall do in the next chapter. In this section we shall interrupt our logical sequence and probe backwards in time into the evolutionary phases of a star prior to its main-sequence phase.

The pre-main-sequence phases of stellar evolution represent the contraction of a stellar mass from the state of an interstellar cloud to the initial main-sequence state. By definition, the latter state represents the phase in which for the first time nuclear hydrogen burning provides the main energy source. Accordingly, hydrogen burning is negligible during the pre-main-sequence phases. Regarding other nuclear energy sources, we have seen in §10 that processes using deuterium, lithium, beryllium, or boron can proceed at the relatively low interior temperatures we should expect for the pre-main-sequence phases. But barring any unforeseen high abundances of these normally rare elements, these nuclear processes have an energy store only sufficient to retard somewhat the pre-main-sequence contraction but not to alter it substantially.

Accordingly, the contraction phases we are now considering are characterized by a negligible rate of nuclear energy generation ε, so that the entire energy expenditure must be covered by the gravitational energy set free by the contraction. Thus we can not use Eq.(5.8) to represent the thermal equilibrium condition as before, but we must use the more general Eq.(5.10), with $\varepsilon = 0$.

This switch in the equation for thermal equilibrium, however, does not alter the character of the over-all problem discussed in §12; given one particular state of a star one can derive all subsequent evolutionary states uniquely one after the other. The only problem is: what is the state with which we shall begin?

Ideally, we should like to start with a model simulating a dense interstellar cloud. But such a starting model has two formidable difficulties. First, what is a proper model for an interstellar cloud? And second, what is the opacity of stellar matter at the low temperatures and densities characteristic for the very early phases following the cloud state? To avoid these difficulties for the time being, investigations have largely concentrated on the later contraction phases, starting with the states in which the stellar radius is not more than, say, ten times larger than in the main-sequence state. There remains the problem of what is an appropriate model for this more moderate starting state.

The general character of the over-all problem for the contraction phases makes it appear probable that the influence of the choice of a particular starting model on the subsequent evolution will last only for a comparatively short time interval, that is that the evolution tracks starting with various possible models for a star will all converge fairly fast to one and the same track. This estimate of the situation has one unfortunate and one fortunate consequence. On the unfortunate side, if the estimate is right we cannot hope to learn anything about the early contraction phases of a star by observing its later states. On the fortunate side, we may begin our computations of the later contraction phases by choosing a starting model reasonably arbitrarily.

Our problem then reduces to this. Choose a star of fixed mass and composition. Select arbitrarily a starting model with a radius of, say, ten times the expected main-sequence radius. Follow the evolution by solving the over-all problem of §12, time step after time step. Terminate the computation when the internal temperature has risen to the value necessary for the hydrogen burning to provide the main energy supply. You will then have arrived at the main-sequence state which we have discussed in the preceding sections.

The computations necessary to carry out this program are rather laborious, mainly because of the occurrence of a time derivative in Eq.(5.10), vital to the contraction phases. This problem, however, does not exceed the capabilities of the larger existing electronic computers, and as a matter of fact a number of evolutionary model sequences for the pre-main-sequence contraction phases have been obtained by using an electronic computer not only for carrying out the numerical integrations but also, for the first time, for the fully automatic determination of the eigen-values.

The evolution track of one model sequence obtained in this manner is shown in the Hertzsprung-Russell diagram of Fig. 19.1. The track starts at the right with a phase of low effective temperature and large radius. It proceeds to the left and ends at the main-sequence state.

To gain further insight into the physical character of the pre-main-sequence contraction we may either study in more detail the numerical results of the computations carried out with the help of electronic computers, or we may investigate the problem independently by an approximate but simpler method which shows up the essential phenomena somewhat more directly. Here we shall follow the second alternative.

Homologous Contraction

Let us assume for the pre-main-sequence contraction that in first approximation the star reduces its size but does not alter the mass distribution throughout its interior. This assumption is well substantiated by the detailed numerical computations.

Under this assumption of a homologous contraction the density, the pressure, and the temperature do not change their relative distributions but do change in their absolute values. The rate of change with time of density, pressure, and temperature at every layer is completely determined by the rate of change in the radius of the star. More precisely, the hydrostatic equilibrium condition (12.1) and the equation of state of an ideal gas give the proportionalities

$$\rho \propto R^{-3}, \ P \propto R^{-4}, \ T \propto R^{-1}. \tag{19.1}$$

Thus when the radius contracts by a factor 10 the temperature increases by a factor 10, the density by a factor 1,000, and the pressure by 10,000.

The proportionalities (19.1) permit a great simplification in the formulation of the contractional energy release, that is in the time-derivative term of the basic thermal equilibrium condition (5.10):

$$\frac{3}{2} \rho^{\frac{2}{3}} \frac{d}{d\tau} \left(\frac{P}{\rho^{\frac{5}{3}}} \right) = \frac{3}{2} \rho^{\frac{2}{3}} \frac{P}{\rho^{\frac{5}{3}}} \left(-\frac{4}{R} \frac{dR}{d\tau} + \frac{5}{R} \frac{dR}{d\tau} \right) = +\frac{3}{2} \frac{P}{\rho} \frac{1}{R} \frac{dR}{d\tau} \tag{19.2}$$

This relation shows that in a homologous contraction gravitation provides an energy release per gram of matter equal to the thermal energy content per gram multiplied by the logarithmic time derivative of the radius.

Construction of Contraction Model

The assumption of a homologous contraction and the consequent equa-(19.2) permit now the construction of a contraction model in much the manner we have used for the main-sequence models.

The four basic differential equations are given by Eqs.(12.1) and (12.2) for the hydrostatic equilibrium conditions, by the first of Eqs.(12.4) for the radiative transfer condition, and by Eq.(5.10) for the thermal equilibrium condition, in which we shall introduce $\varepsilon = 0$ and the expression (19.2) for the gravitational energy liberation. Since the gravitational energy source according to Eq.(19.2) is not at all highly concentrated toward the center, we can be sure that the contraction model will not have a convective core. Furthermore, we shall here ignore the possibility of a deep convective zone under the surface. This neglect may affect the resulting radii fairly seriously but should not vitiate the general character of the contraction which is the aim of this study. For the radiative opacity let us use Kramers' law as a fair approximation.

As usual, we may apply the transformation (13.2) to our basic equations. Thus the hydrostatic equilibrium conditions take the form of Eqs.(13.5) and (13.6) (with $1 = 1$ since we here consider homogeneous models) while the radiative transfer and the thermal conditions take the form

$$\frac{dt}{dx} = -C \frac{p^2}{t^{8.5}} \frac{f}{x^2}, \quad \frac{df}{dx} = +D\ px^2 \tag{19.3}$$

$$\text{with } C = \frac{3}{4ac} \left(\frac{k}{HG}\right)^{7.5} \left(\frac{1}{4\pi}\right)^3 \frac{\varkappa_0}{\mu^{7.5}} \frac{LR^{0.5}}{M^{5.5}} \tag{19.4}$$

$$\text{and } D = \frac{3}{2} \frac{GM^2}{L} \left(-\frac{1}{R^2} \frac{dR}{d\tau}\right). \tag{19.5}$$

The boundary conditions are as before given by Eqs.(13.11) and the first version of (13.12). The two quantities C and D are the eigenparameters of the problem.

If we want to solve this problem numerically, we have to divide the star at an arbitrarily chosen fitting point and carry out numerical integrations both from the center and from the surface to this fitting point. Because of the occurrence of C and D in the equations our problem has a two-parameter family of solutions from the surface. Accordingly, for any particular numerical integration from the surface we have to choose trial values for both C and D. For the integrations from the center it is again advantageous to apply the supplementary transformation (13.13) and thus to eliminate C and D from the equations. The boundary conditions at the center now take the form of Eq.(13.19) and $p_c{}^*$ is the only free parameter for which a trial value has to be chosen for each numerical integration from the center.

The construction of the complete contraction model may then be carried out as follows. Choose a set of trial values for C and D and carry out the corresponding integration from the surface to the value x_f chosen for the fitting point. Plot this integration in the UV plane. Find by trial-and-error integrations that value of $p_c{}^*$ which gives an integration from the center that, in the UV plane, passes through the termination point of the integration from the surface. For the fitting point thus fixed read the values of the non-asterisked variables from the integration from the surface and the values of the asterisked variables from the integration from the center. Introduce all these quantities into the supplementary transformation equations (13.13) and thus determine the constants with zero subscripts. Introduce these constants into the third and fourth of Eqs.(13.14) and see whether these two conditions are fulfilled (the first and the second of these equations are automatically satisfied by the fitting in the UV plane). If these two conditions are not fulfilled choose another set of trial values for C and D and repeat the entire procedure. Repeat as often as is necessary to fulfill the two conditions. Thus C and D are determined as well as $p_c{}^*$, and all the other non-dimensional quantities of the contraction model follow in the usual manner.

Results for the Contraction Model

A computation like the one just described has given the following numerical values for the four most essential non-dimensional characteristics of the contraction model:

$$\log C = -5.725, \quad \log D = +0.278 \tag{19.6}$$

$$\log p_c = +2.163, \quad \log t_c = -0.060. \tag{19.7}$$

Let us compare these data with our results for the main-sequence models, for example with those of Table 16.1 for the lower main sequence. If we first consider the quantities p_c and t_c, which characterize well the over-all pressure, temperature, and density distributions within the model, we find by comparing values (19.7) with the corresponding values in Table 16.1 that the internal structure of the contraction model does not differ greatly from that of the main-sequence models. Indeed, various main-sequence models—for example with different depths of the outer convection zone—can differ more from one another than the contraction model differs from some of them.

If we next consider the quantity C, which through Eq.(19.4) characterizes the mass-luminosity relation, we may compare the value of Eq.(19.6) with the corresponding values in Table 16.1. If we extrapolate in this table to $\alpha = 0$, which corresponds to the unmodified Kramers' opacity law as used for the contraction model, we find that in the mass-luminosity law, just as in the internal structure, the contraction model does not greatly differ from the main-sequence models. A comparison in terms of the quantity D, defined by Eqs.(16.10) and (19.5), would not be sensible since this quantity refers directly to the energy sources, which are nuclear for the main-sequence models and gravitational for the contraction model.

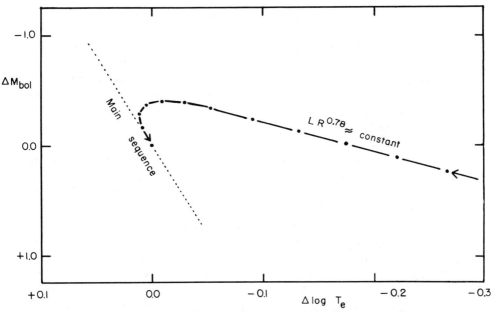

Fig. 19.1. Evolution track in Hertzsprung-Russell diagram for pre-main-sequence contraction. (Henyey, Levee, and LeLevier, *Publ. Astr. Soc. Pacific*, Vol. 67, No. 396, 1955)

We may conclude that a contracting star is quite similar to a main-sequence star, both in its over-all internal structure and its mass-luminosity relation.

We can now interpret the evolution track in the Hertzsprung-Russell diagram of Fig. 19.1, which shows the results of detailed computations with a large electronic computer, with the help of our approximate homologous contraction model. The key equation (19.4) gives directly the evolution track for the homologous contraction model. During this contraction the mass and the composition do not change, nor does the structure of the model change. In particular, the parameter C maintains the fixed value listed in Eq.(19.6). Thus nothing changes in the key equation (19.4) except for L and R. This equation then says that during the pre-main-sequence contraction the luminosity should increase proportionally to the square root of the reciprocal of the radius. Does this relation fit the evolution track obtained with the help of the electronic computations? Yes, in fair approximation, if we ignore for the moment that part of the track which is close to the main sequence. The slight difference in the LR relation of Eq.(19.4) and that of Fig. 19.1 arises simply from the fact that the electronic computations were based on detailed opacity tables while Eq.(19.4) is based on the unmodified Kramers' law. If we had modified this law by introducing α according to Eq.(16.2) so that the key equation (19.4) took the form of Eq.(16.9) and if we had chosen a value for α of 0.09 we would have obtained exact agreement with Fig. 19.1.

Thus we see that the evolution track in the Hertzsprung-Russell diagram for the pre-main-sequence contraction is essentially determined by the mass-luminosity relation. If the opacity is given by Kramers' law then the theoretical mass-luminosity relation as given by Eq.(19.4) involves the square root of the radius and the luminosity will slowly increase during the contraction. If the opacity is dominated by electron scattering—as may be the case for the latest contraction phases of the heaviest stars—the mass-luminosity law does not involve the radius and the luminosity will remain essentially constant during these phases. If however, the guillotine factor in the opacity law is highly sensitive to the density (large α value), as is the case in the less massive stars, the radius appears in the theoretical mass-luminosity relation with a power higher than the square root and correspondingly the luminosity will increase during the contraction by an appreciable amount.

There remains to be interpreted the peculiar behavior of the evolution track of Fig. 19.1 just before the main sequence is reached. During the final phases of the contraction the gravitational energy release does not stop suddenly, nor does the nuclear energy production start suddenly. Rather, as the temperature and the density increase in accordance with Eqs.(19.1) the nuclear energy generation will rise while the contraction will slow down, and thus the nuclear sources will replace the gravitational sources in a smooth and continuous transition.

During this transition period the stellar model will change from the contraction model to the main-sequence model. This change, as we have seen, is not large. Nevertheless, the mass-luminosity relations for these two models are not identical, as can be seen if their log C values are accurately compared. In fact this difference gives just the drop in luminosity which is shown by the left-hand end of the evolution track in Fig. 19.1.

Thus we see that during the very last phases of the pre-main-sequence contraction the evolutionary track in the Hertzsprung-Russell diagram is dominated by the change from the contraction to the main-sequence model, small though this change is.

Rate of Contraction

We still have to investigate one more essential question regarding the pre-main-sequence contraction: how much time does this contraction take? An order-of-magnitude answer to this question we have already given in §5 where we estimated the total gravitational energy of a star. Now, however, we can give a somewhat more precise answer.

The speed of evolution during the contraction phases is governed by the rate of liberation of gravitational energy, and this in turn can be expressed in terms of the rate of change of the radius according to Eq.(19.2), if we use again the approximation of homology during the contraction. For the construction of the contraction model we have represented the rate of change of the radius by the parameter D defined by the second key equation (19.5). Since the value of D is determined by the solution for the model, as listed in Eq.(19.6), the key equation (19.5) permits us to determine the rate of change of the radius.

If we multiply the two key equations (19.4) and (19.5) by each other and thus eliminate L we obtain a differential equation for the radius as a function of the time τ. The solution of this differential equation gives

$$R = \left(\frac{2}{CD}\right)^2 \left[\frac{3}{4ac}\left(\frac{k}{HG}\right)^{7.5}\left(\frac{1}{4\pi}\right)^3 \frac{\varkappa_0}{\mu^{7.5}}\right]^2 \left[\frac{3}{2}G\right]^2 \frac{1}{M^7}\frac{1}{\tau^2}. \qquad (19.8)$$

Hence τ is counted from the time when the radius was very large compared with the radius the star has when it reaches the main sequence. If this solution for the radius is introduced back into the first key equation (19.4) we obtain for the luminosity

$$L = \frac{C^2 D}{2}\left[\frac{3}{4ac}\left(\frac{k}{HG}\right)^{7.5}\left(\frac{1}{4\pi}\right)^3 \frac{\varkappa_0}{\mu^{7.5}}\right]^{-2}\left[\frac{3}{2}G\right]^{-1}M^9\,\tau. \qquad (19.9)$$

Eqs.(19.8) and (19.9) give the radius and the luminosity as functions of time during the pre-main-sequence contraction.

Unfortunately our solution for R and L is fairly sensitive to the opacity law assumed. Had we used electron scattering or a modified Kramers' law in place of the unmodified Kramers' law, Eqs.(19.8) and (19.9) would have been changed sensibly, both in the exponents of τ and in the coefficients. This sensitivity is not surprising since, as we have discussed above, the variation of the luminosity during the contraction depends on the opacity law.

There exists, however, a combination of Eqs.(19.8) and (19.9) which has a simple physical meaning and for which the changes for various opacity laws can be estimated fairly directly. By multiplying Eq.(19.8) by Eq.(19.9) and by then solving for the contraction time τ we obtain

$$\tau = \frac{3}{D}\frac{GM^2}{LR}. \qquad (19.10)$$

This equation gives the contraction time in terms of the gravitational energy of the star divided by its luminosity, exactly the form which we had used in §5 to estimate the length of time a star can live from its gravitational energy.

Even though, because of its physical nature, the form of Eq.(19.10) is entirely general the numerical coefficient of this equation does depend on the opacity law. Had we used electron scattering we would have obtained a coefficient half as big as the one we obtained for Kramers' law. Had we used a modified Kramers' law with $\alpha = 0.25$ we would have found that the time for contraction from a larger and larger initial radius does not converge, owing to the very low luminosity at the earliest phases. If we then define τ as the time it takes the radius to decrease by a factor 100, we find for this modified opacity law a contraction time approximately four times larger than that given by Eq.(19.10).

Duly aware of the uncertainty in the numerical coefficient of Eq.(19.10), we may now apply this equation to main-sequence stars in order to compute the time each of them spent in the pre-main-sequence contraction. To start with the sun, Eq.(19.10) gives from the solar values of L, M, and R

$$\tau_\odot = 5 \times 10^7 \text{ yrs.} \qquad (19.11)$$

For other main sequence stars we may write Eq.(19.10) by referring to the solar values in the form

$$\tau_* \approx \tau_\odot \frac{M_*}{M_\odot} \frac{L_\odot}{L_*} \qquad (19.12)$$

where we have used the fact that M/R does not vary much along the main sequence. From this equation we obtain for the contraction time of a B0 star on the upper main sequence approximately 100,000 years.

We shall see in the next chapter that the time a star spends in the pre-main-sequence contraction phase is shorter by two or three powers of ten than the time it takes the star to evolve substantially beyond the main-sequence phase. This great difference in the speed of evolution is not caused by a difference in the rate of energy emission; the luminosity of a star is substantially the same in the contraction phases and in the main-sequence phase. The large difference in the evolution speed is caused solely by the fact that the gravitational energy store which the star uses during the contraction is enormously smaller than the nuclear store which the star begins to use in the main-sequence state.

CHAPTER V
EARLY EVOLUTIONARY PHASES

20. *Simplified Example of Evolution*

In the preceding chapter we have built and tested the homogeneous models which describe the stars in their initial state when they first start burning their hydrogen fuel. In this chapter we shall follow the stars through their early evolutionary phases, during which they burn up the hydrogen in their cores. In §22 and §23 we shall follow this evolution for particular stars of given mass and given initial composition. As a preparatory step, we shall consider in this section the early evolutionary phases under special circumstances. These special circumstances have the disadvantage of not applying accurately to any type of star. But they have the advantage of leading to very simple stellar models which show in a particularly simple form the essential phenomena occurring during the early evolutionary phases.

Simplifying Circumstances and Basic Equations

These special circumstances arise if the star consists of a convective core and a radiative envelope, if practically all the nuclear energy generation occurs within the core so that composition changes by transmutations are negligible in the envelope, and if the convective core contains the same fraction of the stellar mass throughout the evolutionary phases under consideration so that the transition from the convective core to the radiative envelope always occurs at the same mass layer. Under these circumstances the evolutionary changes consist of nothing but a continuous change in time of the composition of the core. The core maintains homogeneity of composition by convective mixing, while the envelope retains its initial homogeneous composition. Thus a steadily growing discontinuity in composition develops at the edge of the core. Accordingly, the models which represent this evolution will have to consist of a homogeneous convective core and a homogeneous radiative envelope with a discontinuity of composition at the interface.

The special circumstances which we here require turn out to be realized if we select for the gas characteristics relations the equation of state of an ideal gas, Kramers' opacity law, and the carbon cycle as energy source. We have already used these choices for the gas characteristics relations as an example in our discussions of §13. They are represented by Eqs.(13.3).

In Eqs.(13.3), which define the composition functions l, j, and i, reference is made to a standard composition. Let us here choose for this standard composition the composition of the envelope, which does not change with time. Hence we get for the composition functions in the envelope

$$l_e = j_e = i_e = 1. \tag{20.1}$$

In the core, however, the composition will differ from that of the envelope—except in the initial state—and hence we shall have for the three composition functions in the interior

$$l_i = \frac{\mu_i}{\mu_e}, \quad j_i = \frac{1 + X_i}{1 + X_e} l_i, \quad i_i = \frac{X_i}{X_e} l_i. \tag{20.2}$$

These relations follow from the definitions (13.3), from the explicit equations (9.16) for Kramers' law, and from the explicit equation (10.15) for the carbon cycle, where we have to remember that, since the transmutation of hydrogen into helium does not alter the abundance of the heavier elements, $Z_i = Z_e$ holds throughout the early evolutionary phases. Thus the composition factors in the three gas characteristics relations are completely defined, and the only item still undefined in these relations is the temperature exponent in the equation for the carbon cycle, for which we shall choose

$$\nu = 16, \tag{20.3}$$

a value which we have already used for the upper main-sequence stars.

If we now apply again our standard transformation (13.2) we get the basic equilibrium conditions in the form of Eqs.(13.5) to (13.8) with the two parameters C and D defined by Eqs.(13.9) and (13.10). In the envelope nuclear energy generation is negligible and hence D does not play a role in this region. The required envelope solutions are therefore represented by a one-parameter family of integrations depending only on C. For the convective core we may use the supplementary transformation (15.17) which we have already applied to the upper main sequence models. Here, however, we have to replace β by l in this transformation. Thus we obtain again the polytropic differential equation (15.20), of which the unique solution from the center is available in tabular form. We still have to take care of the thermal equilibrium condition (13.8) in the core. If we introduce the supplementary transformation (15.17) into this equation and then integrate it over all layers, we obtain

$$1 = D \frac{i_i}{l_i^2} (2.5)^{\frac{3}{2}} p_c^{\frac{1}{2}} t_c^{17} \int_0^R t^{*19} x^{*2} dx^*, \tag{20.4}$$

where the integral is a known constant.

Fitting Conditions with Discontinuity of Composition

Up to this point we could handle our present problem in much the same manner in which we have previously handled the homogeneous models. Now, however, we come to the fitting at the interface between the core and the envelope where the discontinuity in composition introduces a new feature. If the star is to be in equilibrium the five basic physical variables r, P, T, M_r, and L_r must be continuous at the interface in spite of the discontinuity in composition. Hence the basic fitting conditions (13.1) still hold. Instead of these basic conditions, however, we would like to use the fitting conditions (13.30) for the invariants, and these have to be altered here as follows.

Since the pressure and the temperature are to be continuous, the equation of state requires that the number of free particles, i.e. ρ/μ, must be continuous as well. Since the molecular weight has a discontinuity at the interface, it follows that the density must have a compensating discontinuity. Thus we have at the fitting point

$$\frac{\rho_{fi}}{\rho_{fe}} = \frac{\mu_i}{\mu_e} = l_i.$$

$$(20.5)$$

The invariants U and V, according to their definitions (13.24), are proportional to the density, and hence we obtain for their fitting conditions

$$\frac{U_{fi}}{U_{fe}} = l_i, \quad \frac{V_{fi}}{V_{fe}} = l_i.$$

$$(20.6)$$

These two conditions replace the first two of the former conditions (13.30). We still have to derive a substitute for the third of these equations.

If radiative equilibrium held on both sides of the interface, we would compute the third invariant $n + 1$ according to its definition (13.24) by dividing the hydrostatic equilibrium condition (12.1) by the radiative equilibrium condition (12.4). In this division the density cancels out and the only remaining discontinuous factor is the absorption coefficient. Thus with the help of Eq.(9.16) for Kramers' opacity we would obtain for the third fitting condition

$$\frac{(n + 1)_{fe}}{(n + 1)_{fi}} = \frac{\varkappa_{fi}}{\varkappa_{fe}} = \frac{\rho_{fi}(1 + X_i)}{\rho_{fe}(1 + X_e)} = j_i.$$

$$(20.7)$$

Eq.(20.7) expresses a relation between the radiative temperature gradients at both sides of the interface. In reality, however, convective equilibrium holds inside the interface. In order to formulate the condition for the transition from convective to radiative equilibrium, do we have to state that the hydrogen-rich mixture just outside the interface must be in neutral equilibrium, i.e. $(n + 1)_{fe} = 2.5$, or rather that the hydrogen-poorer mixture just inside the interface is in neutral equilibrium, i.e. $(n + 1)_{fi} = 2.5$?

In the first alternative the following situation arises. The molecular weight inside the interface is slightly higher than that outside, and the same holds for the density according to Eq.(20.5). Because of the density factor in Eq.(20.7) the opacity is also higher inside the interface than outside, and hence the radiative temperature gradient is steeper inside than outside. If now, according to the first alternative, the temperature gradient outside the interface is just in neutral equilibrium, then the radiative temperature gradient inside the interface must be unstable by a finite margin. In the consequent convective equilibrium the radiation flux will fall short of the total flux by a finite amount. Hence the convective flux cannot go to zero as one approaches the interface from the inside. This means, according to Eq.(7.7), that the super-adiabatic temperature gradient $\Delta \nabla T$ also does not go to zero, and it follows, according to Eq.(7.5), that the average convection velocity remains finite to the very edge of the convective core. Under these circumstances it seems certain that every so often a convective element will overshoot and will mix in with the radiative layers just outside the interface. This will lower the hydrogen content of these layers and will increase their opacity. Since previously they were only in neutral equilibrium they will then become convectively unstable. Thus these layers will join the convective core and the core will grow. Our first alternative therefore does not lead to a stable situation.

In the second alternative, i.e. with

$$(n + 1)_{fi} = 2.5, \tag{20.8}$$

the outermost layers of the convective core will just reach neutral equilibrium. The average convective velocity will therefore go to zero as the edge of the convective core is approached and overshooting of convection elements will be unimportant. Besides, the layers just outside the interface are now in stable radiative equilibrium by a finite margin—not just in neutral equilibrium—so that a minor mixture of the hydrogen-poorer gases from the core will not affect their stability. The second alternative thus leads to a persistent situation and consequently is the one to be used. We therefore shall employ Eq.(20.8), which together with Eq.(20.7) gives us for the third fitting condition

$$(n + 1)_{fe} = 2.5 j_i. \tag{20.9}$$

Construction of Models

In the construction of the models we shall have to make use of the thermal equilibrium condition in the form of Eq.(20.4) and of the fitting conditions (20.6) and (20.9). In these equations occur the three quantities l_i, j_i, and i_i. These quantities are not independent of one another. If the initial composition, that is the envelope composition, is given, and if l_i is fixed, then the first of Eqs.(20.2) gives μ_i, this in turn fixes X_i (since $Z_i = Z_e$), and j_i and i_i are determined by the last two of Eqs.(20.2). Hence we may consider j_i and i_i as functions of l_i. These functions can be represented by the following interpolation formulae

$$j_i = l_i^{+0.3}, \quad i_i = l_i^{-1.5}. \tag{20.10}$$

These formulae represent the exact functions with fair accuracy for a wide range of initial compositions and for values of l_i from 1.0 to 2.3.

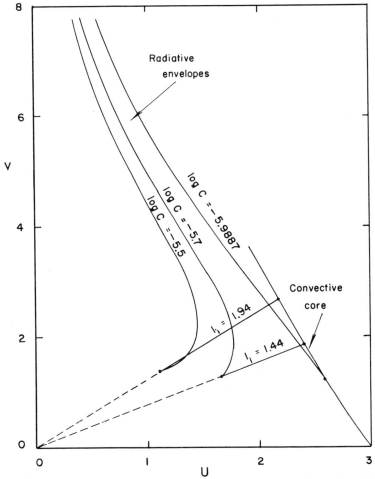

Fig. 20.1. Evolutionary sequence of simplified models in the UV plane.

The use of these formulae has the advantage that we do not have to pin ourselves down from the outset to a specific initial composition.

Our models can now be constructed by the following recipe. Choose a value of C. Obtain the corresponding envelope solution by numerical integration. Plot this solution in the UV plane as shown in Fig. 20.1. Plot in the same graph the unique solution for the convective core. Choose arbitrarily a termination point U_{fe}, V_{fe} on the envelope solution. Draw a straight line through this point and through the origin of the UV plane. Where this line intersects the core solution you have, according to Eqs.(20.6), the termination point U_{fi}, V_{fi} on the core. Determine the value of l_i from Eqs.(20.6) and the corresponding value of j_i from Eqs.(20.10). Read the value of $(n + 1)_{fe}$ from the envelope solution at the chosen termination point and check whether this value fulfills the fitting condition (20.9). If not, choose another trial for the termination point on the envelope—without, however, changing the value of C and the corresponding numerical integration—and repeat the whole procedure. After you have found the termination point which leads to the fulfillment of condition (20.9), read off all the non-dimensional variables without asterisks from the envelope solution at the termination point and the values of the asterisked variables from the core solution at its termination point. Of these values introduce the relevant ones into the first two of the

TABLE 20.1

Evolutionary sequence of simplified models with composition discontinuity between convective core and radiative envelope.

$\log C$	-5.9887	-5.9	-5.8	-5.7	-5.6	-5.5	-5.25
U_{fe}	2.584	2.280	1.958	1.658	1.372	1.108	0.488
V_{fe}	1.218	1.237	1.258	1.288	1.325	1.372	1.615
$(n + 1)_{fe}$	2.500	2.581	2.679	2.789	2.913	3.052	3.547
l_i	1.000	1.112	1.261	1.440	1.663	1.942	3.210
x_f	0.170	0.156	0.141	0.125	0.110	0.094	0.051
q_f	0.146	0.147	0.148	0.149	0.149	0.149	0.144
$\log p_f$	$+1.735$	$+1.826$	$+1.939$	$+2.063$	$+2.204$	$+2.365$	$+2.967$
$\log t_f$	-0.151	-0.118	-0.078	-0.035	$+0.012$	$+0.064$	$+0.244$
x^*_f	1.191	1.263	1.352	1.456	1.578	1.721	2.275
t^*_f	0.787	0.769	0.733	0.697	0.653	0.600	0.395
$\log p_c$	$+1.995$	$+2.111$	$+2.275$	$+2.454$	$+2.668$	$+2.919$	$+3.976$
$\log t_c$	-0.047	-0.004	$+0.056$	$+0.121$	$+0.197$	$+0.285$	$+0.648$
$\log D$	$+0.320$	-0.307	-1.233	-2.222	-3.399	-4.786	-10.717
$\log R/R_0$	0.000	$+0.029$	$+0.074$	$+0.122$	$+0.180$	$+0.250$	$+0.556$
$\log L/L_0$	0.000	$+0.075$	$+0.152$	$+0.228$	$+0.299$	$+0.364$	$+0.461$
$\log T_e/T_{eo}$	0.000	$+0.004$	$+0.001$	-0.004	-0.015	-0.034	-0.163
X_i	1.000	0.838	0.668	0.511	0.362	0.224	(-0.10)
$\overline{X}$	1.000	0.976	0.951	0.927	0.905	0.884	—

transformation equations (15.17) and thus determine p_c and t_c. In turn, introduce these into Eq.(20.4) and obtain D from it.

The results of computations according to this recipe are shown in Table 20.1 for seven different values of C. The different values of C give different values of l_i and hence correspond to different compositions of the convective core. Thus the one-parameter family of models represented by Table 20.1 is exactly the desired evolutionary sequence of models in which the composition of the convective core varies from hydrogen-rich to hydrogen-poor.

The first model of Table 20.1 has $l_i = 1$ and therefore corresponds to the initial homogeneous state. To obtain this particular model the appropriate C value had to be found by a series of trial integrations. For each of the subsequent models, for which no specific values of l_i were required, one numerical integration for the envelope was all that was needed.

Evolutionary Track in Hertzsprung-Russell Diagram

We still have to determine the luminosities and the radii of our model stars. If we were to do this in absolute units, we should now have to choose a particular mass and initial composition. Instead we may choose the luminosity L_0 and the radius R_0 of the initial homogeneous state as units. In this manner we will not obtain the absolute luminosities and radii of our model stars, but we will obtain the variation of the luminosity and of the radius of a given star during its evolution. We may introduce these new units by the following procedure. If we form the ratios C/C_0 and D/D_0 (where the subscript zero refers to the initial homogeneous model) with the help of the key equations (13.9) and (13.10), all the constant factors on the right-hand side of these equations cancel out. This includes the composition factors since they refer only to the standard composition for which we have chosen the envelope composition and the latter does not vary during the evolution of the star. Similarly the mass cancels out and we have left on the right-hand side only the two ratios L/L_0 and R/R_0. If we solve the two equations for the latter two ratios, we obtain

$$\frac{R}{R_0} = \left(\frac{C}{C_0}\frac{D}{D_0}\right)^{-\frac{1}{18.5}}, \quad \frac{L}{L_0} = \frac{C}{C_0}\left(\frac{R}{R_0}\right)^{-\frac{1}{2}}. \tag{20.11}$$

Thus the relative radii and luminosities are easily computed from the C and D values listed in Table 20.1 and the results are given in the same table, together with the corresponding values for the relative effective temperatures.

The luminosities and the effective temperatures of our model sequence are shown in Fig. 20.2. This figure represents a small portion of the Hertzsprung-Russell diagram on a greatly magnified scale. The zero

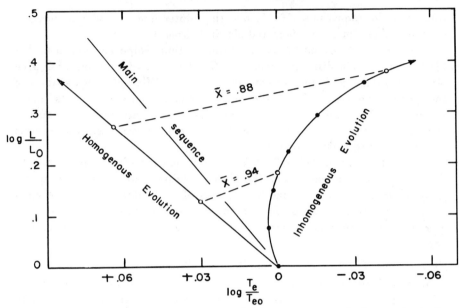

Fig. 20.2. Comparison of homogenous and inhomogenous evolution tracks in the Hertzsprung-Russell diagram for simplified models.

point of the coordinates refers here to the homogeneous model, which corresponds to the initial stellar state on the main sequence. The dots in this figure represent the models of Table 20.1. The curve through the dots gives us our first theoretical evolutionary track in the Hertzsprung-Russell diagram for the hydrogen-burning phases. We must remember that this theoretical track is based on simplifying approximations which do not hold accurately for any star. Nevertheless the general character of this track can hardly be wrong: the track turns off the main sequence to the right and points into the region of the red giants.

This preliminary result is doubly encouraging. First, it indicates that a star during its early evolutionary phases does not move far from its initial point in the Hertzsprung-Russell diagram. Even when the hydrogen content of the core has dropped down to 10 percent—a state represented by the tip of the right-hand arrow in Fig. 20.2—the brightness of the star has increased only by one magnitude (according to our present simplified models) and the effective temperature has dropped by less than 15 percent. Thus a star will probably be classified as a "main-sequence star" during most of the phases in which it consumes the hydrogen of its core. If we had found that stars move more rapidly out of the main-sequence band it would have been hard to see why the main sequence is as well populated as we observe it to be.

The second encouraging feature of our present result is the indication that the evolutionary track turns to the right of the main sequence. In this direction we find in the Hertzsprung-Russell diagram the well-populated area of the red giants and subgiants. These types of stars cannot, it seems, be explained in terms of initial stellar states since, as we have seen in the preceding chapter, all homogeneous models fall on or to the left of the main sequence. Our present result raises the hope that we will be able to interpret the red giants and subgiants in terms of later evolutionary states.

Comparison of Inhomogeneous and Homogeneous Evolutions

In the construction of the present models we have made the tacit assumption that there exists no mechanism which will mix the gases of the core with the gases of the envelope during the evolution of the star. The question arises whether this assumption is essential or whether much the same evolutionary track would have been obtained even if an effective mixing mechanism existed. To answer this question we may for the sake of comparison follow the evolution of a star under the extreme assumption that all the layers within the star from the center to the surface will be completely mixed throughout the evolution.

Under this new assumption the star will have a homogeneous composition not only in its initial state but also in the subsequent evolutionary states. If we choose the same forms for the gas characteristics relations, the structure of the star will correspond in all states to the first model of Table 20.1 since this model represents the unique homogeneous solution under the chosen circumstances. This constancy of the model throughout the evolution does not mean, however, constancy of luminosity or of radius. As before, L and R have to be determined from the key equations (13.9) and (13.10). The parameters C and D which occur in these equations are now constant since the model does not vary. But the composition factors in these equations which refer to the composition of the envelope are now varying since the assumed complete mixing will cause the nuclear transformations to change the composition of the envelope just as much as that of the core. If we again use as units the quantities of the initial state designated by subscript zero, the key equations (13.9) and (13.10) become

$$1 = \frac{1+X}{1+X_0}\left(\frac{\mu}{\mu_0}\right)^{-7.5}\frac{L}{L_0}\left(\frac{R}{R_0}\right)^{\frac{1}{2}}, \quad 1 = \frac{X}{X_0}\left(\frac{\mu}{\mu_0}\right)^{16}\frac{L_0}{L}\left(\frac{R}{R_0}\right)^{-19}. \qquad (20.12)$$

For the first factor in each of these two equations we may use approximate relations equivalent to Eqs.(20.10). If we then solve Eqs.(20.12) for L and R we obtain

$$\frac{R}{R_0}=\left(\frac{\mu}{\mu_0}\right)^{0.286}, \quad \frac{L}{L_0}=\left(\frac{\mu}{\mu_0}\right)^{8.06}, \quad \frac{T_e}{T_{e0}}=\left(\frac{\mu}{\mu_0}\right)^{1.87}. \qquad (20.13)$$

We have thus derived under the new assumption of complete mixing the variation of the radius, the luminosity, and the effective temperature during the evolution as a function of the change in the molecular weight caused by the nuclear transmutations.

The evolutionary track in the Hertzsprung-Russell diagram which corresponds to the last two of Eqs.(20.13) is represented by the straight line at the left-hand side of Fig. 20.2. This homogeneous evolutionary track certainly differs greatly from the inhomogeneous evolutionary track derived before. We must conclude that the existence or non-existence of a mechanism which keeps the gases of the radiative envelope mixed with those of the core is of decisive influence on the evolutionary track.

Only one such mixing mechanism appears possible for normal stars, that of rotation. We shall discuss this particular mechanism in full detail in the next section.

For the present let us continue our comparison between homogeneous and inhomogeneous evolutionary tracks a little further. To make possible a comparison in terms of states of equal hydrogen depletion we have to compute the average hydrogen content $\bar{X}$ of our inhomogeneous models. The average hydrogen content is the mean of the hydrogen content of the envelope and that of the core, weighted according to the relative masses of these two parts of the star. Accordingly the average hydrogen content is given by

$$\bar{X} = X_e (1 - q_f) + X_i q_f. \tag{20.14}$$

Let us here assume $X_e = 1$ for the initial composition, that is for the envelope composition. This means that the initial content in helium and the heavier elements is taken to be small. The hydrogen content of the core X_i can be determined from the molecular weight of the core μ_i with the help of Eq.(8.2). For small values of Z this gives

$$X_i = \frac{0.8}{\mu_i} - 0.6 = \frac{1.6}{l_i} - 0.6. \tag{20.15}$$

In the last part of this equation we have replaced μ_i by l_i according to the first of Eqs.(20.2) with $\mu_e = 0.5$. We can now compute for each model of Table 20.1 the hydrogen content of the core from Eq.(20.15) and the average hydrogen content from Eq.(20.14). The results are listed at the bottom of Table 20.1. Thus we know the average hydrogen content for each point along the inhomogeneous evolutionary track of Fig. 20.2.

In the homogeneous evolution the average hydrogen content is equal to the actual hydrogen content throughout the star. This hydrogen content is related to the molecular weight by Eq.(20.15), without the subscripts. Since, furthermore, the molecular weight is related to the position on the homogeneous evolutionary track by Eqs.(20.13), it is easy to derive for each point on the homogeneous track the average hydrogen content.

We may now compare points of equal average hydrogen content on the two evolutionary tracks. Two such pairs of points, corresponding to remaining hydrogen contents of 94 percent and 88 percent respectively, are indicated by the four circles in Fig. 20.2 and by the two dashed lines which connect the comparable points. These dashed lines show that points of equal average hydrogen content on the two evolution tracks differ little in luminosity and much in effective temperature. For example, in the state when 12 percent of the total hydrogen has been burned up, i.e. $\bar{X} = 0.88$, the inhomogeneously evolving star is brighter than the homogeneous star by only a quarter of a magnitude, but their effective temperatures differ by about 25 percent; the difference in absolute magnitude would be hard to determine observationally while the difference in spectral type would be obvious.

We may then conclude from this preliminary investigation that the variation of the luminosity during the early evolutionary phases does not depend sensitively on the distribution of composition through the star, but that, if we are to predict the evolutionary changes of the radii and effective temperatures, it is essential for us to know whether there do or do not exist mechanisms capable of effective mixing even in regions of radiative equilibrium.

21. Rotational Mixing

Rotation is in itself not a motion which mixes up the various layers within a star. It was discovered, however, that in a star which deviates from the spherical form because of rotation the radiative flux alone is in general not capable of fulfilling the thermal equilibrium condition everywhere. This deficiency of the radiative flux causes a slow circulation, inwards in the equatorial plane and outwards along the rotation axis; and it is this circulation in meridional planes which is, in principle, capable of mixing the interior with the exterior layers of a star. The essential question is: will the circulation be fast enough to achieve mixing in time intervals in which the nuclear processes alter the core composition appreciably?

Hydrostatic Equilibrium in a Rotating Star

To solve this problem we shall have to consider the full set of the basic equilibrium conditions (12.1) to (12.4). If we restrict ourselves to regions in radiative equilibrium, and if we exclude the central regions in which the nuclear energy generation occurs, we may write the four basic equations in the vectorial form

$$\nabla P = -\rho \nabla V_s + \rho \Omega^2 \, \vec{a} \tag{21.1}$$

$$\nabla^2 V_s = 4\pi G \rho \tag{21.2}$$

$$\vec{H} = - \frac{4\,ac}{3}\, \frac{T^3}{\varkappa\rho}\, \nabla T \qquad\qquad (21.3)$$

$$\nabla \cdot \vec{H} = 0 . \qquad\qquad (21.4)$$

In the hydrostatic equilibrium condition (21.1) we have added a term corresponding to the centrifugal force. Here Ω represents the angular velocity of rotation and a the distance from the axis of rotation, as indicated in Fig. 21.1.

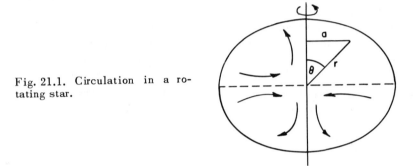

Fig. 21.1. Circulation in a rotating star.

Let us now restrict our investigation to the case of solid-body rotation, i.e. to $\Omega = $ const. The centrifugal force can be expressed in this case as the gradient of a distorting potential V_d which is given by

$$V_d = - \tfrac{1}{2}\, \Omega^2\, a^2 . \qquad\qquad (21.5)$$

If we add this distorting potential to the gravitational potential V_s, we can rewrite the hydrostatic equilibrium condition

$$\nabla P = - \rho \nabla V \qquad\qquad (21.6)$$

$$\text{with } V = V_s + V_d . \qquad\qquad (21.7)$$

Thus we have obtained a form for the hydrostatic equilibrium condition identical with Eq.(18.1), which we used when we investigated the distortion by a close companion. By the same method as we used there we may deduce for a rotationally-distorted star that the pressure and the density are constant on equipotential surfaces, i.e.

$$P = P\,(V), \quad \rho = \rho\,(V), \quad T = T\,(V) . \qquad\qquad (21.8)$$

The last of these equations states that the temperature also must be constant on equipotential surfaces. This conclusion follows from the corresponding statements regarding the pressure and the density through the equation of state as long as the composition is constant on equipotential surfaces.

Deficiency of Radiation Flux and Meridional Circulation

Eqs.(21.8), which follow directly from the hydrostatic equilibrium condition (21.1), have highly restrictive consequences for the radiative equilibrium. According to the last of Eqs.(21.8), we may express the temperature gradient in the radiative equilibrium condition (21.3) by the potential gradient. Thus we obtain in place of Eq.(21.3)

$$\vec{H} = \left(-\frac{4\,ac}{3}\,\frac{T^3}{\varkappa\rho}\,\frac{dT}{dV} \right) \nabla V .\qquad(21.9)$$

All factors within the parentheses depend only on V, that is are constant on equipotential surfaces. Eq.(21.9) therefore shows that the radiation flux not only must be parallel to the potential gradient everywhere, but also that its absolute value must be proportional to the absolute value of the potential gradient for all points of one equipotential surface. By adjusting its internal temperature distribution the star can therefore only affect the average absolute value of the radiation flux at each equipotential surface; it cannot alter the direction of the radiation flux anywhere, nor can it change the relative amounts of this flux at different points of the same equipotential surface.

This severe restriction on the radiation flux has serious consequences regarding the fulfillment of the thermal equilibrium condition (21.4). If we use Eq.(21.9) to compute the divergence of the radiation flux, we obtain

$$\nabla \cdot \vec{H} = \frac{d}{dV} \left(-\frac{4\,ac}{3}\,\frac{T^3}{\varkappa\rho}\,\frac{dT}{dV} \right) |\nabla V|^2 + \left(-\frac{4\,ac}{3}\,\frac{T^3}{\varkappa\rho}\,\frac{dT}{dV} \right) \nabla^2 V .\qquad(21.10)$$

The last factor in the last term of this equation can be expressed with the help of Eq.(21.2) and (21.5) by

$$\nabla^2 V = 4\pi\,G\,\rho - 2\Omega^2 .\qquad(21.11)$$

Thus we see that all quantities occurring on the right-hand side of Eq.(21.10) are constant on equipotential surfaces with the sole exception the potential gradient $|\nabla V|$. This quantity cannot possibly be constant on equipotential surfaces in a rotating star since in such a star these surfaces are more widely spaced in the equatorial plane than along the rotation axis and thus the absolute value of the potential gradient is smaller in the equatorial regions than in the polar regions. Hence, according to Eq.(21.10), as long as the factor by which the square of the potential gradient is multiplied is different from zero, as is actually the case everywhere, the divergence of the radiation flux cannot be constant on equipotential surfaces. In particular, it cannot be zero everywhere as required by the thermal equilibrium condition (21.4). Thus in a rotationally-distorted star the radiation flux cannot fulfill the thermal equilibrium condition.

What will happen under such conditions? The star will adjust its temperature gradient so that on the average over each equipotential surface the divergence of the radiation flux is zero. As we have seen, this average adjustment cannot prevent the divergence of the radiation flux from being slightly positive in certain regions and slightly negative in other regions. The first regions will lose energy by the radiation flux and will therefore cool slowly while the latter regions will gain energy from the radiation flux and will therefore heat up slowly. The cooling regions will start sinking and the heating regions will start rising. Thus, a slow circulation in meridional planes will commence. In this circulation the sinking regions will be heated by compression and the rising regions will cool by expansion. The speeds of these motions will adjust themselves automatically so that in the sinking elements the heating by compression exactly compensates the cooling by the radiation flux, and in the rising elements the cooling by expansion will exactly compensate the heating by the radiation flux. This condition determines the speed of the circulation. It remains our problem to compute this speed so that we can judge whether or not the circulation is sufficiently fast to mix a star in the required time intervals.

Determination of Rotational Distortion

We start again with the hydrostatic equilibrium condition. We can write the distorting potential (21.5), according to the geometry indicated in Fig. 21.1, in the form

$$V_d = \left[-\tfrac{1}{3}\Omega^2 r^2\right] + \tfrac{1}{3}\Omega^2 r^2 P_2(\theta) \tag{21.12}$$

where P_2 designates the second Legendre polynomial. Here we have put the first term into brackets since it is spherically symmetrical. In our discussion of Eq.(21.10) we have seen that the non-vanishing of the divergence of the radiation flux is caused by the deviations from spherical symmetry of the equipotential surfaces in a rotating star. We are therefore not interested here in spherical distortions and will disregard the bracketed term. With this elimination the distorting potential (21.12) is identical with the distorting potential caused by a close companion as given by Eq.(18.4) if we make the replacement

$$-\frac{GM_2}{D^3} \longrightarrow +\frac{1}{3}\Omega^2. \tag{21.13}$$

Hence our present hydrostatic problem is the same as that which we have already discussed for the apsidal motion test. We may again describe the form of the equipotential surfaces by

$$r = \bar{r}[1 + c_2(\bar{r})P_2(\theta)] \tag{21.14}$$

where $\bar{r}$ designates the mean distance from the center of an equipotential surface and c_2 measures the deviation from spherical symmetry of this surface. This distortion function c_2 is again determined, in first order approximation, by the differential Eq.(18.11), by the initial condition (18.12), and by the normalization condition (18.15) (in which, of course, the replacement (21.13) has to be used). This solves our present hydrostatic problem and we can now turn to the radiative problem.

Determination of the Divergence of the Radiation Flux

To compute the divergence of the radiation flux from Eq.(21.10) we may express the potential gradient, according to Eq.(21.14), in first order approximation by

$$| \nabla V | = \frac{dV}{d\bar{r}} \bigg/ \frac{dr}{d\bar{r}} = \frac{dV}{d\bar{r}} \left[1 - \frac{d(\bar{r} c_2)}{d\bar{r}} P_2(\theta) \right]. \qquad (21.15)$$

If we introduce this as well as the expression (21.11) into Eq.(21.10), we find that the right-hand side of this equation contains a number of terms which depend only on $\bar{r}$, as well as one term which, because of the P_2 factor in Eq.(21.15), depends also on θ. All the terms which depend only on $\bar{r}$ have to add up to zero since, as we have mentioned before, the star will adjust its temperature distribution so that the divergence of the radiation flux is zero on the average over each equipotential surface. There remains on the right-hand side of Eq.(21.10) the one term which depends on θ. This term contains the factor c_2, which is already of first order. As long as we aim only at a first order approximation we need to compute all the other factors in this term only to zero order accuracy. These factors reduce accordingly to the simple expression

$$\frac{d}{dV} \left(-\frac{4ac}{3} \frac{T^3}{\varkappa\rho} \frac{dT}{dV} \right)\left(\frac{dV}{d\bar{r}}\right)^2 = \frac{d}{dr} \left(\frac{L}{4\pi r^2} \frac{r^2}{GM_r} \right)\frac{GM_r}{r^2} = -\frac{L\rho}{M_r},$$

and we obtain for the remaining term of Eq.(21.10)

$$\nabla \cdot \vec{H} = \frac{L\rho}{M_r} \times 2 \frac{d(\bar{r} c_2)}{d\bar{r}} P_2(\theta). \qquad (21.16)$$

Since the first factor on the right-hand side of this equation is determined by the unperturbed stellar model and since c_2 is determined by the solution of the hydrostatic problem, Eq.(21.16) completely determines the non-vanishing divergence of the radiation flux.

Speed of the Meridional Circulation

Next we have to derive the rate of heating or cooling caused by the meridional circulation. If $\vec{v}$ represents the velocity vector of the circulation, the rate of energy loss per cubic centimeter is given by

$$-\frac{dE}{dt} = + P \nabla \cdot \vec{v} + \nabla \cdot \left(\frac{3}{2} \frac{k}{\mu H} T \rho \vec{v} \right), \qquad (21.17)$$

where the first term represents the work done against the surrounding and the second the divergence of the convective heat transport. With the help of the continuity equation

$$\nabla \cdot (\rho \, \vec{v}) = 0$$

we may transform the expression for the rate of energy loss to

$$- \frac{dE}{dt} = - \frac{P}{\rho} \, \nabla \rho \cdot \vec{v} + \frac{3}{2} \, \frac{P}{T} \, \nabla T \cdot \vec{v} .$$

Since the velocity of the circulation is already a first order quantity, we need to compute all the other factors on the right-hand side of this equation only in zero order approximation. If we express the density gradient in terms of the pressure and temperature gradients according to the equation of state of an ideal gas, and if we express the temperature gradient in terms of the pressure gradient with the help of the invariant $n + 1$ as defined by Eq.(13.24), we obtain for the rate of energy loss

$$- \frac{dE}{dt} = + \left(1 - \frac{2.5}{n+1} \right) \rho \, \frac{GM_{\bar{r}}}{\bar{r}^2} \, v , \qquad (21.18)$$

where v stands for the radial component of $\vec{v}$.

The speed of the circulation will adjust itself so that the energy losses caused by the circulation exactly compensate the energy gains caused by the non-vanishing divergence of the radiation flux. Hence, we have

$$\nabla \cdot \vec{H} = \frac{dE}{dt} . \qquad (21.19)$$

If we introduce into this condition on the left-hand side the expression (21.16) for the divergence of the radiation flux and on the right-hand side the expression (21.18) for the energy losses by the circulation, and if we then solve for the radial velocity of the circulation, we obtain

$$v = \frac{n+1}{n+1-2.5} \, \frac{L \bar{r}^2}{G M_{\bar{r}}^2} \, 2 \, \frac{d (\bar{r} \, c_2)}{d \bar{r}} \, P_2 \, (\theta) . \qquad (21.20)$$

This is the equation for the circulation speed that we have been aiming for. Before we apply it let us bring it into a more convenient form, first, by introducing our usual non-dimensional variables x and q defined by

$$\bar{r} = x R , \qquad M_{\bar{r}} = q M ;$$

secondly, by using the equatorial velocity of rotation W in place of the angular velocity Ω according to the relation

$$W = \Omega R ;$$

and finally, by employing explicitly the normalization condition (18.15) modified according to the replacement (21.13). Thus we obtain

$$v = + W^2 \frac{L R^3}{G^2 M^3} \times v^* (x) \times P_2 (\theta) \qquad (21.21)$$

$$\text{with } v^* (x) = \frac{10}{3} \; \frac{n + 1}{n + 1 - 2.5} \; \frac{x^2}{q^2} \; \frac{c_2 + x \dfrac{dc_2}{dx}}{\left(2 c_2 + x \dfrac{dc_2}{dx}\right)_R}. \qquad (21.22)$$

Eq.(21.21) gives the radial component of the circulation velocity for any point within a rotating star as a function of the position coordinates x and θ and as a function of the equatorial rotation velocity W.

The function v^* can be computed for any star for which an unperturbed model has been derived, giving $n + 1$ and q as functions of x, and for which the differential Eq.(18.11) has been integrated, giving c_2 as a function of x. The results of such computations are shown in Table 21.1 for the simplified homogeneous model of §20. We shall use the numbers of this tabulation as roughly representative for main sequence stars.

Numerical Results

As a first example let us apply Eq.(21.21) to the sun. We have for the sun

$$W_\odot = 2 \; \frac{km}{sec} \quad \text{and} \quad \left(W^2 \frac{L R^3}{G^2 M^3}\right)_\odot = 1.46 \times 10^{-9} \; \frac{cm}{sec} .$$

Furthermore, according to Table 21.1, the function v^* as well as the Legendre polynomial P_2 are of the order of unity. We therefore find the velocity of circulation in the sun to be of the order of 10^{-9} cm/sec. It follows from this extremely low circulation velocity that it would take several times 10^{12} years for matter to rise from the interior regions of the sun to its surface. Since this time interval exceeds the age of the sun by about a factor 1000, we conclude that the circulation in the sun caused by its rotation can have produced practically no mixing whatsoever.

Does the same conclusion hold also for the faster rotating stars of the upper main sequence? To answer this question in a general manner let us define the mixing time t_{mix} as the time required by an element to rise along the axis of rotation from the core to the surface. According to this definition we obtain for the mixing time

$$t_{mix} = \int_{r_f}^{R} \frac{dr}{v} = \frac{G^2 M^3}{W^2 L R^2} \; \frac{1}{v^*} \qquad (21.23)$$

$$\text{with } \frac{1}{v^*} = \int_{x_f}^{1} \frac{dx}{v^*} = 2.20 \,. \tag{21.24}$$

Here we have used Eq.(21.21) for the radial circulation velocity along the axis, i.e. with $P_2 = 1$ at $\theta = 0$. The numerical value of Eq.(21.24) is based on the values of v^* given in the last column of Table 21.1, which we may consider as approximately representative.

TABLE 21.1

Rotational distortion and circulation velocity for simplified model (c_2 normalised to 1 at center). (Härm and Rogerson, *Ap.J. 121*, 439, 1955.)

x	q	U	$n+1$	c_2	$\dfrac{dc_2}{dx}$	v^*
0.2	0.22	2.42	2.85	1.09	1.01	0.61
0.3	0.51	1.69	3.62	1.24	2.04	0.15
0.4	0.76	1.02	3.98	1.52	3.63	0.16
0.5	0.90	0.54	4.14	1.99	5.89	0.27
0.6	0.97	0.25	4.21	2.72	8.91	0.53
0.7	0.99	0.10	4.24	3.80	12.70	1.09
0.8	1.00	0.02	4.25	5.29	17.19	2.08
0.9	1.00	0.00	4.25	7.25	22.28	3.78
1.0	1.00	0.00	4.25	9.76	27.89	6.43

We may compare this mixing time with a critical time, t_{crit}, defined as the time during which 5 percent of the stellar mass is transmuted from hydrogen into helium. We have chosen this value of 5 percent since Fig. 20.2 shows that the evolutionary changes of a star over time intervals in which several per cent of the mass are transmuted depend significantly on the existence or non-existence of effective homogenizing mixing currents. According to our definition of the critical time, we have

$$t_{crit} = 6 \times 10^{18} \times \frac{0.05 M}{L} \,, \tag{21.25}$$

where we have used the value given by Eq.(10.21) for the number of ergs liberated per gram of hydrogen burned.

We may now ask: what is the value of the critical equatorial rotation velocity for which the mixing time is equal to the critical time? By equating Eq.(21.23) to Eq.(21.25) we find for the critical rotation velocity

$$W_{crit} = 52 \frac{M}{M_\odot} \frac{R_\odot}{R} \frac{\text{km}}{\text{sec}} \,. \tag{21.26}$$

By using in this equation the average masses and radii as functions of spectral type for upper main sequence stars, we find the critical rotation velocities listed in Table 21.2. For comparison the average observed rotational velocities are also given in this table.

From the data of Table 21.2 we conclude that the majority of the main sequence stars of type F and later have rotation velocities well below the critical rotation velocity. This means that in these stars the circulation caused by the rotation is too slow to produce effective mixing throughout the radiative layers. For the majority of these stars, therefore, the "inhomogeneous evolution" appears to be a better approximation than the "homogeneous evolution"—a welcome conclusion since the inhomogeneous evolution leads into the red giant area while the homogeneous one does not.

On the other hand, Table 21.2 gives for the early main-sequence stars of type A and earlier an average observed rotation velocity definitely

TABLE 21.2

Average observed rotation velocities and critical velocities (km/sec) for upper main-sequence stars. (Slettebak, *Ap.J. 121*, 653, 1955.)

Spectrum	W_{obs}	W_{crit}
B	210	100
A	170	60
F	30	55

larger than the critical velocity, suggesting the possibility of effective mixing. But even in these stars mixing by circulation caused by rotation is probably not effective, as is shown by the following additional consideration.

Effects of Inhomogeneities in Composition

The values for the critical rotation velocity listed in Table 21.2 are based on a number of approximations. Specifically, we have used a simplified stellar model and we have assumed solid body rotation. There does not appear to be much danger that these two approximations have caused serious errors. More important appear to be the effects which we have omitted by assuming that the composition is homogeneous on each equipotential surface. This assumption is fulfilled in rotating stars in the absence of meridional circulation since nuclear burning causes inhomogeneities along the radius vector but not along an equipotential surface. Let us now, however, consider the effects of composition inhomogeneities caused by the very circulation which we have been discussing. Assume that a slow circulation sets in as discussed before and starts carrying material of heavier molecular weight upwards along the rotation axis. Now the molecular weight is not any longer constant on equipotential surfaces. We still can deduce from the hydrostatic equilibrium condition that the pressure and density must be constant on equipotential surfaces. On the other hand, we must now conclude from the equation of state that the higher molecular weight in the regions near the axis has to be compensated by slightly higher temperatures in these regions as compared with points that are on the same equipotential surfaces but located near the equatorial plane. Hence there will now exist a temperature gradient parallel to the equipotential surfaces, and this will cause an additional radiation flux along the equipotential surfaces from the axial regions to the equatorial regions. This additional radiation flux will cause an additional term for the divergence of the radiation flux, positive near the axis and negative in the equatorial plane. This new term has therefore the opposite sign from the original term given by Eq.(21.16) since c_2 is negative, according to the normalization Eq.(18.15) with the replacement (21.13), and P_2 is positive along the axis and negative in the equatorial plane. Thus the net divergence of the radiation flux will be reduced, the speed of the circulation will be diminished, and the critical rotation velocity will be raised.

In fact, quantitative estimates indicate that the inhomogeneities on equipotential surfaces caused by the circulation itself increase the critical rotation velocity quite appreciably, so that, contrary to Table 21.2, the critical rotation velocity will exceed the observed average rotation velocity even for the early main-sequence stars.

Thus we conclude that rotation does not cause effective mixing in the vast majority of stars.

We shall therefore proceed in the following sections under the assumption that no mising occurs in the radiative layers of a star—while, of course, we continue to assume that turbulent mixing is highly effective in convective layers and keeps these layers homogeneous in composition.

Even though our present conclusion appears well founded for most normal stars, it does seem likely that mixing mechanisms of the type here discussed will eventually have to be taken into account in those exceptional cases in which stars suffer extraordinarily large distortions. These exceptional cases may include not only those early-type stars with extremely high rotation velocities but also perhaps such contact binaries as the W Ursae Majoris stars in which the strong mutual gravitational distortion may cause a circulation much like the one here discussed.

22. *Evolution of Upper Main-Sequence Stars*

We might be tempted to assume now that we can compute the accurate evolutionary model sequences for upper main sequence stars by exactly the same method which we have employed for the simplified models of §20. If this were correct, all we would have to do would be to determine more accurately the appropriate gas characteristics relations and then to repeat the computations of §20 with these new relations. That this procedure, however, would lead to incorrect results will be seen in the following paragraphs.

Occurrence of Inhomogeneous Zone

For the simplified models we had assumed that the star consisted of a convective core, in which the hydrogen abundance was steadily diminished by the nuclear transmutations, and of a radiative envelope, in which the composition remained unchanged—in complete agreement with the results of the last section. Furthermore we had assumed that the interface between the core and the envelope remained at the same layer throughout the evolutionary phases considered, that is that the mass fraction contained in the convective core did not vary. The validity of this last assumption for the simplified models of §20 is verified by the nearly complete constancy of the q_f values of Table 20.1. It turns out, however, that this lucky circumstance depends sensitively on the law for the absorption coefficient. For example, if we had not chosen Kramers' opacity but rather electron scattering, we would have found that the mass fraction in the convective core was steadily decreasing during the early evolutionary phases and that therefore it would have been wrong to assume simply the development of a sharp discontinuity in composition at a fixed interface between core and envelope.

How then does the early evolution of upper main sequence stars proceed in reality? From the initial models of §15 we know that in these stars electron scattering plays a major role. We therefore have to expect that the convective core will contain a smaller and smaller mass fraction as the evolution proceeds. Consequently, there will appear an intermediate zone between the core and the envelope. The layers of this zone are initially contained in the convective core, but later on will lie outside the core. The composition of any layer in this zone is determined by the moment when the layer ceases to belong to the convective core. After this moment the composition of the layer will not change any more—at least not during the early evolutionary phases we are considering—since the nuclear transformation occur only in the central portion of the core and their effect can reach out only as far as convective mixing will carry them. In consequence, the lower layers of the intermediate zone, which belonged to the convective core for a long time, will have a lower hydrogen abundance than the higher layers of this zone, which remained in the convective core only for a short time. Thus the composition of the intermediate zone will be a continuous function of the distance from the center, starting at the inner edge with the hydrogen-poor composition of the core and ending at the outer edge with the hydrogen-rich composition of the envelope.

We may conclude that we now have to construct for each upper main sequence star an evolutionary sequence of models consisting of a convective core which steadily decreases in mass, an inhomogeneous radiative zone which steadily increases in mass, and a homogeneous radiative envelope. Before we start the actual construction, let us assemble the necessary equations.

Basic Equations and Definitions

For the gas characteristics relations we shall clearly choose again Eqs.(15.1) to (15.5), which we have found appropriate for the initial state of upper main sequence stars. When we apply the standard transformation (13.2) we shall use the non-varying composition of the envelope as a standard, as we have done for the simplified models. Similarly, we shall define the parameters A, B, C, and D again by Eqs.(15.9), (15.11), (15.13), and (15.16), but we shall use the envelope composition in these equations wherever the composition occurs. Furthermore, we shall use the numerical values of Table 15.1 for the parameters in the absorption law and for A and B, as before.

The composition functions l, j, and i are given by Eq.(20.1) for the envelope and (20.2) for the core, as was the case for the simplified models. In the new intermediate zone we may use once more Eqs.(20.2) if we drop the subscript i in these equations. For the relation between j and l we shall here use the approximate formula.

$$j = l^{+0.285} \tag{22.1}$$

This formula is a little more accurate than Eq.(20.10), which we used for the simplified models.

With these choices and definitions we now obtain the basic equilibrium conditions for the envelope and the intermediate zone in the form

$$\frac{dp}{dx} = -l \frac{p}{t} \frac{q}{x^2} \beta, \quad \frac{dq}{dx} = +l \frac{p}{t} x^2 \beta \tag{22.2}$$

$$\frac{dt}{dx} = -C j^{1-\alpha} l \frac{p^{2-\alpha}}{x^2 t^{8.5-\alpha}} \delta \beta^{2-\alpha} \tag{22.3}$$

The two auxiliary relations (15.10) and (15.12), which fix the contribution of the radiation pressure to the total pressure and of the electron scattering to the opacity, now become

$$\beta = 1 - B \frac{t^4}{p}, \quad \delta = 1 + AB \frac{j^\alpha}{l} \frac{t^{4.5-\alpha}}{(\beta p)^{1-\alpha}}. \tag{22.4}$$

For the convective core the hydrostatic equilibrium conditions are given by Eqs.(22.2), but the radiative equilibrium condition (22.3) must be replaced by the adiabatic relation (15.14). If we again neglect the slight variation of β in the core, as well as the small effect of the radiation pressure on γ, we may apply the supplementary transformation (15.17) as we did for the initial models, although in the present case we have to replace β by $l_i \beta_i$ in these equations. With this transformation we again obtain the polytropic equation (15.20) and hence may use again the unique and available core solution of this equation.

Finally we have to modify the thermal equilibrium condition (15.15) in accordance with the composition functions of our present case, and we obtain

$$\frac{df}{dx} = +D \; i_i \, l_i \; p^2 \, t^{14} \, x^2 \, \beta_i^2. \tag{22.5}$$

Since we are not interested in the run of f through the core we may integrate Eq.(22.5) as before. Thus, with due regard to the boundary conditions for f and with the help of the supplementary transformations (15.17), we obtain

$$1 = D \; \frac{i_i}{l_i^2} \; \frac{(2.5)^{\frac{3}{2}}}{\beta_i} \; p_c^{\frac{1}{2}} \, t_c^{17} \int_0^R t^{*19} \, x^{*2} \, dx^*. \tag{22.6}$$

This completes our assembling of the necessary equations.

Construction of the "First Model"

Let us consider in detail the construction of the "first model," which follows the initial model in the evolution of a star of, say, 10 solar masses. We could start by choosing a definite value for the time interval which we suppose to have elapsed between the initial and the first models. Next we could compute the changes in the composition caused by the nuclear transmutations during this time interval, and last we could solve the over-all boundary value problem for this new composition. This last step would then give the changes of all the stellar properties during the time interval considered; in particular it would give the change in the parameter C.

We may, however, invert this procedure and, instead of choosing a definite value for the elapsed time interval, choose a definite value for the parameter C for the new model. In this latter inverted procedure the choice of C fixes the entire new model, including the run of the new composition, whereas in the direct physical procedure the run of composition is determined by the rate of nuclear transmutations in the convective core and by the reduction rate of the core mass. The inverted procedure has the advantage of leading to a relatively simple computing scheme, as we shall see.

We start then by selecting a value of C for the first model, for example $\log C = -6.500$ as compared with $\log C = -6.579$ for the initial model. Next we obtain the solution for the envelope without any arbitrariness by a single numerical integration of the Eqs.(22.2) to (22.4) from the surface inwards.

We have to terminate this envelope solution when the interior mass fraction q reaches the value q_1 which is equal to the mass fraction q_f at the interface between core and envelope in the initial model. This follows from the fact that the envelope is defined as that part of the stellar mass which is at no time contained in the convective core so that the mass of the envelope for the subsequent phases is fixed from the outset by the mass of the envelope of the initial model.

If we want to continue the envelope solution beyond q_1 into the intermediate zone, we have to take into account the variation of the composition in this zone. Since the intermediate zone will still be shallow in the first model, it will suffice to represent the variation of the composition function l by the simple formula

$$l = l_0 \, q^{-m}. \tag{22.7}$$

Here we can determine the factor l_0 immediately by the condition that the composition must not change discontinuously at the interface between the envelope and the intermediate zone. The exponent m, however, which characterizes the steepness of the composition variation from the outer edge of the zone to the inner edge, can be determined only with the help of the fitting conditions at the edge of the convective core. For the time being, therefore, we have to assume a trial value for m. With this trial value we may introduce Eqs.(22.7) and (22.1) into the main Eqs.(22.2) to (22.4) and then by numerical integration of these equations obtain the continuation of the envelope solution through the intermediate zone.

We still have to fit the intermediate zone solution to the core solution —with the help of the UV plane, as always. Fig. 22.1 shows a section of the UV plane in which is drawn the envelope solution of the initial model as well as the convective core solution. This figure also shows the envelope solution for the first model and its termination point. The intermediate zone solution is shown by the solid line which starts at this termination point and leads sharply to the right.

For the fitting of the intermediate zone solution to the core solution we must remember that in our present models—in contrast to the simplified models of §20—the composition is continuous at the edge of the core so that the three invariants U, V, and $n + 1$ must be continuous as well. The termination point of the intermediate zone solution is therefore that point where this solution crosses the core solution. At this cross point U and V are automatically continuous and we have only to check on the continuity of $n + 1$. We know that $n + 1$ has the value 2.5 in the convective core, and we can read its value at the termination point of the intermediate zone from the numerical integration. If these two values do not agree we have chosen a wrong trial value of the exponent m, we have to choose a new trial value, and we have to repeat the numerical integration for the intermediate zone (the envelope solution remains unchanged). When we have found that value of m which leads to the fulfillment of the fitting condition for $n + 1$ we have completed the integration of the model.

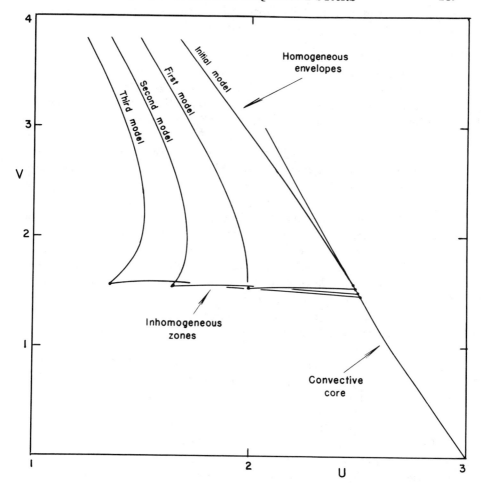

Fig. 22.1. Section of the UV plane showing the evolutionary model sequence for a star of 5 solar masses. (R. S. Kushwaha, *Ap.J.* *125*, 242, 1957)

We can then determine the various properties of this model as follows. If we apply Eq.(22.7) to the edge of the core it gives us l_i. From this in turn we obtain μ_i with the help of Eq.(20.2). For this last step we have to assume a definite envelope composition, i.e. initial composition, for which we shall use consistently here the composition (15.22) already used for the corresponding initial models. With the value of μ_i and with $Z_i = Z_e$ the new composition of the core is entirely determined, i.e. X_i can be computed.

The remainder of the determination of the model properties runs much like that for the initial models. By applying the first two of the transformation equations (15.17) to the fitting point we determine p_c and t_c. D follows from Eq.(22.6), for which i_i can be obtained from the last of

Eqs.(20.2). Finally, the luminosity and the radius of the model are computed from the key equations (15.9) and (15.16).

Construction of the Subsequent Models

After this discussion of the "first model" let us have a short look at the construction of the "second model." We again start by choosing a definite value for C and obtaining the corresponding envelope solution by numerical integration of Eqs.(22.2) to (22.4). We terminate this solution as before at q_1. Next we have to determine the solution for the intermediate zone. In the second model this zone consists of two shells. The first shell contains those layers which already belonged to the intermediate zone in the first model. The second shell contains those layers which still belonged to the convective core in the first model, but which emerged from the convective core during the time interval in which the star evolved from the first to the second model. In the first shell no further changes of composition can have occurred since the time of the first model. Hence in this shell we may use Eq.(22.7) with the value for the exponent m which we have determined for the first model. We can therefore continue the numerical integration of the envelope through this first shell without any arbitrariness. In contrast, we do not know the run of the composition in the second shell since the layers of this shell still belonged partially to the convective core during the time interval between the first and the second models and thus still suffered composition changes. We may use the approximate formula (22.7) again and determine the value of m for this shell by trial and error integrations until we have obtained the fulfillment of the fitting condition for $n + 1$ at the edge of the convective core. In this way the second model is completely determined.

The construction of the subsequent models follows the same pattern. The intermediate zone of any model is divided into shells, the number of which exceeds by one the number of shells of the preceding model. The run of the composition is known from the preceding model for all the shells except the innermost one, which has just emerged from the convective core. The run of the composition in the innermost shell, which is characterized by the value of m for this shell, always has to be determined by trial and error integrations, which luckily are restricted to this one shell.

Results for Hydrogen Depletion

The construction procedure which we have here described has been applied to the three upper main sequence stars for which we have discussed the initial models in §15. The results of these computations are listed in Table 22.1 and shown in Fig. 22.2. In this figure each of the three columns refers to one of the three stars, and the four graphs in a column depict the hydrogen depletion in four successive phases of evolution. In each graph the hydrogen content is plotted as a function of the mass

TABLE 22.1.

Properties of models for three upper main-sequence stars in successive phases of evolution. The corresponding initial states of these stars are given in Table 15.2. (Kushwaha, *Ap.J.* 125, 242, 1957.)

Mass =	$10M_\odot$			$5M_\odot$			$2.5\,M_\odot$		
Model =	1st	2nd	3rd	1st	2nd	3rd	1st	2nd	3rd
$\log C$	− 6.500	− 6.400	− 6.300	− 6.000	− 5.900	− 5.815	− 5.600	− 5.500	− 5.440
$\log D$	+ 0.213	− 1.267	− 3.393	− 1.291	− 2.746	− 4.414	− 2.741	− 4.187	− 5.258
$\log p_c$	+ 1.958	+ 2.204	+ 2.673	+ 2.250	+ 2.505	+ 2.878	+ 2.572	+ 2.847	+ 3.103
$\log t_c$	− 0.033	+ 0.064	+ 0.238	+ 0.054	+ 0.151	+ 0.287	+ 0.136	+ 0.236	+ 0.326
x_1	0.213	0.188	0.160	0.165	0.145	0.127	0.124	0.108	0.098
q_1	0.244	0.244	0.244	0.201	0.201	0.201	0.162	0.162	0.162
$T_1 \times 10^{-6}$	19.3	18.7	17.0	17.1	16.4	15.2	14.8	14.1	13.5
ρ_1	3.87	2.81	1.70	9.28	6.85	4.67	23.3	17.3	13.8
$1 - \beta_1$	0.029	0.035	0.044	0.008	0.010	0.012	0.002	0.003	0.003
δ_1	5.17	6.51	8.48	3.09	3.71	4.45	2.07	2.37	2.61
x_f	0.197	0.146	0.079	0.147	0.108	0.067	0.108	0.077	0.056
q_f	0.205	0.146	0.072	0.155	0.111	0.065	0.119	0.084	0.058
$T_f \times 10^{-6}$	20.4	22.0	26.3	18.3	19.5	22.1	16.0	16.9	18.4
ρ_f	4.73	5.11	7.70	12.7	14.0	19.3	35.0	39.8	50.8
$1 - \beta_f$	0.031	0.045	0.078	0.009	0.013	0.019	0.002	0.003	0.004
δ_f	5.13	6.42	8.54	2.99	3.55	4.37	1.97	2.23	2.50
$T_c \times 10^{-6}$	28.5	30.1	35.1	24.8	26.2	29.5	21.1	22.4	24.3
ρ_c	7.80	8.15	11.88	19.9	21.7	29.7	53.0	60.4	77.0
m	0.6820	0.6639	0.6680	0.8003	0.7001	0.6684	0.9265	0.7178	0.6685
X_i	0.729	0.461	0.061	0.620	0.370	0.077	0.526	0.280	0.090
$\log L/L_\odot$	3.552	3.635	3.718	2.543	2.611	2.664	1.422	1.443	1.466
$\log R/R_\odot$	0.603	0.677	0.784	0.452	0.524	0.610	0.303	0.378	0.433
$\log T_e$	4.347	4.330	4.298	4.170	4.151	4.122	3.965	3.932	3.910
$\bar{X}$	0.8617	0.8149	0.7720	0.850	0.817	0.792	0.848	0.823	0.809
τ_*	0.0350	0.0710	0.0983	0.0455	0.0711	0.0880	0.0466	0.0661	0.0762
$\log \tau$ (yrs)	7.067	7.374	7.515	7.894	8.088	8.181	8.741	8.893	8.954

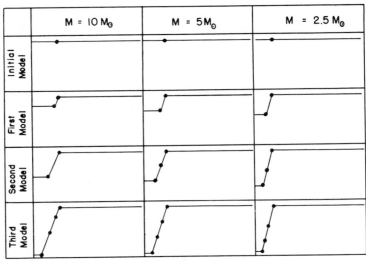

Fig. 22.2. Hydrogen depletion in the core of three upper main-sequence stars in successive phases of evolution. In each of the twelve small graphs the abcissa is the mass fraction q (center at the left, surface at the right) and the ordinate is the hydrogen abundance X (0 at the bottom, 1 at the top). Initially X is taken to be 0.9.

fraction, with the center at the left and the surface at the right. The high horizontal line on the right-hand side of each graph represents the high and invariable hydrogen content of the envelope while the left-hand horizontal line shows the steadily diminishing hydrogen content in the convective core. Between the envelope and the core the graphs show the development of the steadily increasing inhomogeneous zone.

Later Phases with Hydrogen Burning in the Intermediate Zone

Before we plot the results of these evolution computations in the Hertzsprung-Russell diagram let us try to press these computations a little further. Fig. 22.2 shows how the hydrogen abundance of the convective core steadily decreases and how simultaneously the mass fraction contained in the core steadily shrinks during the evolution. There must come a time in the evolution when these two factors together force the energy generation to reach beyond the edge of the core. From then on the energy generation in the intermediate zone will add a substantial contribution to that of the core.

This change has two consequences. First, the energy generation has to be taken into account in the intermediate zone by applying there the thermal equilibrium condition (22.5). This step necessitates a modification of the radiative equilibrium condition (22.3) by multiplying its right-hand side by f. Second, the composition of a layer in the intermediate

TABLE 22.2.

Properties of an upper main-sequence star in two evolutionary phases (following those of Table 22.1) in which hydrogen burning outside the core becomes important. (Kushwaha, *Ap.J.* *125*, 242, 1957.)

Mass =	10 $M_\odot$	
Model =	4th	5th
$\log C$	$-\ 6.293$	$-\ 6.286$
$\log D$	$-\ 3.330$	$-\ 2.53$
$\log p_c$	$+\ 2.731$	$+\ 2.784$
$\log t_c$	$+\ 0.258$	$+\ 0.258$
q_f	0.061	0.034
$T_f \times 10^{-6}$	28.2	33.7
ρ_f	9.34	16.6
$1 - \beta_f$	0.084	0.084
$T_c \times 10^{-6}$	37.0	40.8
ρ_c	14.1	22.1
X_i	0.024	0.002
L_{core}/L	0.947	0.554
$\log L/L_\odot$	$+\ 3.736$	$+\ 3.742$
$\log R/R_\odot$	$+\ 0.781$	$+\ 0.738$
$\log T_e$	$+\ 4.304$	$+\ 4.327$
$\bar{X}$	0.7694	0.7683
τ^*	0.09977	0.09983
$\log \tau$ (yrs)	7.5216	7.5218

zone, which had remained constant with time from the moment the layer emerged out of the convective core, will now start changing owing to the nuclear transmutations within the layer itself.

These changes, though simple in principle, greatly complicate the computations. It does not appear practical to continue with the procedure which we have described for the earlier evolution phases. Instead we have to fall back essentially on the general method described for the over-all problem in §12. This means in terms of the non-dimensional variables that for each evolutionary model we have to determine C and D by trial integrations for the envelope and the intermediate zone. In fact we even have to add a third parameter, m, which has to be determined by trial and error, if we want to use Eq.(22.7) to represent as well as before the run of composition in the shell which has just emerged from the core.

Not only do we have to solve this problem with three trial parameters for each evolutionary model but between every two consecutive models we also have to compute the change in the composition with the help of Eqs.(12.8), (10.22), and (10.23).

Such computations have been carried out for two models which correspond to evolutionary phases following the three models given in Table 22.1 for the evolution of a star of 10 solar masses. The main numerical results for these two additional models are listed in Table 22.2. The successively increasing hydrogen depletion is shown in Fig. 22.3 for

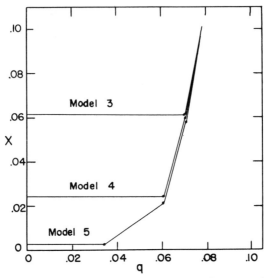

Figure 22.3. Hydrogen depletion in the central portion of an upper main sequence star ($M = 10M_\odot$) in the evolutionary phases in which nuclear burning outside the convective core becomes important.

these two models as well as for the model just preceding, which was already represented by the bottom left-hand graph in Fig. 22.2. The new figure indicates that the hydrogen content is rapidly decreasing not only in the convective core but now also in the inner portions of the intermediate zone.

Evolutionary Tracks in Hertzsprung-Russell Diagram

Let us now plot our evolutionary model sequences in the Hertzsprung-Russell diagram. Fig. 22.4 shows the evolutionary tracks for our three upper main sequence stars according to the data for the luminosities and effective temperatures of Tables 22.1 and 22.2. The three tracks are very similar in character to the inhomogeneous evolutionary track which we have already derived for the simplified models in §20. The tracks turn from the main sequence to the right, the luminosity increases, the effective temperature decreases, the tracks point towards the red giants.

To make possible a more detailed comparison of the three evolutionary tracks they have once more been plotted in Fig. 22.5. For this figure the luminosities and effective temperatures have been expressed in units of the initial luminosity and effective temperature of each star. Thus the three tracks have been moved to one and the same origin, which represents their respective initial models.

Fig. 22.5 shows that the three tracks are very similar in form. This similarity permits us to use graphical interpolation to draw with good ac-

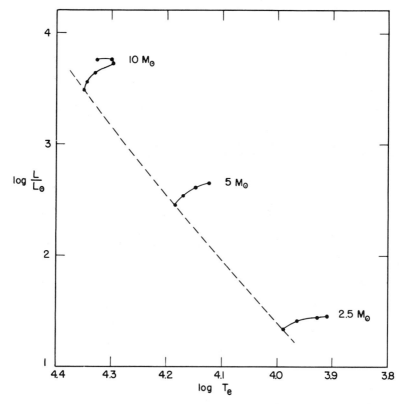

Fig. 22.4. Evolutionary tracks in the Hertzsprung-Russell diagram for three upper main-sequence stars of 10 $M_\odot$, 5 $M_\odot$ and 2.5 $M_\odot$ respectively. (Data from Table 15.2, 22.1 and 22.2).

curacy the evolutionary track of any upper main sequence star with a mass between 2.5 and 10 solar masses.

Time Scale of Evolution

We have this far completely ignored one essential quantity, the time. The inversion of general procedure which we have used for the construction of all the models of Table 22.1 made it unnecessary to refer to the time explicitly. But clearly the evolutionary models which we have derived correspond to very definite times in the life of the respective stars, and we shall need to know these times if we are to interpret the observed Hertzsprung-Russell diagrams. We may obtain these times by the following short calculation.

The time enters the evolutionary phases we are here considering essentially through the rate of change of the hydrogen content, which is given by Eq.(10.22). We may formulate this equation in a more convenient form for our present purposes. Let us define the average hydrogen content, $\overline{X}$, by

$$\overline{X} = \int_0^1 X \, dq. \tag{22.8}$$

The total amount of hydrogen in a star is then represented by $M\overline{X}$. The time derivative of this quantity gives the rate of hydrogen burning throughout the star. If we multiply this rate by the conversion factor E_{cc}^*, which according to its definition (10.21) converts the number of grams burned into ergs, we obtain the total luminosity of the star. Thus we have

$$L = -M \, E_{cc}^* \, \frac{d\overline{X}}{d\tau}. \tag{22.9}$$

This equation could also have been derived by integrating Eq.(10.22) over the whole star.

Let us define the age of the star as the time elapsed since the star first started to burn its hydrogen fuel, that is since the moment when the star was constructed according to its initial model. The age thus defined may be determined for any evolutionary model from Eq.(22.9) by first dividing this equation by L and then integrating it over time from the initial phase to the evolutionary phase in question. In this manner we obtain

$$\tau = \int_{\overline{X}}^{X_0} \frac{M \, E_{cc}^*}{L} \, d\overline{X} \tag{22.10}$$

where the zero subscript refers to the initial state. We can write this equation in the following more convenient form:

$$\tau = \tau_0 \, \tau^* \tag{22.11}$$

$$\text{with } \tau_0 = \frac{ME_{cc}^*}{L_0} \quad \text{and} \quad \tau^* = \int_{\overline{X}}^{X_0} \frac{L_0}{L} \, d\overline{X}. \tag{22.12}$$

We may call τ_0 the "expected life" since this quantity represents the time a star would live if it started with one hundred per cent hydrogen, if it were capable of completely burning up all its hydrogen fuel, and if it maintained its luminosity equal to its initial luminosity throughout its life. On the other hand we may call τ^* the "relative age" since this quantity gives the fraction of the expected life which a star has actually lived. The relative age will be much smaller than 1 throughout the evolutionary phases here considered, first because only a small fraction of the total hydrogen content is burned during these phases, and second because the luminosities increase from their initial values during the evolution.

It is easy to compute the average hydrogen content for all our models, according to Eq.(22.8), from the data on the hydrogen content represented in Figs. 22.2 and 22.3. Next we can plot for each evolutionary sequence

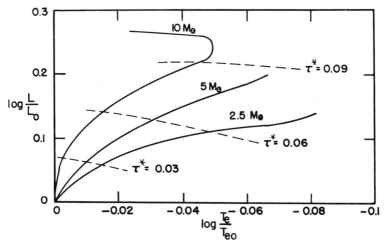

Fig. 22.5. Relative evolutionary tracks in the Hertzsprung-Russell diagram for upper main-sequence stars, in terms of their initial luminosities and effective temperatures. The dashed lines connect points of equal relative age, τ^*. (Data from Tables 15.2, 22.1, and 22.2.)

the reciprocal of the luminosity against the average hydrogen content and then obtain the relative age according to the second of Eqs.(22.12) by graphical integration. Finally we may compute the actual age for each model from Eq.(22.11).

The data resulting from this calculation are given in the bottom section of Tables 22.1 and 22.2. The ages in the last line of these tables are quite of the expected order of magnitude, about a billion years for the lightest of the three stars, but only several times ten million years for the heaviest one.

Apsidal Motion Test for Evolutionary Models

Clearly the most important application of our present theoretical results will be for the interpretation of the observed Hertzsprung-Russell diagrams of galactic clusters. This weighty topic we shall postpone until §29. Now we shall restrict ourselves to the comparison of our evolutionary models with the observed apsidal motion in binaries, the same critical test to which we have subjected the initial models for upper main sequences stars in §18.

As we have seen in §18, this test is centered around the apsidal motion coefficient k. This coefficient can be computed according to Eq.(18.17) for any model for which the differential equation (18.11) has been solved by a numerical integration. These computations have been carried out for several of the evolutionary models considered in this section and the resulting values for k are listed in Table 22.3. This tabulation shows that the apsidal motion coefficient varies significantly during the early

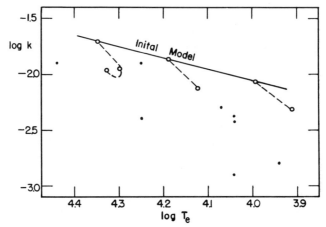

Fig. 22.6. The apsidal motion coefficient as a function of effective temperature (see Fig. 18.3). The dots represent the observed binaries of Table 18.1 and circles the evolutionary model sequences of Table 22.1 to 22.3.

evolution phases of a star and thus provides—at least in principle—the possibility of determining the ages of the binaries in which apsidal motion is observed.

The comparison between theory and observations is given by Fig. 22.6, in which—just as in Fig. 18.3—the apsidal motion coefficient is plotted against the effective temperature. Again the observations of Table 18.1 are represented by dots and the theoretical models by circles. The new figure contains not just three circles for the three initial models but rather three sequences of circles corresponding to the three evolutionary model sequences derived in this section.

Fig. 22.6 shows that the apsidal motion coefficients k for the evolutionary phases here considered agree somewhat better with the observed values than did those for the initial models. There is still left an average discrepancy of about a factor 2 which needs explanation. In view of

TABLE 22.3.

Apsidal motion coefficient of evolutionary model sequences for upper main sequence stars. (Kushwaha, *Ap.J.* 125, 242, 1957.)

Mass	Model	log k
10 $M_\odot$	Initial	−1.70
,,	Third	−1.94
,,	Fifth	−1.96
5 $M_\odot$	Initial	−1.87
,,	Third	−2.11
2.5 $M_\odot$	Initial	−2.07
,,	Third	−2.31

the present theoretical and observational inaccuracies, however, the remaining discrepancy is not large and should not distract us from our conclusion that the apsidal motion observations confirm the over-all structure of our theoretical models.

23. Evolution of the Sun

In principle it is just as easy to compute models for the early evolutionary phases of lower main-sequence stars as we have found it for upper main-sequence stars in the preceding section. In practice, however, these computations are complicated by two circumstances. First, the nuclear transmutations occur in a radiative, not a convective, core and hence one has to compute explicitly from the outset the decrease of the hydrogen abundance at every point in the core for each evolutionary model. Second, lower main-sequence stars have convective envelopes which affect the radii significantly, and the depth of such an envelope can be determined only by the derivation of a model stellar atmosphere for each evolutionary phase.

On the other hand, the calculations for lower main-sequence stars are made less laborious by the fact that we may restrict ourselves to stars with masses close to that of the sun; stars with appreciably smaller masses use up their nuclear fuel so slowly that even the oldest of them in our galaxy have hardly evolved appreciably beyond the initial main sequence state. Accordingly we shall limit ourselves here to the discussion of the evolution of the sun itself.

Depletion of Hydrogen in the Sun

Let us assume that five billion years ago the sun was in its initial homogeneous state, and let us derive the present inhomogeneous model of the sun as it must have evolved in consequence of the nuclear transmutations in the time interval elapsed. To save labor, let us bridge the entire five billion years in a single step. Such a coarse time step can hardly lead to serious inaccuracies since the evolutionary changes in the internal structure of the sun can hardly have been so very great up to now.

We shall start then with one of the homogeneous models which we have constructed in §16 for lower main-sequence stars. The series of models shown in the upper half of Table 16.1—with $\alpha = 0.25$ for the sun—differ from each other in the depth of the convective envelope, i.e. in the value of E. Which of these models is the correct one for the initial state of the sun? We might think of answering this question by using an E value as indicated in Table 16.2. The data of this table, however, were derived under the assumption that the present luminosity and radius of the sun could be applied to its initial homogeneous state. From the improved point of view of the present section we must consider this earlier procedure as inaccurate since the luminosity and the radius of the sun may well have changed from the initial to the present state. We shall therefore not use any of the solar data of Table 16.2, but shall consider the value of E for the initial state of the sun as undetermined.

Luckily this indeterminacy does not at all affect our results for the present state of the sun. The initial value of E, that is the initial depth of the convective envelope, affects only the outer layers but not the structure of the interior portion, at least not as long as the convective envelope does not reach much deeper than it appears to do in the sun. For our present purposes we are interested only in the interior portion of the initial model, since it is only there that hydrogen burns and causes the present inhomogeneities of composition. For the following calculations we may therefore choose any one of the models in the upper half of Table 16.1—for example the one with $E = 1.68$. It is easy to check at every step of the following calculations that another choice would have given us the same results.

To bridge the interval of five billion years we have to compute at every point in the sun the amount of hydrogen consumed. The rate of change of the hydrogen content is given by the transmutation equation (10.22). In this equation we need consider here only the term arising from the proton-proton reaction. The energy generation by this reaction in the sun is represented by Eq.(16.3). We may put this rate of energy generation into a more convenient form by expressing it in terms of its value at the center of the sun, ε_c, and by representing the variation of the energy generation from the center outwards in terms of the asterisked variables defined by the transformation (13.13) and (13.14). Thus we obtain for the transmutation equation

$$\frac{dX}{d\tau} = -\frac{\varepsilon_c}{E_{pp}^*} \times \frac{p^*}{p_c^*} t^{*3.5}. \tag{23.1}$$

We can simplify the transmutation equation still further. According to the definition (16.10) of the parameter D we can express the central rate of energy generation by

$$\varepsilon_c = \frac{L}{M} \times D p_c t_c^{3.5}. \tag{23.2}$$

The numerical values of the three non-dimensional factors in this equation can be read from Table 16.1, giving

$$\varepsilon_c = \frac{L}{M} \times 9.9. \tag{23.3}$$

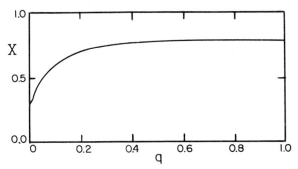

Fig. 23.1. Distribution of the hydrogen abundance X in the sun as a function of the mass fraction q. (Schwarzschild, Howard, and Härm, *Ap.J.* 125, 233, 1957)

If we introduce the last relation into the transmutation equation (23.1), if we ignore any time variations of the right-hand side of this equation, and if we integrate over time from the initial state, with the hydrogen abundance X_e, to the present state, we find

$$X = X_e - \frac{\tau}{\tau_0} \times 9.9 \times \frac{p^*}{p_c^*} t^{*3.5} \qquad (23.4)$$

Here we have introduced again the expected life τ_0 which we defined in the preceding section and which has for the sun the value

$$\tau_0 = E_{pp}^* \frac{M}{L} = 10^{11} \text{ yrs}.$$

If we introduce this value for τ_0 into Eq.(23.4), as well as five billion years for τ and 0.80 for X_e, we finally obtain

$$X = 0.80 - 0.50 \frac{p^*}{p_c^*} t^{*3.5}. \qquad (23.5)$$

With the help of this formula we can compute the hydrogen abundance at every point in the present sun if we read the values of the asterisked variables from the core solution of the initial model. Fig. 23.1 shows the resulting distribution of the hydrogen abundance throughout the sun.

Construction of Inhomogeneous Solar Model

To make the subsequent integrations easier we may represent the run of X as a function of q by the following formulae

$$\log (X/X_e) = -0.416 \qquad \text{for} \qquad \log q < -3.2436$$
$$\log (X/X_e) = -0.1565 + 0.08 \log q \quad \text{for} -3.2436 < \log q < -2.0436$$
$$\log (X/X_e) = +0.0887 + 0.20 \log q \quad \text{for} -2.0436 < \log q < -0.4436 \qquad (23.6)$$
$$\log (X/X_e) = \quad 0.000 \qquad \text{for} -0.4436 < \log q.$$

These formulae represent the computed run of X with an error nowhere exceeding 3 percent.

The composition enters our basic equations through the three composition functions l, j, and i. We may again represent these three functions by simple exponential formulae similar to Eqs.(20.10) which we used for the simplified models. Here we shall use

$$l = \frac{\mu}{\mu_e} = (X/X_e)^{-0.46}$$

$$j = \frac{1+X}{1+X_e} \frac{\mu}{\mu_e} = (X/X_e)^{-0.11} \qquad (23.7)$$

$$i = \left(\frac{X}{X_e}\right)^2 \frac{\mu}{\mu_e} = (X/X_e)^{+1.54}$$

These formulae represent the composition functions with a maximum error of 3 percent for the range of X from 0.3 to 0.8. Eqs.(23.6) and (23.7) together give us the three composition functions for every value of q throughout the present sun.

We may now perform the integrations for the present inhomogeneous model with much the same equations as we have used for the homogeneous models for lower main sequence stars. For the gas characteristics relations we will again employ Eqs.(16.1) to (16.3) with the parameters of the opacity law given by the first line of Eq.(16.4). With the help of the standard transformation (13.2) we can write the basic equations for the convective envelope as before in the form of Eqs.(16.5). In these equations the parameter E is defined by Eq.(16.6), with μ replaced by μ_e.

Continuing as for the homogeneous models, we go on to the radiative interior and apply the supplementary transformation (13.13) with its conditional equations (13.14) modified by (16.8). The parameters C and D are again defined by Eqs.(16.9) and (16.10), in which the composition factors are now defined by the constant envelope composition. The basic equations for the radiative interior then take the form of Eqs.(16.11). Now, however, we have to multiply the first two of these equations by l, the third by $j^{1-\alpha}l$ and the fourth by il. It is through these factors, which we compute according to Eqs.(23.6) and (23.7), that the composition inhomogeneities enter the new models.

The construction of the models proceeds as before: a solution for the interior characterized by a specific value of p_c^* is fitted in the UV plane, shown by Fig. 23.2, to the appropriate envelope solution characterized by a specific value of E. Thus a series of models is obtained in which all the models have the same composition inhomogeneity in the interior but differ from each other in the value of E, i.e. in the depth of the convective envelope.

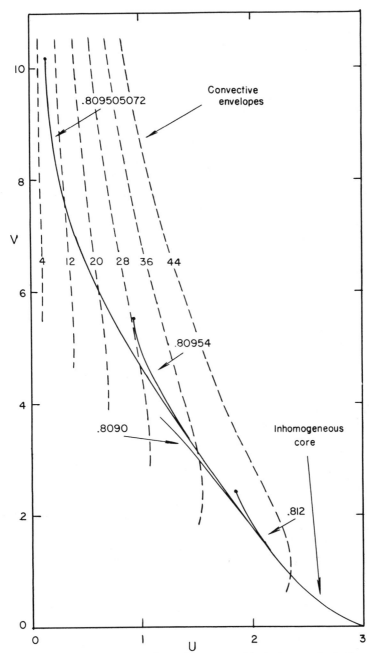

Fig. 23.2. Solutions for the solar core with hydrogen depletion represented in the UV plane. The core solutions (solid curves) are labelled with the p_c^* values while the convective envelopes (dashed curves) are designated by their E values. (Schwarzschild, Howard, and Härm, $Ap.J.$ 125, 233, 1957)

One complication is encountered in this model construction. When one integrates the differential equations (16.11), appropriately multiplied by the composition factors, to obtain the solution for the interior, one needs to know the composition functions l, j, and i as functions of the asterisked variable q^*. But our equations (23.6) and (23.7) give these three functions only in terms of the non-asterisked variable q. Therefore we have to make at the outset a guess at the value of q_0 which relates q with q^* according to the transformation (13.13). With such a guessed value for q_0 we then have to carry out an interior integration. The fitting of this integration to the envelope solutions gives us the accurate value of q_0. If our guessed value turns out to be in error by more than one or two percent, we have to repeat the procedure. This complication therefore increases the amount of numerical work necessary, but the increase is not very great since the homogeneous models of §16 give us good first estimates for the proper q_0 values.

The results of these computations are listed in Table 23.1, which gives the essential mathematical characteristics for three models of this new inhomogeneous series, corresponding to three specific values of E.

Results for the Present State of the Sun

We are ready to apply the new inhomogeneous models to the sun. As with the homogeneous models, we may use the key equations (16.9) and (16.10) to determine the initial helium content Y_e and the depth parameter of the convective envelope, E, for the sun. This determination is shown in graphical form in Fig. 23.3, in which the left-hand curve shows the relation between the parameters C and D for the inhomogeneous models of Table 23.1. For comparison, the curve to the right represents the homogeneous models of Table 16.1 which we have already used in Fig. 16.1. The two horizontal curves are also the same as those in Fig. 16.1; they were computed by introducing the solar values of L, M, and R into

TABLE 23.1

Mathematical properties of lower main-sequence models ($\alpha = 0.25$) with hydrogen depletion in the core as given by Eq.(23.6). (Schwarzschild, Howard, and Härm, *Ap.J. 125*, 233, 1957.)

$p_c{}^*$	0.809505072	0.8095050581	0.8095050576
E	7.58	3.90	1.76
U_1	0.152	0.061	0.020
V_1	10.22	12.80	16.92
q_0	0.070	0.070	0.070
x_f	0.759	0.805	0.852
q_f	0.980	0.994	0.999
$\log p_f$	-1.367	-1.950	-2.657
$\log t_f$	-0.898	-1.016	-1.160
$\log p_c$	$+2.237$	$+2.428$	$+2.582$
$\log t_c$	-0.015	$+0.032$	$+0.072$
$\log C$	-5.424	-5.369	-5.315
$\log D$	-0.831	-1.186	-1.474

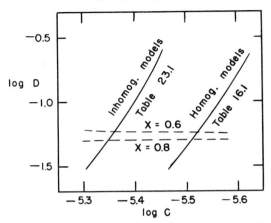

Fig. 23.3. Interrelation of the parameters C and D for solar models.

the key equations (16.9) and (16.10) and by determining from these equations the relation between C and D for any chosen value of X_e, with Z as running parameter. The intersections of these horizontal curves with the curve for the inhomogeneous models give us the C and D values of the new solutions for the sun. With the help of these values we can obtain the various physical properties of interest by interpolation between the models of Table 23.1. The results are given in Table 23.2 for three values of X_e.

Table 23.2 represents our final results for the sun, derived with due regard to the inhomogeneities in composition caused by the nuclear transmutations during the past five billion years. Table 16.2 gave the corresponding data computed neglecting the inhomogeneities. This neglect is well justified for faint main sequence stars such as Castor C since even in the center of such stars the nuclear transmutations can have changed the composition very little. For the sun, however, we see by comparing the two tables that the introduction of the inhomogeneities has had noticeable effects on some of the results.

We notice from Table 23.2 that the abundance of the heavier elements Z has the value 0.02 suggested by the spectroscopic observations of the solar atmosphere when the hydrogen content X_e has the value 0.78, which corresponds approximately to the left-hand column of the table. We shall therefore consider this column as our final representation of the sun in its present evolutionary phase.

There are two values in the left-hand column of Table 23.2 which we can check with other sources. First, the tabulated value of E falls well within the range of E values given in Eq.(16.7) which were derived from model solar atmospheres (the E values here discussed all refer, of course, to the present state of the sun; the E value of the initial state remains undetermined, but has hardly any influence on the determination of the present state). Second, the tabulated value for the temperature T_f at the

TABLE 23.2

Physical properties of the sun as deduced from models with hydrogen
depletion in the core, for various assumed initial hydrogen contents.
(Schwarzschild, Howard, and Härm, *Ap.J. 125*, 233, 1957.)

X_e	0.8	0.7	0.6
Y_e	0.185	0.26	0.30
Z	0.015	0.04	0.10
E	2.95	3.25	3.61
x_f	0.824	0.818	0.810
q_f	0.997	0.996	0.995
$T_f \times 10^{-6}$	1.12	1.27	1.46
ρ_f	0.035	0.041	0.049
$T_c \times 10^{-6}$	14.8	15.8	17.1
ρ_c	132	127	132

bottom of the convective envelope is well below the maximum value of
2.5 million degrees which we had deduced in §10 from the observed
existence of lithium in the solar atmosphere. The satisfactory outcome
of these two qualitative checks may justifiably increase our confidence
in the essential correctness of our latest solar model.

Changes in the Solar Luminosity and Radius in the Past

We have finished our investigation of the present evolutionary state
of the sun. But there still remains the question: how much has the sun
changed in luminosity and radius during the past five billion years?

We can compute these changes from the key equations (16.9) and
(16.10) which apply both to the homogeneous initial models and to the
inhomogeneous present models. For each of the two states we may solve
the key equations for L and R. Then we may form the ratios of present
luminosity to initial luminosity and of present radius to initial radius.
Finally, into the resulting equations we may introduce the values of C
and D of the model with $E = 3.90$ from Table 23.1 for the present state
and the values of the model with $E = 4.14$ of Table 16 1 for the initial
state. Thus we obtain

$$\frac{L \text{ present}}{L \text{ initial}} = \frac{(C^{1.20}D^{0.20}) \text{ inhomog.}}{(C^{1.20}D^{0.20}) \text{ homogen.}} = 1.6,$$

$$\frac{R \text{ present}}{R \text{ initial}} = \frac{(C^{0.16}D^{0.16}) \text{ homogen.}}{(C^{0.16}D^{0.16}) \text{ inhomog.}} \approx 1.04, \qquad (23.8)$$

$$\frac{T_e \text{ present}}{T_e \text{ initial}} \approx 1.1.$$

Here we have included the change in the effective temperature, which results from the changes in luminosity and radius according to the definition (1.3) for the effective temperature.

Before we accept these results let us inquire into the effect of the value of E assumed for the initial state—for which the true E-value is as yet quite undetermined. If we had chosen a very different value for the initial E from that used above—for example $E = 0.00$ instead of $E = 4.14$ —we would have found hardly any effect in the luminosity but a very noticeable effect in the radius and hence in the effective temperature. This situation corresponds exactly to what we have already found in our discussions of the subdwarfs: the depth of the convective envelope has little effect on the deep interior and hence hardly any effect at all on the luminosity, but it does affect the radius noticeably. We must therefore consider the changes in radius and effective temperature given by (23.8) with great caution but may accept the change in the luminosity with good confidence.

We may thus conclude that the solar luminosity must have increased by about a factor 1.6 during the past five billion years. Can this change in the brightness of the sun have had some geophysical or geological consequences that might be detectable?

CHAPTER VI
ADVANCED EVOLUTIONARY PHASES

24. Growth of Isothermal Core

To the best of our present knowledge the basic considerations which we need for the advanced evolutionary phases do not differ seriously from those which we have applied to the early evolutionary phases. Nevertheless this chapter on the advanced phases will read very differently from the preceding one. This difference arises from the circumstance that we could base our discussion of the early phases on fairly detailed and rather accurate computations, while for the advanced phases only the most approximate survey computations are available. This situation will change rapidly since large electronic computers are already at work at several places to fill this gap. In spite of this fast-moving situation it is intriguing to discuss the advanced evolutionary phases even now on the basis of the available preliminary computations, which appear to be sufficiently detailed to indicate the main developments occurring during these phases.

Stars with Cores Exhausted of Hydrogen

What do we expect to have happen when a star has nearly burned up all the hydrogen in its core? We have already seen in §22 that by and by the burning in the core will be replaced by burning in a shell further out which is still hydrogen-rich. When the hydrogen in the core is completely exhausted all the energy generation by nuclear processes will occur in the shell. The star will then consist of a helium core (with an admixture of heavy elements as determined by the initial composition), an energy-producing shell and a hydrogen-rich envelope.

Let us assume that expansions or contractions and the corresponding temperature variations with time do not play a role in the evolutionary phases we are now considering. Then the core must soon reach an isothermal state since any temperature gradient would cause an energy flux for which there exist no sources.

Let us furthermore permit ourselves the following simplifying approximations. The thickness of the energy-producing shell can be estimated to be small in comparison with the stellar radius; for the present rough purposes we may therefore represent the shell by a layer of zero thickness. The envelope is not strictly homogeneous in composition since the composition of the layers just outside the burning shell may have been affected by the hydrogen burning during the earliest evolutionary phases, as we have seen in the preceding chapter. Since these inhomogeneities, however, can hardly have decisive effects we may here ignore them without serious worry.

We are thus led to a model consisting of a homogeneous isothermal helium core, an interface which represents the energy-producing shell, and a homogeneous hydrogen-rich envelope.

In the construction of such models a great difference is encountered between the massive stars—with more than, say, two solar masses—and the less massive stars. The densities in the massive stars are substantially lower than those in the less massive stars, as is shown in detail in Tables 22.1 and 23.2. Thus the ideal gas law will continue to hold in the massive stars, while partial degeneracy is encountered in the less massive stars during the advanced evolutionary phases. We shall see that the occurrence or non-occurrence of degeneracy has a decisive effect on the evolution.

Dilemma of Massive Stars with Isothermal Cores

Let us start with the massive stars. The construction of a non-degenerate isothermal core is particularly simple. Since temperature gradient and energy flux are both zero we have to consider only the hydrostatic equilibrium conditions, which in the non-degenerate case can be used in the non-dimensional form of Eqs.(13.5) and (13.6). If as usual we employ the supplementary transformation (13.13), but this time fix the constants with zero subscript by the conditions

$$l_i \frac{q_0}{t_0 x_0} = 1, \quad l_i \frac{p_0 x_0^3}{t_0 q_0} = 1, \quad p_0 = p_c, \quad t_0 = t_c, \tag{24.1}$$

then the hydrostatic equilibrium conditions take the form

$$\frac{dp^*}{dx^*} = -p^* \frac{q^*}{x^{*2}}, \quad \frac{dq^*}{dx^*} = +p^* x^{*2} \tag{24.2}$$

and the boundary conditions at the center become

$$\text{at } x^* = 0: \quad p^* = 1, \quad q^* = 0. \tag{24.3}$$

The differential equations (24.2) do not contain any parameters, and the boundary conditions (24.3) completely determine the starting values. The solution we are looking for is therefore unique—and has long ago been obtained by numerical integration.

For the envelope we may utilize the same one-parameter family of solutions which we have already used for the early evolutionary phases. These envelopes, specifically for stars of five solar masses, are shown in the *UV* plane of Fig. 24.1, in which the solution for the isothermal core is also represented.

Before we can combine these solutions into a definite model we still have to fix two items. The first is the composition of the envelope and

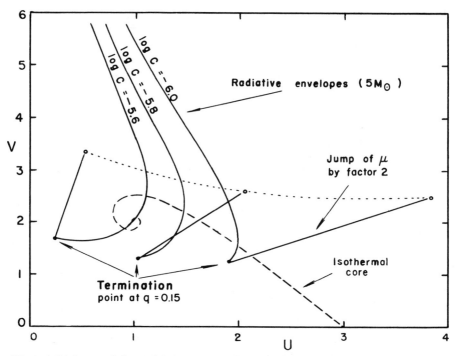

Fig. 24.1. Impossibility of extensive isothermal cores in heavy stars, as shown in UV plane.

the core. Let us use for the envelope a composition similar to that discussed for massive stars at the end of §15, with, say, $\mu_e = 0.67$. For the core we must assume the same abundance of the heavier elements, but zero hydrogen content. Thus we obtain $\mu_i = 1.35$. Since we are representing the energy-burning shell by a sharp interface the molecular weight must therefore jump discontinuously by a factor 2 at this interface. We have already investigated the consequences of such a discontinuity in §20. According to the fitting conditions (20.5) and (20.6) U and V must also jump by a factor 2 at the interface.

The second item we have to fix is the fraction of the mass contained in the core. This mass fraction is clearly an increasing function of time since the hydrogen-burning shell will move further and further out as the evolution progresses. For our present calculations we may use the value $q_1 = 0.15$ which according to Fig. 22.2 appears to be representative for a phase immediately following the early evolutionary phases discussed in §22.

Our fitting problem is now completely defined and may be handled as shown in Fig. 24.1. For each trial value of C we have a definite envelope. We have to terminate this envelope at the point where q reaches the value 0.15. This termination point is marked in Fig. 24.1 by a dot on each envelope. Because of the change in μ at the interface we now have to jump from the termination point on the envelope by a factor of 2 in U and V. The end point of this jump is marked in Fig. 24.1 by a circle. If this end point falls on the solution for the isothermal core we have completed the construction of a consistent model. We do not expect this last condition to be fulfilled for every arbitrarily chosen trial value of C. But is it fulfilled for any value of C? According to Fig. 24.1 the answer is clearly no; the locus of the end-point (dotted line) nowhere crosses the isothermal core (dashed line). We may conclude that under the specific circumstances here considered an equilibrium model with a non-degenerate isothermal core does not exist. Fig. 24.1 suggests that we could have avoided this dilemma either by considering a smaller value of q_1 or by using a smaller jump in μ. In reality, however, neither of these two alterations appears to provide a way out of the difficulty since the mass fraction contained in the core and the difference in molecular weight between core and envelope are both fixed by the previous evolution and are therefore not freely disposable in the search for equilibrium configurations for subsequent phases.

The character of the dilemma into which we have led the star becomes clearer if we make the following thought experiment. Let us assume for the moment that there exists at the interface between core and envelope not only a jump in μ but also a jump in the pressure. If one derives the fitting conditions for U and V at such a double discontinuity, with the help of the defining equations (13.24), one finds that although as before the jump in V is proportional to that in μ the jump in U is proportional not only to that in μ but also to that in P. Accordingly, if one assumes that the pressure in the innermost layers of the envelope is higher than the pressure in the outermost layers of the core, one can shift the locus of the jump end-point in the UV plane (dotted line in Fig. 24.1) to the left and thus enforce an intersection with the core solution (dashed line). But of course such a solution is not stable; the higher pressure in the envelope will force the core to contract. This contraction will raise the temperature in the core and may thus provide the increase in pressure necessary for the support of the envelope.

We had started this section with the assumption that contraction did not play a decisive role during the evolutionary phases here considered. Now we see that this assumption may be incorrect, at least for the massive stars in which the ideal gas law holds. We shall postpone the problem of core contraction in massive stars to the next section and now turn to the problem of isothermal cores in less massive stars.

Stars of Moderate Mass with Partially Degenerate Cores

The essential feature which we have to introduce for the less massive stars is the possibility of partial degeneracy in the central portion of the helium core. We have already discussed the mathematical character of partially degenerate isothermal cores in §13. We found that these cores can be represented by a family of solutions which is defined by Eqs.(13.20) to (13.23) and which depends only on the one parameter p_c^*. These solutions are shown in the UV plane of Fig. 13.2. We see that the solutions reach higher and more to the right in the UV plane the higher the value of p_c^*, that is the higher the degree of degeneracy at the center. The comparison of Fig. 13.2 for the partially degenerate cores with Fig. 24.1 for the non-degenerate massive stars strongly suggests that partial degeneracy will save us from our previous dilemma—provided of course that our final models for the less massive stars will give us an appreciable degree of degeneracy at the center in the evolutionary phases we are here considering.

The envelope solutions we may again derive in terms of the non-dimensional variables from our basic equations (13.5) to (13.10). As generally for stars with moderate mass, we have to expect that the outer portion of the envelope will be convective, and hence we have to use the second alternative of the surface condition (13.12). Thus the envelope solutions depend on the three parameters C, D, and E.

The value of E has to be determined for each stellar model by the construction of an appropriate model atmosphere. We shall postpone the discussion of the atmospheres until towards the end of this section and for the time being consider E a free and undetermined parameter. We shall eliminate the parameter D by the following approximate procedure. Instead of integrating the nuclear energy production according to Eq.(13.8) throughout the energy-producing shell and thus insuring that the total nuclear production equals the luminosity of the star, we shall assume that we can fulfill this latter condition by simply choosing an appropriate temperature for the shell. The specific value which we shall use for the subsequent computations will be $T_1 = 2 \times 10^7$. This procedure introduces the danger of a poor guess for the temperature necessary in the shell to fulfill the thermal equilibrium condition (13.8) accurately, but it greatly simplifies the computations. With this disposal of D the envelope solutions, for an arbitrarily but definitely chosen value of E, depend only on the one parameter C.

Before we can construct a definite model we have to prescribe the stellar mass, the stellar composition, and the specific evolutionary phase. For the stellar mass we shall use $1.2\,M_\odot$ for reasons which will become apparent in the discussion of the Hertzsprung-Russell diagram of globular clusters in §29. For the composition—which is not of critical importance here—we shall use for the envelope 90 percent hydrogen and a negligible amount of heavier elements and for the core accordingly practically pure helium. These data lead to a jump of the molecular weight at the interface between the envelope and the core by a factor of 2.5. The evolutionary phase, finally, may be fixed by the mass fraction contained in the core, the larger this fraction the later the phase. The following computations have been carried out for two phases, corresponding to $q_1 = 0.23$ and $q_1 = 0.25$ respectively.

The models can now be constructed according to the following recipe. Choose a value of E arbitrarily. Determine by numerical integration several envelope solutions corresponding to several values of C. Plot these solutions in the UV plane and terminate each solution at the point where q reaches the prescribed value of q_1. Jump from the termination point to a point in the UV plane with U and V values 2.5 times larger. Connect the several end points corresponding to the several envelope solutions by a curve. Next turn to the core solutions. From the given values of M and q_1 compute M_{r_1}. Introduce this last value into the first of Eqs.(13.21) and determine the corresponding value of q_1^*. Mark in the UV plane on each core solution the point where q^* reaches the value q_1^*. Connect these points on the several core solutions by a curve. Find the intersection of this curve with that one which had previously been drawn through the jump end-points. This intersection fixes the entire model since it fixes the particular envelope solution and the particular core solution which lead to the intersection point. The various physical quantities follow from the non-dimensional quantities in the usual way. The only modification is that while we can use the key equation (13.9) involving C we cannot use the second key equation (13.10) involving D since the latter parameter has been eliminated. Instead of the second key equation we have to use the last of Eqs.(13.2) applied to the interface for which the temperature T_1 was fixed at the outset.

The results of computations according to this recipe are given in Table 24.1. Six models are given in this table corresponding to three arbitrary values of E for each of the two selected values of q_1. The data of Table 24.1 are not accurate since in the application of the above recipe a number of additional simplifying approximations have been em-employed. Nevertheless it appears probable that these data are not misleading regarding the main phenomena occurring during the evolutionary phases we are here surveying.

The group of data at the bottom of Table 24.1 shows the extent of the convection zone in these evolutionary phases. The small values of x_2 and q_2, which refer to the interface between the convective and radiative parts of the envelope, indicate that now the convection zone not only covers from one half to three quarters of the radius but even contains a substantial fraction of the entire stellar mass. The values of T_2 show that the temperature at the bottom of the convection zone is of the order of a million degrees, much as in the initial models for lower main-sequence stars.

The group of data on log ρ in Table 24.1 makes it clear that the central densities are indeed high enough to provide a large degree of degeneracy but that the density drops outwards so rapidly that at the energy-producing shell just outside the core degeneracy has already completely disappeared.

Finally, the group of data for R, L, and T_e in Table 24.1 give the radii and luminosities of the models which have been plotted in the Hertzsprung-Russell diagram of Fig. 24.2. In this diagram the six models fall above

TABLE 24.1.

Models with partially degenerate isothermal helium cores (subscripts c to 1), radiative intermediate zones (1 to 2), and convective envelopes (2 to surface). ($M = 1.2 M_\odot$, $T_1 = 20 \times 10^6$). (Hoyle and Schwarzschild, *Ap.J.* Suppl. No. 13, 1955.)

		0.23			0.25	
$q_1 =$ $q_1^* =$		25.0			27.5	
$E =$	25.3	20.0	17.0	23.9	20.0	18.9
$\log C$	-5.00	-4.60	-4.50	-5.00	-4.75	-4.50
$\log p_c^*$	$+2.11$	$+2.09$	$+2.08$	$+2.20$	$+2.19$	$+2.19$
U_{1e}	0.031	0.045	0.049	0.022	0.027	0.033
V_{1e}	3.06	3.26	3.31	3.38	3.50	3.61
$(n+1)_{1e}$	3.93	3.93	3.93	3.96	3.96	3.96
$\log x_1$	-1.75	-1.93	-2.01	-2.02	-2.34	-2.45
$\log q_1$	-0.64	-0.64	-0.64	-0.60	-0.60	-0.60
$\log t_1$	$+0.61$	$+0.75$	$+0.80$	$+0.84$	$+1.11$	$+1.28$
$\log p_1$	$+3.72$	$+4.64$	$+4.94$	$+4.49$	$+5.90$	$+6.55$
U_2	0.53	0.70	0.59	0.34	0.58	0.60
V_2	1.91	3.19	3.68	1.90	2.28	2.88
$\log R/R_\odot$	$+0.48$	$+0.62$	$+0.67$	$+0.71$	$+0.98$	$+1.15$
$\log L/L_\odot$	$+1.68$	$+1.44$	$+1.01$	$+1.98$	$+1.67$	$+1.47$
$\log T_e$	$+3.94$	$+3.81$	$+3.68$	$+3.90$	$+3.69$	$+3.55$
$\log \rho_c$	$+4.46$	$+4.46$	$+4.44$	$+4.52$	$+4.51$	$+4.51$
$\log \rho_{1e}$	$+1.42$	$+1.78$	$+1.88$	$+1.27$	$+1.60$	$+1.57$
$x_2 = r_2/R$	0.21	0.40	0.52	0.16	0.26	0.35
$q_2 = M_{r2}/M$	0.33	0.63	0.72	0.32	0.45	0.59
$T_2 \times 10^{-6}$	4.1	1.7	1.0	3.0	1.1	0.6

and to the right of the main sequence. This suggests that in these more advanced phases the evolutionary track in the Hertzsprung-Russell diagram will continue the same trend away from the main sequence and towards the red giants that we had already found for the earlier evolutionary phases. These six models, however, do not permit the deduction of a definite evolutionary track as long as we have not determined the appropriate E values by consideration of the atmosphere. For the time being we may note that the range of E values required by the stellar interior to give reasonable radii and effective temperatures is very narrow.

Stars with Large Partially Degenerate Cores

Before we turn to the atmosphere and the determination of E we should obviously try to push our present derivation of stellar models to higher values of q_1, that is to later evolutionary phases. But the computations according to the above recipe for models with larger q_1 values become increasingly laborious since longer and longer numerical integrations are required for the core as well as for the envelope. Furthermore, it is not possible to make offhand the necessary guess for the temperature in the energy-producing shell, particularly since this temperature will keep increasing as the luminosity of the star grows. Luckily, appreciable in-

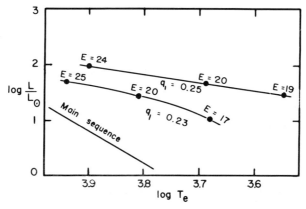

Fig. 24.2. Hertzsprung-Russell diagram representing two evolutionary phases with helium cores of moderate size, for various trial values of E (data from Table 24.1).

sight into the structure of the core during the more advanced evolutionary phases can be obtained by the following trick.

We see from Table 24.1 that the invariants V and $n + 1$ at the inner edge of the envelope approach the values

$$V_{1e} \approx 4, \quad (n + 1)_{1e} \approx 4. \tag{24.4}$$

These values are found to be the limiting values to which V and $n + 1$ converge on radiative solutions with electron scattering if these solutions are followed inwards to small fractions of the radius. Since in reality electron scattering provides the dominant opacity in the inner portions of the envelopes here considered, and since we have to follow these radiative solutions to quite small fractions of the radius—the smaller the more advanced the phase—we may use these values with good confidence.

According to the definitions (13.24) for the invariants V and $n + 1$ the above numerical values fix the relative pressure and temperature gradients in the innermost layers of the envelope, in the energy-producing shell. Thus from Eqs.(13.24) and (24.4) we obtain for the pressure and temperature run through the energy-producing shell

$$P = P_1 \left(\frac{r}{r_1}\right)^{-4}, \quad T = T_1 \left(\frac{r}{r_1}\right)^{-1}. \tag{24.5}$$

The second of these formulae is not very accurate since the relative temperature gradient must vary through the shell from the value given here at the outer edge of the shell to a value of 0 at the inner edge. Such a correction, however, would hardly alter the following estimates.

We may now use Eqs.(24.5) for the pressure and temperature run through the shell to determine the total nuclear energy production within the shell. If we use Eq.(10.15) for the carbon cycle together with the appropriate coefficients from Table 10.1 we obtain

$$L = \int_{r_1}^{\infty} \epsilon \rho \, 4\pi r^2 \, dr = 10^{-15.6} \, \rho_{1e}^2 \, X_e \, X_{CN} \left(\frac{T_1}{10^6}\right)^{15} \frac{4\pi r_1^3}{18} . \qquad (24.6)$$

This equation relates the total luminosity as produced by the carbon cycle to the physical conditions in the shell.

On the other hand, the luminosity may also be determined through the condition of radiative equilibrium. If we use the definition (13.24) for the invariant $n + 1$ and if we replace the pressure and temperature gradients in this definition by the conditions (12.1) and (12.4) for the hydrostatic and radiative equilibrium, and if we finally express the opacity coefficient by Eq.(9.19) for electron scattering, we then find

$$(n + 1)_{1e} = 4 = \frac{4 \, ac}{3} \, \frac{4\pi GM_{r1}}{0.19(1 + X_e)L} \, \frac{T_1^4}{P_1} . \qquad (24.7)$$

This equation represents a second relation between L and the physical conditions in the shell. We may eliminate L between Eqs.(24.6) and (24.7). We may furthermore eliminate P_1, ρ_1, M_{r1}, and r_1 with the help of Eqs.(13.20) and (13.21). After these eliminations we end up with an equation for the temperature in the shell T_1 in terms of the values of the asterisked variables at the shell.

Our new recipe for constructing model cores for more advanced evolutionary phases goes as follows. Take a particular solution for a partially degenerate isothermal core corresponding to a particular value of p_c^*. Follow this solution over its maximum in the UV plane and down to the point where V is equal to 10. At this point you have reached the shell since, because of the jump in μ, the value of V at the interface must be 2.5 times larger in the core than the value in the envelope given by Eq.(24.4). Read the values of all the asterisked variables from your core solution at this point. Introduce these values into the equation for T_1, the derivation of which we have just described. With the value of T_1 thus obtained compute the other physical quantities for the shell with the help of Eqs.(13.20) and (13.21). Finally, derive the luminosity from Eq.(24.7).

Table 24.2 gives the results of such computations for a sequence of values of p_c^*. The top group of data shows the asterisked variables at the shell. The middle group of data gives the quantities most essential for us: the shell temperature, the mass fraction of the core and the luminosity. We see that the core models of Table 24.2 cover a range of evolutionary phases in which the mass fraction of the core increases from about 30 percent to nearly 60 percent. During these phases the internal temperature steadily rises, reaching values above forty million degrees (retrospectively we notice that the temperature guess of twenty million

TABLE 24.2.

Core characteristics for red-giant models with large isothermal helium cores ($M = 1.2 M_{\odot}$, $X_{CN} = 0.001$). (Modified data from Hoyle and Schwarzschild, *Ap.J.* Suppl. No. 13, 1955.)

$\log p_c^* =$	$+2.36$	$+2.43$	$+2.48$	$+2.54$	$+2.58$	$+2.63$
$\log s_c^*$	$+1.58$	$+1.62$	$+1.65$	$+1.68$	$+1.72$	$+1.74$
q_1^*	27.0	28.0	29.0	30.0	31.0	32.0
$\log x_1^*$	$+0.44$	$+0.45$	$+0.46$	$+0.48$	$+0.49$	$+0.50$
$\log s_1^*$	-0.98	-1.21	-1.49	-1.75	-2.01	-2.25
$T_1 \times 10^{-6}$	28.9	31.9	36.1	40.7	45.7	50.9
$q_1 = M_{r1}/M$	0.32	0.36	0.41	0.47	0.53	0.59
$\log L/L_{\odot}$	$+1.80$	$+2.15$	$+2.56$	$+2.96$	$+3.34$	$+3.71$
$\log r_1$	$+9.47$	$+9.47$	$+9.48$	$+9.47$	$+9.47$	$+9.46$
$\log \rho_c$	$+4.86$	$+4.96$	$+5.08$	$+5.19$	$+5.30$	$+5.39$
$\log \rho_{1e}$	$+2.30$	$+2.13$	$+1.94$	$+1.75$	$+1.57$	$+1.40$

degrees which we have used in the preceding discussion for the somewhat earlier evolutionary phases may be correct for $q_1 = 0.23$ but is most likely somewhat too low for $q_1 = 0.25$). The luminosity steadily increases during these evolutionary phases, covering the range from about a hundred times to over a thousand times the solar luminosity—a range fitting well for the red giants.

The last group of data in Table 24.2 gives some further information regarding the core. Its radius turns out amazingly constant, amounting to about 4 percent of the solar radius, in spite of the substantial increase in the core mass. The central density increases slowly and reaches values above 10^5, well inside the degenerate range. On the other hand, the density at the interface remains relatively low, definitely in the nondegenerate range.

The Surface Layers of Stars with Isothermal Cores

We have followed the evolution of a star of moderate mass up to the phase when its helium core contains about one half of the stellar mass. We have determined the increase of the luminosity as a function of the growth of the core. But we still have to determine the behavior during these phases of the radius, which is determined not so much by the inner portion of the star on which we have concentrated thus far in this section but rather by the outermost parts to which we now turn.

We may consider the parameter E as the connecting link between the interior and the atmosphere. The determination of E from the interior is graphically shown in Fig. 24.2. The determination of E from the atmosphere may be formulated as follows. To start with, we select a definite star, that is one with a definite mass and a definite envelope composition. Next we choose arbitrary values for L and R, which fix the effective temperature and the surface gravity. Then we compute the photospheric density according to Eq.(11.3). From there we integrate inwards, first through the layers of the lower photosphere in radiative equilibrium. When we reach the onset of convective instability we add to the radiation flux the convective heat transport according to Eqs.(7.4) to (7.7). We continue the integration inwards through the hydrogen and helium ionization zones and finally into the deep convective layers in which the simple adiabatic relation (11.7) holds. The value of the coefficient K in this relation determines E according to its definition (13.12). Thus we have determined E as required by the atmosphere for a definite star and an arbitrarily chosen set of L and R values.

If we repeat this computation for a number of sets of L and R values, we can determine E for the star as a function of L and R, as a function of the position in the Hertzsprung-Russell diagram. On the other hand, our earlier discussion, represented by Fig. 24.2, shows that for a definite star in a definite phase of evolution the analysis of the stellar interior gives a line in the Hertzsprung-Russell diagram and that for every point on this line the interior requires a definite value of E. Combining the interior and the atmosphere analyses we find that there exists only one point on such a line in which the interior E value and the atmospheric E value agree. This point in the Hertzsprung-Russell diagram must then represent the star in the particular evolutionary phase considered. If we determine this point for a number of phases, that is for a number of q_1 values, we obtain the evolutionary track of the star for the advanced phases here considered.

The computations which we have just sketched have thus far been carried out only in the roughest of approximations. We shall not reproduce them here in detail, but just emphasize the major points of uncertainty and then indicate the results.

The worst uncertainties arise from the convection in the subphotospheric layers. In the outermost convective regions the convective energy transport efficiency is so low that the radiative temperature gradient holds in good approximation. In the deeper convective layers the convective energy transport efficiency is so high that the adiabatic temperature gradient is an excellent approximation. But in the transition region, which has an appreciable influence on the pressure-temperature relation further in, the exact values of the quantities occurring in the convective equations (7.4) to (7.6) play a serious role. In particular, the characteristic length l for the convective elements has to be known. Since in these outer layers the density drops by large factors over distances small compared with the radius it seems unlikely to be correct to take l equal to a moderate fraction of the stellar radius as we have done for the deep interior. Instead it appears more likely that at every point l will be equal to a moderate multiple of the scale height (i.e. the distance in which the pressure drops by a factor e). Exactly which multiple to take appears, however, rather uncertain.

Another uncertainty arises when the convective velocities computed from Eq.(7.5) turn out to be supersonic—as is frequently the case for the red giants. Under these circumstances one can certainly not use Eq.(7.5) any further, and it may be best for the time being simply to replace v by the sound velocity in Eq.(7.4) for the convective energy transport.

Results for Stars of Various Compositions

With these uncertainties duly in mind, we may now look at the results of the approximate computations thus far available. Fig. 24.3 gives the evolutionary tracks in the Hertzsprung-Russell diagram obtained by combining the interior analysis discussed in detail in this section with the atmospheric analysis just sketched. The two evolutionary tracks shown in this figure both refer to stars of 1.2 solar masses. The left-hand track is derived for a composition with very low metal abundance, appropriate for Population II stars. We see that this track covers the evolutionary phases in which the helium core grows from 25 percent of the mass to 50 percent. The evolutionary track for these phases agrees most satisfactorily with the red-giant branch observed in globular clusters.

Similarly encouraging are the preliminary results for Population I stars with higher metal abundance, as represented by the right-hand track. This evolutionary track leads to lower effective temperatures, just as is observed for Population I giants.

The effect of the difference in the metal abundance, which causes the difference between the two evolutionary tracks, can be understood in terms of the temperature-density diagram of Fig. 11.1. An increased metal abundance provides a relatively large number of free electrons and

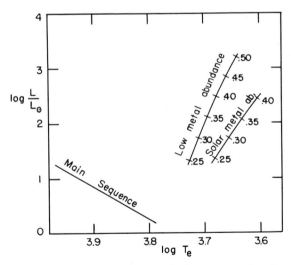

Fig. 24.3. Approximate evolutionary tracks in Hertzsprung-Russell diagram for stars of 1.2 $M_{\odot}$ in phases with large helium cores. The numbers give the mass fraction of the core, q_1, as it increases during the evolution. (Hoyle and Schwarzschild, *Ap.J.*, Supplement No. 13, 1955)

thus increases the opacity in the photosphere. This increase has to be compensated by a lower photospheric density to preserve the proper optical depth. Thus the point representing the photosphere moves to the left in the diagram. Furthermore, the lower densities will cause the convective transport to be ineffective rather far in, so that the steep radiative temperature gradient will hold for an appreciable distance inwards. The curve in the temperature-density diagram which represents these outer layers will therefore rise steeply from the photospheric point and will eventually turn over into an adiabat relatively high in the diagram, that is at relatively high temperatures. Finally, high temperatures produce low pressure gradients according to the hydrostatic equilibrium condition and thus cause an extended envelope. It is this chain of effects that appears to explain the relatively more extended radii of Population I giants compared with Population II giants.

Onset of Fast Core Contraction

Again we postpone the detailed comparison of the theoretical evolutionary tracks with the observed Hertzsprung-Russell diagrams until §29. Let us finish this section by answering the question why we have not pushed the evolutionary tracks with the same technique beyond the last phases indicated in Fig. 24.3. For the right-hand track representing Population I giants the answer is that a continuation of the track leads to such low photospheric temperatures that a more detailed analysis of the photospheric state of ionization is first needed. For the left-hand track representing Population II giants the most direct answer is that a continuation of the track would have led to higher luminosities than are ever observed among Population II giants.

Happily a good theoretical reason also offers itself for stopping where we have. Consider the rate of evolution of a star during the interval in which the helium core increases its mass fraction from 0.45 to 0.50. According to Fig. 24.3, during this phase the star will have a luminosity approximately a thousand times larger than that of the sun. Correspondingly one finds from Eq.(22.10) that the star will need only six million years to run through this evolutionary interval. We see that because of the high luminosity the rate of evolution is enormously increased compared with the early evolutionary phases. This large increase makes it practically certain that the thermal effects of the core contraction, which is evidenced by the steadily increasing central density, can no longer be ignored.

We may conclude that sooner or later the contraction of the core has to be taken into account in the evolution of any star—for the massive stars immediately after the convective core has exhausted its hydrogen, as we have deduced early in the section, and for the less massive stars only after an extensive development of a partially degenerate isothermal helium core.

25. *Heating of Core by Contraction*

We have followed the evolution of stars, both massive and less massive ones, to the phase when the contraction of the core starts to play a decisive role. Shall we pursue stellar evolution still further, into the short but critical phases in which the contraction of the core raises the central temperature to more than a hundred million degrees so that helium burning commences? These phases introduce nothing that we have not already included in our discussion of the over-all problem in §12. To follow the evolution through these phases in a completely determined manner all we have to do is to replace the thermal equilibrium condition (12.3) by the more general equation(12.10). This replacement, however, greatly increases the necessary computational effort. It is not surprising then that only the roughest surveys have been made for these contraction phases until very recently when the large electronic computers have been set to work attacking this problem.

It would thus clearly be safer if we stopped our discussion of stellar evolution here and waited for the results from the big computers, which we may expect in the nearest future. But for those whose curiosity is stronger than their wish for safety we shall go on—fully aware of the risk.

Contracting Cores in Massive Stars

For the massive stars we have found that contraction becomes important virtually as soon as the hydrogen is exhausted in the center. For the subsequent phases we must therefore consider models which consist of a helium core in which gravitational energy is released by contraction, of a hydrogen-burning shell and of a hydrogen-rich envelope. Such models would be easy to construct if it were not for the unknown distribution throughout the core of the energy set free by the contraction. This distribution is determined by the laborious application of Eq.(12.10). We can circumvent this complication, however, by the following crude approximation.

Let us assume that the net energy release by contraction per gram per second, that is the difference between the gravitational energy set free and the thermal energy consumed for increasing the temperature, is constant throughout the core. This simplifying assumption relieves us from having to apply Eq.(12.10) point by point throughout the core. It does not, however, relieve us from having to compute the total amount of energy released by the contraction in the core as a whole, as we shall see below.

Constant energy release per gram means that the total energy release within a sphere must be proportional to the mass contained in the sphere, that is that L_r must be proportional to M_r throughout the core for any given phase. If in addition to this proportionality we assume that degeneracy is negligible and that electron scattering provides the main opacity—two fair assumptions for the cores of massive stars—and if we apply our standard transformation (13.2) and the supplementary transformation (13.13), and if we dispose of the constants with zero subscripts in a manner similar to Eq.(13.14), we obtain for the basic conditions

$$\frac{dp^*}{dx^*} = -\frac{p^* q^*}{t^* x^{*2}}, \quad \frac{dq^*}{dx^*} = \frac{p^* x^{*2}}{t^*}, \quad \frac{dt^*}{dx^*} = -\frac{p^* q^*}{t^{*4} x^{*2}}, \quad f^* = q^* \qquad (25.1)$$

and for the boundary conditions at the center

$$\text{at } x^* = 0: f^* = q^* = 0, \ t^* = 1. \qquad (25.2)$$

Thus we find again that the core solution is determined by a set of differential equations which contain no extra parameter and by a set of boundary conditions which leave indeterminate only p_c^*, so that altogether p_c^* is the sole free parameter—exactly as in the solutions for non-contracting cores governed by Eqs.(13.15) to (13.19).

The single-parameter family of core solutions defined by Eqs.(25.1) and (25.2) are represented by the solid curves in the UV plane of Fig. 25.1. The character of these solutions is strikingly similar to the partially degenerate isothermal cores depicted in Fig. 13.2. This similarity in the solutions is caused by a similarity in the pressure-density relation; in the inner portions of both types of cores the pressure rises somewhat more steeply than the density—in the contracting cores because of the increase of temperature towards the center and in the partially degenerate cores because of the degeneracy—while in the outer portions both types of cores approach nondegenerate isothermal conditions. The dilemma of Fig. 24.1 in which we found ourselves when we tried to fit a non-degenerate isothermal core into the appropriate envelopes is clearly avoided by the contracting cores of Fig. 25.1 for the massive stars, just as we found that it was avoided by the partially degenerate cores of Fig. 13.2 for the less massive stars.

Gravitational Energy Release in a Contracting Core

To fit the contracting cores to the proper envelopes we may again approximate the thin hydrogen-burning shell by a sharp interface at which the composition jumps discontinuously from the helium of the core to the hydrogen-rich mixture of the envelope. As before, the discontinuity in composition causes a discontinuity in U and V and we may write the fitting conditions for these invariants according to Eqs.(20.5) and (20.6) in the form

$$\frac{U_{1i}}{U_{1e}} = \frac{V_{1i}}{V_{1e}} = \frac{\mu_i}{\mu_e} \qquad (25.3)$$

The fitting condition for the third invariant $n + 1$ we may derive from its definition (13.24), in which we may replace the pressure gradient with

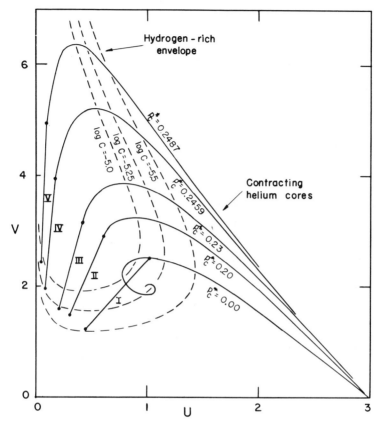

Fig. 25.1. *UV* plane showing models with nondegenerate, contracting helium cores. (Sandage and Schwarzschild, *Ap.J. 116,* 463, 1952)

the help of the hydrostatic equilibrium condition (12.1) and the temperature gradient with the help of the radiative equilibrium condition (12.4). After these eliminations we find an expression for $n + 1$ in terms of various physical quantities, all of which are continuous at the interface with the exception of the absorption coefficient $\varkappa$ and the energy flux L_r. The discontinuity in $\varkappa$ is directly caused by the discontinuity in composition, while the discontinuity in L_r arises from the fact that, whereas just inside the interface the energy flux represents only the gravitational energy released by the core contraction, outside the interface it represents also the nuclear energy released in the shell. Our third fitting condition, therefore, is

$$\frac{(n + 1)_{1e}}{(n + 1)_{1i}} = \frac{\varkappa_{1i} L_{1i}}{\varkappa_{1e} L_{1e}}. \tag{25.4}$$

Strictly speaking we should already have applied this third condition when we were constructing models with isothermal cores. But in a core which is isothermal because it lacks energy sources, L_{1i} is zero and

$(n + 1)_{1i}$ is by definition infinite. Hence condition (25.4) was automatically fulfilled by the models we considered in the previous section. Now, on the other hand, L_{1i}, though small, is not zero because of the gravitational energy released, and $(n + 1)_{1i}$, though large, is not infinite in a contracting core. We must now, therefore, take explicit care of this third fitting condition.

To be able to fulfill condition (25.4) we have to compute L_{1i}, and it is for this reason that we have to determine the total amount—not the distribution—of the gravitational energy release in the core. Let us consider two definite neighboring phases of evolution. The energy δE_g set free by the contraction during the time interval between the two phases is given by the difference between the two phases in the excess of the gravitational energy over the thermal energy throughout the core and by the work done by the pressure at the surface of the core owing to the change in volume of the core. Thus we have

$$\delta E_g = \delta \left[\int_0^{M_{r_1}} \left(\frac{GM_r}{r} - \frac{3}{2} \frac{kT}{\mu H} \right) dM_r \right] - P_1 \delta \left[\frac{4}{3} \pi r_1^3 \right]. \qquad (25.5)$$

Had we wanted to stress the relation between the new Eq.(25.5) and our basic conditions we could have derived it from the basic energy balance condition (5.9) by integrating this condition over the entire core, by performing a partial integration, and by employing the formulae

$$\frac{dP}{dM_r} \times \delta \left[\frac{4}{3} \pi r^3 \right] = + \delta \left[\frac{GM_r}{r} \right], \quad \frac{1}{\rho} = \frac{d}{dM_r} \left(\frac{4}{3} \pi r^3 \right),$$

which follow from the hydrostatic equilibrium conditions (12.1) and (12.2). We may thus consider Eq.(25.5) just as an integrated form of the basic condition (5.9). We should note that the differences on the right-hand side of Eq.(25.5) are to be taken at a fixed value of M_r. It may sometimes be more convenient to use different values of M_{r_1} for the two phases, as, for example, the actual values of the core, which will grow in mass a little from the first to the second phase. In such a case we have to replace the simple difference operator δ by the modified operator

$$\delta - \delta[M_{r_1}] \times \frac{d}{dM_r}.$$

While the gravitational energy released between the two phases is given by Eq.(25.5), the nuclear energy released in the same time interval is determined by the mass increment of the helium core since this mass increment represents the amount of hydrogen which has been transmuted into helium in this time interval. Thus, according to the transmutation coefficient (10.21), we obtain

$$\delta E_n = E_{cc}^* \, \delta [M_{r_1}].$$ (25.6)

Since the two energy releases δE_g and δE_n occur over the same time interval their ratio gives directly the ratio of the two corresponding energy fluxes L_g and L_n. Accordingly we can express the jump in the energy flux at the interface between core and envelope by the relation

$$\frac{L_{1i}}{L_{1e}} = \frac{L_g}{L_n + L_g} = \frac{\delta E_g}{\delta E_n + \delta E_g}.$$ (25.7)

This relation in conjunction with Eqs.(25.5) and (25.6) finally enables us to apply the fitting condition (25.4).

Evolution of Massive Stars During Core Contraction

How do we now construct an evolutionary sequence of models with contracting cores? To start the sequence let us use a model with a core which is not yet seriously contracting and is hence still isothermal and which contains the maximum mass fraction that still avoids the dilemma of Fig. 24.1. With this choice for the first model we can be sure that the star is forced into a core contraction during the subsequent phases. This starting model is represented in the UV plane of Fig. 25.1 by the lowest of the core solutions and by the jump line labelled I.

For the second model in the sequence let us use the second core solution shown in Fig. 25.1. Let us terminate this core solution at an arbitrary trial point. From this trial termination point we have to take a jump in the UV plane in accordance with the fitting conditions (25.3). The end point of this jump uniquely determines the envelope which we have to fit to the core. In this manner we have completely determined a trial model for the second phase and we can compute all the various physical quantities in the usual manner. This trial model fulfills by its construction all the necessary equations and conditions except for the fitting condition (25.4). We can check on the fulfillment of this condition by using Eqs.(25.5) to (25.7), where we can compute the differences occurring on the right-hand sides of Eqs.(25.5) and (25.6) from the differences in the physical quantities between the definitively chosen first model and the trial model for the second phase. If we find that condition (25.4) is not fulfilled, we deduce that we have chosen a wrong termination point on the core solution. By trial and error we eventually find the correct termination point which leads to the fulfillment of condition (25.4). We see that nothing is left arbitrary in the construction of the second model and that this second model is tied to the first one through the differences occurring in Eqs.(25.5) and (25.6).

By the same method of construction we can derive the subsequent models of the evolutionary sequence, with each model uniquely determined through its relation to the preceding one.

Such calculations have been carried out for the sequence of five models indicated in Fig. 25.1. The results of this computation must be considered rather uncertain, for the following reasons. First, we used only an approximation for the distribution of the gravitational energy release throughout the core. Second, the envelope solutions used take account of the transition from Kramers' opacity in the outer parts to electron scattering in the inner parts only in a most approximate manner. Third, the jump in opacity needed for Eq.(25.4) was inadvisedly computed for Kramers' opacity instead of for electron scattering. We shall therefore not discuss the results of these computations in detail but only consider the following overall results, in which we may have reasonable confidence.

During the evolution from the first to the fifth model shown in Fig. 25.1 the mass contained in the helium core increases by only a fraction of a percent of the total mass. The smallness of this effect of the hydrogen burning proves that the entire evolutionary period which we here consider must be of very short duration. This is no surprise. We have seen already in §5 that the total gravitational energy available to a star is very small. Hence the period in which gravitational energy release plays a noticeable role can only last a small fraction of the total life of the star.

The computations further show that during the phases considered the luminosity of the star changes little while the radius seems to increase by perhaps a factor 3 and the effective temperature decreases by about 40 percent. The small change in luminosity is again no surprise; we have already found other examples indicating that the luminosity of a star is insensitive to the exact internal structure and depends mainly on the average composition of the star, which changes little during the short evolutionary period of the core contraction. The changes in radius and effective temperature we should consider with great caution since they depend rather sensitively on the detailed structure of the envelopes and cores, which have not been treated with sufficient accuracy in the available computations. In fact the true changes in radius and effective temperature may well differ appreciably between stars of say three and ten solar masses.

Finally—and most important—the computations show that the short but effective core contraction raises the internal temperature from about thirty million degrees, adequate for hydrogen burning in the shell of the first model, to over a hundred million degrees, adequate for helium burning in the center of the fifth model. The most essential effect of the core contraction then is that it leads to the start of helium burning by the triple-alpha process described in Eq.(10.16). After the helium burning has once commenced it will be likely to maintain by itself the temperature it needs.

What will happen next in the evolution of a massive star? Only he who will investigate in detail models with helium-burning centers and hydrogen-burning shells can hope to be able to give a reliable answer.

Core Contraction in Stars of Moderate Mass

For the less massive stars we have seen in the preceding section that core contraction does not begin to play a decisive role as soon as the hydrogen in the center is exhausted. On the contrary, a long sequence of evolutionary phases follows, characterized by a steadily growing partially degenerate isothermal helium core. Eventually, however, when the luminosity of the star has increased to a high value and when therefore the evolutionary changes have become very fast, a short but critical period must occur—just as in an earlier phase of the massive stars—in which the core contraction raises the central temperature to the ignition point of helium. At what phase does this decisive event occur in the evolution of a star of, say, 1.2 solar masses?

We can find an approximate answer to this question by the following considerations. The energy flux L_g caused by gravitational energy release will not produce a steep temperature gradient within the inner degenerate portion of the core since the highly efficient conduction by the degenerate electrons will prevent any appreciable temperature differences there. Similarly, in the outermost portions of the core the density, and hence the opacity, is so low that radiative transfer will eliminate any great temperature gradients. The greatest temperature differences will occur in the layers just outside the degenerate portion since electron conduction there is not efficient whereas the radiative opacity is rather high. Let us designate the layer which is most effective for the buildup of a temperature difference by the subscript g. From the fact that this critical layer must be the innermost layer in which degeneracy is not yet serious we can locate this layer in terms of the asterisked variables defined by Eqs.(13.20) and find that the layer occurs where $s_g^* \approx 6$. With the help of this condition we can determine all the physical characteristics in the critical layer by first reading the values of all the asterisked variables from the core solution (appropriate for the evolution phase) at the point where s_g^* reaches the above number, and then by introducing these values into Eqs.(13.20) and (13.21).

If we are now to estimate the temperature gradient in the critical layer at any evolution phase, we first have to compute the gravitational energy released in the inner portion of the core according to Eq.(25.5). Next we have to compute the energy flux through the critical layer L_g; the relation between L_g and δE_g is most easily established with the help of the total luminosity of the star L and the nuclear energy δE_n freed during the same evolutionary time interval, which together lead to the relation

$$L_g = L \frac{\delta E_g}{\delta E_n + \delta E_g}. \qquad (25.8)$$

Finally, we have to introduce the energy flux into the radiative equilibrium condition (12.4) for the critical layer. If we consider now only those phases in which an appreciable temperature differential has already been produced by the contraction, we may estimate that the central temperature has approximately the value given by the temperature gradient in the critical layer multiplied by the radius of this layer. Thus we obtain from the radiative equilibrium condition

$$T_c{}^4 = \frac{3}{4ac} \varkappa_g \rho_g \frac{L_g}{4\pi r_g} \tag{25.9}$$

At what phase does Eq.(25.9) give us a central temperature sufficient to start the helium burning? We find the answer by applying Eq.(25.9), in conjunction with Eqs.(25.5), (25.6), and (25.8), to each of the models of the evolutionary sequence listed in Table 24.2. We find that in the early models of this sequence the core contraction affects the internal temperatures hardly at all, but that somewhere between the two last models listed in Table 24.2 the central temperature suddenly rises to the critical value. We may conclude that a star of about 1.2 solar masses will commence to burn helium in its center when it has exhausted its hydrogen in somewhat more than 50 percent of its mass and when its luminosity has risen to about

$$L \approx 5000 \ L_\odot. \tag{25.10}$$

This critical value for the luminosity agrees within one magnitude with the luminosity observed for the top of the red-giant branch in globular clusters. It seems likely, therefore, that the onset of the helium burning is in fact the process which terminates the evolution up the red-giant branch.

We have led our star of moderate mass into a precarious situation. The helium burning will tend to increase the central temperature further. Since the central portion of the star is degenerate, the increase in temperature does not lead to a noticeable increase in pressure and hence does not lead to a cooling expansion. It thus appears that the central temperature must continue to rise until the core has become non-degenerate. May the temperature gradient have become so steep by then that convection will mix the star from the center to the surface? May the star by this process again become homogeneous in composition and in consequence return to the main sequence, with the luminosity more than ten times brighter now than its initial luminosity because of the increased helium content? May then the star embark for a second time on an evolutionary development leading from the main sequence towards the right in the Hertzsprung-Russell diagram? Question after question...all unanswered at present, all surely within our grasp.

We have followed stars of various masses in their evolution to the furthest phases analyzed thus far. Ahead of us stretches a large territory of stellar evolution as yet hardly explored. In it we recognize many of the most fascinating types of stars: the supergiants, the pulsating variables, the blue stars of Population II, perhaps the Wolf-Rayet stars and the nuclei of planetary nebulae, surely the novae and supernovae. These are exciting topics for the future. For now we shall pass them by in a big jump to the end point of stellar evolution, the white dwarfs.

CHAPTER VII
FINAL STELLAR STRUCTURE

26. Structure of White Dwarfs

What will be the state of a star after it has exhausted the nuclear energy store accessible to it? The precise answer to this question we can know only after we have solved the over-all problem formulated in §12 successively for all the phases of nuclear burning. Meanwhile we may venture the guess that a star after exhausting the accessible nuclear fuel will commence a profound contraction. For the result of this contraction two alternatives appear plausible. The contraction may lead to very high temperatures, well above the range discussed in the second chapter, so that completely new physical phenomena—as, for example, endothermic nuclear reactions—may completely dominate the final evolution phases. Or the contraction may lead to very high densities so that degeneracy occurs not only in the core but virtually throughout the star, a situation which could permit the star to settle into a final state little affected by the continual loss of residual energy from the surface.

This second evolutionary alternative then suggests that the final state of a star may be represented by the white dwarfs, for whom high mean densities were long ago deduced from observations. We shall therefore now analyze the structure of white dwarfs with this interpretation in mind.

Models for Completely Degenerate Stars

A white dwarf cannot be in a completely degenerate state, first because in the very surface layers the density will be too low for any degeneracy, and second because even in the interior the density suffices for degeneracy of the electrons but not of the nucleii. On the other hand, we shall find in the next section that the non-degenerate surface layers of a white dwarf are very thin indeed. In addition, we may estimate that throughout the main part of a white dwarf the non-degenerate partial pressure of the nuclei is quite small compared with the degenerate partial pressure of the electrons. We shall permit ourselves therefore the simplifying approximation of complete degeneracy, that is of neglecting all non-degenerate pressure contributions and taking into account only the degenerate pressure of the electrons.

In this approximation we can represent the equation of state according to Eqs.(8.4), (8.5), (8.15), and (8.17) by

$$P = \frac{\pi m^4 c^5}{3h^3} f(x), \quad \rho = \mu_E \frac{8\pi H m^3 c^3}{3h^3} x^3 \tag{26.1}$$

$$\text{with } \mu_E = \frac{2}{1+X}.\tag{26.2}$$

Here the function $f(x)$ is completely defined by its definition (8.16). From the discussion of §8 we remember that Eq.(26.1) represents the equation of state properly for the non-relativistic degenerate case at moderate densities as well as for the relativistic degenerate case at the highest densities.

The fact that the equation of state (26.1) does not involve the temperature has a consequence of great convenience; we can completely separate the hydrostatic problem, which we shall analyze in this section, from the thermodynamic problem, which we shall take up in the next section. Accordingly we consider for now only the hydrostatic equilibrium conditions, which as always are given by

$$\frac{dP}{dr} = -\rho \, \frac{GM_r}{r^2}, \quad \frac{dM_r}{dr} = 4\pi r^2 \rho.\tag{26.3}$$

Eqs.(26.1) to (26.3) completely define the hydrostatic problem for white dwarfs.

We can solve this problem in the following direct manner. First let us choose a definite composition for the star. We thus fix by Eq.(26.2) the value of μ_E, which is the only composition-dependent quantity in our problem. Next let us choose arbitrarily a definite value for the central density. With these choices all the starting values at the center are determined and we can derive the corresponding definite solution of the differential Eqs.(26.3) by numerical integration from the center outwards. We continue with the integration to the point where the pressure drops to zero, that is to the surface. The value of M_r at this point gives us the mass of the white dwarf we have just constructed, and the value of r gives us its radius.

If we repeat the procedure for different values of the assumed central density, we obtain a one-parameter family of white dwarf models, corresponding to different values of the mass and radius.

If we now want to change the composition and correspondingly choose a different value of μ_E, do we have to obtain another one-parameter family of solutions by more numerical integrations? No, we can obtain the solutions for the new composition (variables without superscripts) from the solutions for the old composition (variables with zero superscripts) by the simple transformation

$$P = P^0, \quad \rho = \frac{\mu_E}{\mu_E^0}\rho^0, \quad M_r = \left(\frac{\mu_E}{\mu_E^0}\right)^{-2} M_r^0, \quad r = \left(\frac{\mu_E}{\mu_E^0}\right)^{-1} r^0.\tag{26.4}$$

If we introduce this transformation into the basic Eqs.(26.1) and (26.3) we find that the new variables will fulfill these equations for the new value of μ_E just as the old variables did for the old value. We conclude that our one-parameter family of integrations is sufficient to represent all completely degenerate stellar models.

This one-parameter family of completely degenerate models is well represented by the ten available integrations, which correspond to ten different values of the central density for a given composition. The actual numerical integrations were performed not in terms of the physical variables of Eqs.(26.3) but in terms of special non-dimensional variables. The use of these special variables simplified appreciably the necessary numerical work for each integration, but could not reduce the number of necessary integrations in this case. The main results of these integrations are summarized in Table 26.1. The data of this table refer to a composition containing a negligible amount of hydrogen ($\mu_E = 2$). We shall see in the next section that it is exactly this composition which we have to assume for the interior of white dwarfs.

Corrections to the Degenerate Equation of State

Before we consider the results shown in Table 26.1 we have to mention two inaccuracies which we have tacitly committed. Both these inaccuracies occur in the derivation of the degenerate equation of state. They are of importance only at the highest densities and therefore affect only the data in the last three or four lines of Table 26.1.

The first inaccuracy arises as follows. In a degenerate state, whereas the gravitational force acts mainly on the nuclei the pressure acts mainly on the electrons. These opposing forces will therefore slightly separate the nuclei and the electrons, thus producing an electrostatic field. It is just this electrostatic field which provides the linking force that permits the balancing of the gravitational force on the nuclei with the pressure force on the electrons and thus permits us to use the usual equation for hydrostatic equilibrium. On the other hand, this electric field has the

TABLE 26.1.

Mass-radius relation for white dwarfs ($\mu_E = 2$), uncorrected for secondary effects modifying degenerate equation of state. (Chandrasekhar, *Stellar Structure*, p. 427, Univ. of Chicago Press, 1938.)

$\log \rho_c$	$M/M_\odot$	$\log \dfrac{R}{R_\odot}$
5.39	0.22	−1.70
6.03	0.40	−1.81
6.29	0.50	−1.86
6.56	0.61	−1.91
6.85	0.74	−1.96
7.20	0.88	−2.03
7.72	1.08	−2.15
8.21	1.22	−2.26
8.83	1.33	−2.41
9.29	1.38	−2.53
∞	1.44	−∞

secondary consequence of slightly altering the number of quantum-mechanical states available for the electrons. It therefore necessitates a correction in our derivation of the degenerate equation of state in §8.

The second inaccuracy we have committed by ignoring the possibility that at very high degeneracy free electrons might be forced into nuclei. At very high densities this does in fact occur. It reduces the number of free electrons, i.e. increases the value of μ_E.

The effect of these two corrections for the degenerate equation of state on the white dwarf models has been estimated, and it appears that the mass values given in the lowest lines of Table 26.1 have to be reduced somewhat, in the extreme cases probably as much as 20 percent. With these corrections in mind we turn now to the results shown in Table 26.1.

Mass-Radius Relation and Mass Limit

The first main result is the existence of a definite relation between mass and radius for white dwarfs (assuming here always an interior composition with negligible hydrogen content). That such a relation should exist follows directly from the fact that the completely degenerate models for a given composition form a one-parameter family. Even though the data of Table 26.1 are still affected by the inaccuracies which we have just discussed, they suffice to show two main features of the mass-radius relation: first, the larger the mass of a white dwarf the smaller its radius, and second, for white dwarfs with masses comparable to that of the sun the radii are by order of magnitude a hundred times smaller than that of the sun.

The mass-radius relation affords in principle an excellent empirical test for the theory of completely degenerate stellar models. In reality, however, the observational data, as well as the theoretical results, on the masses and radii of white dwarfs are still afflicted by appreciable inaccuracies which prevent a sharp comparison as yet. All one can say at present is that there has not appeared thus far a serious discrepancy between theory and observations.

The second main result shown by Table 26.1 is the existence of an upper limit for the mass of a completely degenerate star. As we increase the central densities to higher and higher values, the mass does not increase indefinitely but approaches a finite limit. The numerical value given for this limit in Table 26.1 still has to be corrected for the secondary effects which we have discussed earlier. We may estimate that the actual value of the limiting mass for completely degenerate stars is approximately 1.2 solar masses.

How can we understand the occurrence of this limit? Let us consider the dependence on mass and radius for those internal physical quantities which are essential for hydrostatic equilibrium. If we ignore for this qualitative investigation any relative changes in the internal structure and consider only the over-all scale changes, we clearly find for the density

$$\rho \propto \frac{M}{R^3} \qquad (26.5)$$

and for the gravitational force

$$\rho \, \frac{GM_r}{r^2} \propto \frac{M^2}{R^5}. \qquad (26.6)$$

If we want to compare the pressure force with the gravitational force, we have to distinguish between non-relativistic and relativistic degeneracy. In §8 we found that non-relativistic degeneracy occurs at the less extreme densities which, according to Table 26.1, are relevant in the less massive degenerate stars, while relativistic degeneracy occurs at the extreme densities found in the more massive degenerate stars. If we use the equations of state (8.6) and (8.7), which are the limiting forms of the general Eq.(26.1) for the cases of completely non-relativistic and completely relativistic degeneracy respectively, we find for the pressure

$$\begin{aligned}
\text{non-relativistic: } & P \propto \rho^{\frac{5}{3}} \propto \frac{M^{\frac{5}{3}}}{R^5} \\
\text{relativistic: } & P \propto \rho^{\frac{4}{3}} \propto \frac{M^{\frac{4}{3}}}{R^4}
\end{aligned} \qquad (26.7)$$

and for the pressure force

$$\begin{aligned}
\text{non-relativistic: } & \frac{dP}{dr} \propto \frac{M^{\frac{5}{3}}}{R^6} \\
\text{relativistic: } & \frac{dP}{dr} \propto \frac{M^{\frac{4}{3}}}{R^5}.
\end{aligned} \qquad (26.8)$$

Now compare the pressure force and the gravitational force for a star of a given mass. For the non-relativistic case you find that the two forces depend on a different power of the radius. Thus the star has the ability of bringing the two forces into balance by adjusting its radius. For example, if the star finds the pressure force to exceed the gravitational force, it will expand, the increase in the radius will reduce the pressure force more than the gravitational force, and expansion will stop when equilibrium is reached.

Now compare the two forces for the relativistic case. You find that the two forces depend on the same power of the radius. Hence the star does not have the ability of bringing the two forces into equilibrium by simply adjusting its radius appropriately. Note also that in the relativistic case the two forces depend on different powers of the mass. Hence there exists a specific mass value, the limiting mass, for which the two forces happen to be exactly in balance. For masses larger than the limiting mass the gravitational force will always exceed the pressure force, whatever the radius. The star cannot find an equilibrium state—unless it manages to reduce its mass by ejection. On the other hand, in a star with a mass smaller than the limiting mass the gravitational force will be smaller than the pressure force and the star will expand. In this expansion the density will decrease until, at least in the outer portions, the degeneracy changes from relativistic to non-relativistic. Now with increasing radius the pressure force decreases faster than the gravitational force so that eventually the two forces will come into balance.

We may conclude that in stars heavier than the limiting mass the force of the degenerate pressure is never sufficient to balance gravity, that a star lighter than the limiting mass is able to balance gravity with degenerate pressure, and that to achieve this balance in the latter case the star has to adjust its radius to the value prescribed by the mass-radius relation for degenerate models.

Location of Degenerate Stars in Hertzsprung-Russell Diagram

To what extent does the analysis of the hydrostatic equilibrium of completely degenerate stars determine the location of white dwarfs in the Hertzsprung-Russell diagram? For a degenerate star of a given mass—less than the limiting mass—hydrostatic equilibrium determines only the radius (by the mass-radius relation), not the luminosity. What can we conclude from the radius alone?

According to the definition of the effective temperature the luminosity, the radius, and the effective temperature are related by Eq.(1.4). If we read the radius of a degenerate star of a definite mass from Table 26.1, correct this value as best as we can for the secondary effects we have discussed earlier, and introduce the corrected value into Eq.(1.4), then this equation gives us a relation between the luminosity and the effective temperature which we can represent by a line in the Hertzsprung-Russell diagram. Four such lines of constant radius are drawn in Fig. 26.1 for four mass values which fairly cover the masses which we might expect for white dwarfs.

The four lines in Fig. 26.1 show that our hydrostatic analysis leads us to the expectation that the majority of the white dwarfs should lie in a fairly narrow strip in the lower left corner of the Hertzsprung-Russell diagram. The only exceptions should be white dwarfs with masses very close to the limiting mass, which should fall to the left of the indicated strip, and white dwarfs with exceptionally low masses, which should fall to the right of the four lines. Comparing the theoretical Fig. 26.1 with Fig. 1.6, which represents the observed luminosities and effective temperatures of white dwarfs, we may conclude that the observational data are in satisfactory accordance with the theoretical expectations.

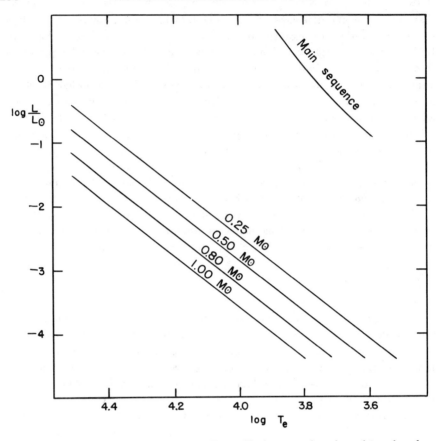

Fig. 26.1. Theoretical Hertzsprung-Russell diagram for the white dwarfs.

27. *Thermodynamics of White Dwarfs*

What are the internal temperatures of white dwarfs and what are the sources of their luminosity? These are two major questions which we could not answer in the preceding section where we analyzed solely the hydrostatic structure of degenerate stars. It is the thermal structure which we have to investigate if we are to answer these questions.

Throughout the degenerate interior of a white dwarf conduction by the degenerate electrons is highly efficient. The degenerate interior must therefore be practically isothermal. This isothermal interior is blanketed by the non-degenerate surface layers with their high opacity. Accordingly, an analysis of these surface layers will lead us to the determination of the internal temperatures.

The Internal Temperatures

According to our discussion of §11 we may assume that the surface layers of the majority of white dwarfs, with their early spectral types, are in radiative equilibrium. Now for radiative envelopes in which the opacity can be approximated by Kramers' law we have found—by a simple integration of the hydrostatic and radiative equilibrium conditions—the pressure-temperature relation (11.5), which we may transform into the density-temperature relation

$$\rho = \left(\mu \frac{2}{8.5} \frac{4ac}{3} \frac{H}{k} \frac{4\pi GM}{\varkappa_0 L} \right)^{\frac{1}{2}} \times T^{3.25} . \tag{27.1}$$

For the coefficient in the opacity law we should use here, according to Eq.(9.16),

$$\varkappa_0 = 4.34 \times 10^{25} \frac{Z(1+X)}{(t/\overline{g})} .$$

Eq.(27.1) gives us the run of the density as a function of temperature in the surface layers of a star of given mass, composition and luminosity.

If we follow this relation further and further inwards through the surface layers of a white dwarf, we eventually reach the point where degeneracy begins. At the transition layer between the degenerate and the non-degenerate state the density and the temperature must fulfill the criterion (8.10), which gives

$$\rho_{tr} = \mu_E \left(\frac{k \ T_{tr}}{H \ K_1} \right)^{\frac{3}{2}} . \tag{27.2}$$

If we apply our general relation (27.1) specifically to the transition layer, we can eliminate ρ_{tr} between Eqs.(27.1) and (27.2) and thus obtain a condition for the temperature at the transition layer, which we may write in the form

$$L = 5.7 \times 10^4 \times \frac{\mu}{\mu_E^2} \frac{(t/\overline{g})}{Z(1+X)} \times \frac{M}{M_\odot} \times T_{tr}^{3.5} \tag{27.3}$$

This equation determines the temperature at the transition layer, that is the temperature of the practically isothermal interior, for any white dwarf with given mass, composition, and luminosity.

Let us assume for the subsequent computations the following composition, mass, and average guillotine factor

$$x = 0, \quad y = 0.9, \quad z = 0.1; \quad \text{i.e. } \mu = 1.38, \quad \mu_E = 2.00$$
$$M = M_\odot, \quad (t/\overline{g}) = 10 . \tag{27.4}$$

With these numerical values Eq.(27.3) becomes

$$L = 2.0 \times 10^6 \times T_{tr}^{3.5} \qquad (27.5)$$

From this formula we can compute directly the internal temperature for any given luminosity. The results of such a computation are shown in Table 27.1 for three values of the luminosity which fairly represent the observed luminosities of white dwarfs.

The main uncertainty in these results arises from our lack of knowledge about the composition of the surface layers of white dwarfs. Yet even if we were to replace the hydrogen-free composition (27.4) by pure hydrogen and thus lower the opacity by about a factor 20, we would decrease the internal temperature according to Eq.(27.3) by only about a factor 2. We may then conclude from the data of Table 27.1 with good certainty that the internal temperatures of white dwarfs must be of the order of several million degrees—quite comparable with the central temperatures of red dwarfs.

Extent of Non-Degenerate Layers

We have answered the first of our two main questions regarding the thermodynamic structure of white dwarfs. Before we turn to the second main question let us fill in a hole which we have left in the preceeding section, namely the justification of the neglect of the non-degenerate surface layers in the analysis of the hydrostatic structure of white dwarfs. To justify this neglect two conditions must be fulfilled: first, the density at the bottom of the non-degenerate layers must be small compared with the internal densities so that zero density is a good approximation for the boundary condition of the degenerate model; and second, the geometrical depth of the non-degenerate layers must be small compared with the radius so that it can be neglected in the computation of the radius.

We can compute the density at the bottom of the non-degenerate layers directly from the degeneracy criterion (27.2) with the help of the temperature values which we have already derived. The resulting density values are shown in Table 27.1. Comparison with the central densities listed in Table 26.1 shows that the densities at the transition level at the bottom of the non-degenerate layers are in fact very much smaller than the internal densities, so that the first condition is fulfilled.

TABLE 27.1.

Temperatures and cooling times of white dwarfs, specified by Eq.(27.4).

L	T_{tr}	$\log \rho_{tr}$	$\dfrac{R - r_{tr}}{R}$	$\tau_{cooling}$
$10^{-2} L_\odot$	17×10^6	3.5	0.011	0.3×10^9 yrs
$10^{-3} L_\odot$	9×10^6	3.1	0.006	1.6×10^9 yrs
$10^{-4} L_\odot$	4×10^6	2.6	0.003	8×10^9 yrs

The depth of the non-degenerate surface layers we can derive with the help of the relation between T and r which we have derived for a radiative envelope by a simple integration of the radiative equilibrium condition and which is represented by the first of Eqs.(14.5). If we apply this equation to the transition level we obtain

$$T_{tr} = \frac{1}{4.25} \frac{\mu H}{k} \frac{GM}{R} \left(\frac{R}{r_{tr}} - 1\right) \tag{27.6}$$

Into this formula we may introduce for the molecular weight and the mass the values listed under (27.4) and for the radius the value $0.005\,R_\odot$ in accordance with the mass-radius relation of the preceding section. If, finally, we use the temperature values listed in the second column of Table 27.1 we obtain from Eq.(27.6) the relative depth of the non-degenerate layers as given in the fourth column of the same table. We see that this depth is very small indeed and contributes extremely little to the radius of a white dwarf.

We may then rest assured that both the conditions under which the neglect of the non-degenerate layers in the construction of models for white dwarfs is permitted are in fact satisfactorily fulfilled.

Proton-Proton Reaction and the Interior Hydrogen Content

There remains the question regarding the source of the luminosity of white dwarfs. Could the energy source lie in some residual nuclear fuel? Of the various energy-liberating processes which we have discussed in §10 only the proton-proton reaction can be effective at the relatively low temperatures of the white dwarfs. For the moment let us postpone consideration of whether the proton-proton reaction is actually the main energy source of white dwarfs, and first derive on the basis of the proton-proton reaction an upper limit to their hydrogen content.

If we select a mean temperature according to Table 27.1 and a mean internal density according to Table 26.1 we can compute the corresponding mean rate of energy generation from the proton-proton reaction with the help of Eq.(10.14) and the numerical parameters of the first line of Table 10.1. We find

$$\text{for } \overline{T} = 6 \times 10^6 \quad \text{and} \quad \overline{\rho} = 10^6 : \overline{\varepsilon}_{pp} \approx 10^4 \times \overline{X}^2. \tag{27.7}$$

On the other hand, we can estimate the mean rate of energy generation also from the luminosity and the mass, obtaining for a typical white dwarf

$$\text{for } L = 10^{-3} L_\odot \quad \text{and} \quad M = M_\odot : \overline{\varepsilon} = \frac{L}{M} = 2 \times 10^{-3}. \tag{27.8}$$

By setting the mean rate of energy production as given by Eq.(27.7) equal to that obtained from the observed luminosity according to Eq.(27.8) we find for the average internal hydrogen content of a typical white dwarf the value

$$\overline{X} \approx 0.0005. \tag{27.9}$$

We may conclude that the residual hydrogen fuel remaining in the interior of the white dwarf must be very low; otherwise, under the prevailing temperatures and densities, the proton-proton reaction would liberate energy at a rate far exceeding the observed luminosities.

This low average hydrogen content in the interior does not, of course, imply necessarily a low value in the surface layers. A high hydrogen content there would not cause an excessive rate of the proton-proton reaction because of the relatively low temperatures and densities.

For the interior, however, even the low hydrogen content indicated by (27.9) is only an upper limit, for two reasons. First, we have assumed that nuclear processes are the only energy source; if other sources—which we shall discuss later—provide appreciable contributions then the hydrogen content necessary to feed the proton-proton portion of the energy liberation will naturally have to be less than our above estimate. The second reason lies in the fact that Eq.(10.14), which we have used for the proton-proton reaction, underestimates the rate of energy liberation for the high densities encountered in white dwarfs. At these high densities the electrons screen the coulomb fields of the protons quite effectively and thus make it much easier for two protons to come close to and react with each other. Correction of this underestimate of the reaction rate reduces still further the permissible hydrogen content. We may therefore conclude with good certainty that the average internal hydrogen content of a white dwarf must be extremely small, even smaller than the value given by (27.9).

This conclusion has one immediate comforting consequence. We can now be entirely sure that μ_E, which is the only composition-dependent parameter occurring in the hydrostatic equilibrium equations for the white dwarfs and which is determined by Eq.(26.2), has the definite value

$$\mu_E = 2, \tag{27.10}$$

a value which in anticipation we have already used throughout the preceeding section. Hence no doubt caused by uncertainty in the composition is left in the analysis of the hydrostatic structure of white dwarfs.

Secular Instability Caused by Nuclear Reactions

As to our main question regarding the source of the luminosity of the white dwarfs, we may now ask ourselves whether the proton-proton reaction fed by a very low residual hydrogen content does provide the source. The answer is probably no, because the star, though in equilibrium, would probably not be in stable equilibrium. We have to watch out for two types of instability, secular and pulsational.

Consider a small perturbation which slightly raises the internal temperature of a white dwarf. Let us start with the simplifying assumption that Eq.(10.14) represents appropriately the rate of the proton-proton reaction even at the high densities here in question, at least as far as the temperature dependence is concerned. We find then from Table 10.1 that in the temperature range in question the nuclear energy production will rise with about the 5th power of the temperature. On the other hand, the luminosity according to Eq.(27.5) increases only with the 3.5th power of the internal temperature. The increase in temperature will therefore cause a larger increase in the energy gained from the nuclear reactions than in the energy lost by the luminosity. Hence there will be a net energy gain, which will cause a further increase in temperature. Thus the temperature will continue to increase indefinitely; the star is secularly unstable.

We have once before encountered this type of instability, in §25 at the top of the red-giant sequence in globular clusters. There we had tentatively concluded that the appearance of this instability might well be the cause of the termination of the red-giant sequence and that the instability might lead to a complete remixing of the stellar interior. For a white dwarf the consequences of such an instability would probably be much more catastrophic because the very high densities within white dwarfs would necessitate a temperature increase by a dangerously large factor before non-degeneracy could be reached with a subsequent cooling expansion.

Before we conclude that any hydrogen burning in the interior of white dwarfs would lead to an instability which is in no way suggested by the observations we have to make a correction to the above argument. As we have already mentioned, Eq.(10.14) does not properly represent the rate of the proton-proton reaction at the high densities of the deep interior of white dwarfs. At these densities the electrons screen the coulomb fields of the protons. This reduction in the coulomb fields permits the participation in the reactions not only of those protons which have exceptionally high energies but also of those with more nearly average energies. While the number of exceptionally energetic protons is highly temperature-sensitive, the number of protons with more nearly average energy is not. In consequence the temperature sensitivity of the proton-proton reaction is much reduced at the high densities. In fact, at densities above 10^5, such as occur in the deep interior of the white dwarfs, the temperature sensitivity of the proton-proton reaction is so low that it probably does not lead to secular instability.

On the other hand, it is not in the deep interior that we should expect the star to have retained some residual fuel. It is in the outer layers that it appears more likely that the temperature has never been high enough to complete the exhaustion of hydrogen by nuclear processes. In the outer degenerate regions, however, with densities below 10^5, electron screening plays only a minor role, and our original argument which indicated secular instability must hold.

Thus far our attempt to provide a white dwarf with a source for its luminosity by burning hydrogen in its degenerate interior has not been successful; in the deep interior hydrogen burning would probably not lead to any instability but there it appears highly unlikely that any hydrogen could have survived the earlier hotter evolution phases, while in the outer portions of the degenerate structure hydrogen burning would lead to secular instability. Might there be a way out by placing the hydrogen burning in the very outermost layers, which are not degenerate?

Pulsational Instability Caused by Nuclear Reactions

Hydrogen burning in the non-degenerate layers would probably not lead to the danger of secular instability because under non-degenerate conditions an increase in the temperature causes an increase in the pressure, so that any accidental heating would right away be compensated by a cooling expansion. Another instability, however, does threaten—pulsational instability.

We may consider any star as an elastic system in equilibrium. Such a system is capable of a large variety of elastic oscillations. The simplest such oscillations for a star are radial pulsations. In general such pulsations will be highly damped. One damping mechanism is heat leakage, which tends to cause a loss of energy for stellar pulsations just as it causes an energy loss for ordinary acoustic oscillations. Another damping mechanism consists of direct loss of pulsational energy through the atmosphere of the star if in the atmosphere the pulsation has the character of outgoing progressive waves.

If there exists under special circumstances an energizing mechanism for the pulsations which is stronger than the damping mechanisms, the pulsations will not be damped and pulsational instability will occur. Much effort has been directed towards finding an energizing mechanism which could explain the observed pulsations of cepheids and other variables—without any success. The only case thus far in which pulsational instability has been found applies to a model for white dwarfs with nuclear energy generation in the non-degenerate outer layers—while observations do not give any indication of pulsations for white dwarfs. In this special case the unwanted energizing mechanism works as follows.

During a pulsation cycle at the phase of highest compression the temperature will be above normal and in consequence the nuclear processes will exceed their average rate. This excess in the energy liberation will cause a slight extra increase of the temperature. This will produce for the subsequent expansion phases a slight pressure excess and hence a small extra acceleration outwards. Accordingly the outward pulsational motion will reach to a somewhat increased expansion amplitude. Similarly, at the phase of greatest expansion the temperature, and hence the nuclear rate of energy liberation, will be below average. The deficiency in the energy production will cause a slight deficiency in the temperature and hence in the pressure during the subsequent contraction phases. Thus there will be a deficient pressure compensation of the gravitational acceleration inwards and the pulsational motion will swing to a higher contraction amplitude. Altogether we see that the variation of the nuclear energy liberation in this way automatically feeds energy into the pulsation in the expansion phase as well as in the contraction phase.

This energizing mechanism for pulsations exists in all stars which contain nuclear energy sources. Normally, however, it is very ineffective, for the following reason. The pulsational amplitude—in particular the amplitude of the temperature variation—is found to be strongly dependent on the distance from the center of the star, in the sense that it is much larger in the outermost layers than in the deep interior. In consequence the damping mechanisms are most active in the outer layers, while in all the models we have discussed in the preceeding chapters the nuclear energy sources which could provide the energizing mechanism are located fairly deep in the interior. With respect to the pulsations, any interior nuclear reaction operates therefore at a highly disadvantageous location. In fact no case has been found in which an interior nuclear reaction can provide an energizing mechanism sufficiently strong to overcome the damping mechanisms.

In the special case, however, of a white dwarf with nuclear energy sources in the non-degenerate outer layers, the nuclear reactions are exceptionally favorably located for energizing the pulsations. In fact the analysis of this situation indicates that the energizing mechanism would be stronger than the damping mechanisms, that is that pulsational instability would occur, and that therefore even a nuclear energy source in the non-degenerate outer layers does not lead to a stable model for the white dwarfs.

The detailed quantitative analysis of the instabilities which we have just discussed in a qualitative manner is complicated and has not as yet led to entirely certain results. Nevertheless we may summarize the discussion by inferring that the luminosity of white dwarfs probably does not stem from the burning of a residual of nuclear fuel.

Residual Thermal Energy and Cooling Time

What other source for the luminosity of a white dwarf is available? The gravitational energy, though large, is not available; after a star has contracted so far that its radius has practically reached the value prescribed by the mass-radius relation of completely degenerate models, no further contraction is possible and hence no further release of gravitational energy can occur. Similarly, the kinetic energy of the electrons is not available; when a white dwarf has reached an essentially completely degenerate state the energy of the electrons cannot be any further reduced because the exclusion principle prohibits any additional crowding into low-energy states. There remains the thermal energy of the non-degenerate nuclei.

The total thermal energy of the nuclei in a white dwarf must be of the same order of magnitude as the thermal energy of the sun since the masses and the temperatures are comparable. For the sun we have seen in §5 that the thermal energy could account for the solar luminosity for about 30 million years. Accordingly for a white dwarf, with a luminosity smaller than that of the sun by two or more powers of ten, the available thermal energy should be an entirely adequate source. Let us investigate this possibility in somewhat more detail.

If the thermal energy of the nuclei is the only source within a white dwarf, then the rate of loss of this energy is given directly by the luminosity. Thus we have

$$L = - \frac{d}{d\tau} \left(\frac{3}{2} \frac{k\,T}{\mu_A\,H} M \right) \tag{27.11}$$

Here the molecular weight of the nuclei μ_A is defined by Eq.(8.9) and has the value 4.44 for the composition listed in (27.4). Now the luminosity of a typical white dwarf is related to the internal temperature by Eq.(27.5). If we use this equation to eliminate the luminosity from Eq.(27.11) we obtain a differential equation for the internal temperature as a function of time. If we integrate this differential equation we obtain for the cooling time

$$\tau_{\text{cooling}} = \frac{1}{2.0 \times 10^6} \; \frac{3}{2} \frac{k\,M}{\mu_A\,H} \; \frac{1}{2.5} \; T^{-2.5} = \left(\frac{3}{2} \frac{k\,T}{\mu_A\,H} M \right) \Big/ (2.5\,L). \tag{27.12}$$

Here we have set the integration constant equal to zero. Strictly speaking, this choice corresponds to counting the cooling time from the moment when the temperature was infinite. Practically, however, we are counting in this way only the interval in which the temperature drops from, say, twice its present value down to its present value since the earlier interval during which the temperature may have dropped from very high values to twice the present value is very short. We may therefore consider the cooling time given by Eq.(27.12) as representing the time taken by a white dwarf to reach its present state since its last burning of nuclear fuel.

If we apply Eq.(27.12) to white dwarfs with mass and composition fixed by (27.4) and with luminosities and temperatures as listed in the two first columns of Table 27.1, we obtain the cooling times listed in the last column of this table. We see that the cooling times which we have derived in this manner are of just the right order of magnitude. The cooling times are sufficiently short so that many stars have had time enough to reach those values of the internal temperature which, according to Eq.(27.3), give the observed low luminosities of white dwarfs. At the same time, the cooling times are sufficiently long so that a cooling star will not become unobservably faint too quickly.

It thus appears probable that the only source for the luminosity of the white dwarfs is the remaining thermal energy. Accordingly, they have reached a state in which no further changes will occur except for a slow and steady decrease of the internal temperature and the luminosity— truly the final state of stellar evolution.

CHAPTER VIII
SUMMARY AND REVIEW

28. *Physical State of the Stellar Interior*

In the preceding four chapters we have constructed laboriously and in detail models for a variety of stars in various evolutionary phases, from their initial states to their final ones. The models are based on the stellar masses and initial compositions for which we have assembled the observational data in Chapter I. On the basis of these fundamental data the models we have constructed follow directly from the physical laws discussed in Chapter II. Our aim for this section is to review the main physical results that can be deduced from all this model construction work.

An over-all consequence of the investigations described in the preceding chapters is that we can now identify the various observed types of stars with the main evolutionary phases of a star's life. These identifications, which now seem fairly certain, go as follows. The initial main sequence, deduced from the observations of stellar clusters, represents that state of stars in which the nuclear hydrogen burning just commences its role as the major energy source. The contraction phase which precedes the main sequence phase and lasts only a very short time for each star must be represented by a very small percentage of the red dwarfs; a definite identification of this contraction phase has been made thus far only in some very young galactic clusters. The main sequence of the nearby stars is made up of a mixture of stars, partly still in the initial state and partly in early evolutionary phases in which the nuclear processes have begun the depletion of the hydrogen in the core. The sub-dwarfs of Population II (at least those among them which do not lie too far below the ordinary main sequence in the Hertzsprung-Russell diagram) appear to correspond to exactly the same evolutionary states as the ordinary main-sequence stars of Population I, and differ from the latter only because of the difference in initial composition between the stellar populations.

The more advanced evolutionary phases, in which the core of the star is exhausted of hydrogen and the nuclear burning occurs in a shell surrounding the core, can be identified with the subgiants and the red giants; during these phases the star moves towards the right in the Hertzsprung-Russell diagram, further and further away from the main sequence. The most advanced phases of stellar evolution, which follow the state when helium burning commences at the center, cannot yet be identified securely since we have thus far not succeeded in constructing the appropriate models. It seems highly likely, however, that among the stars in these late, bright and fast evolutionary phases must belong the stars occupying the horizontal branch in the Hertzsprung-Russell diagram of the globular clusters, all the pulsating stars, the carbon and S stars, many of the supergiants, the novae, and probably the supernovae. Finally, in spite of this gap in our model construction we may be sure that the end state of stellar evolution is represented by the white dwarfs, in which the nuclear energy appears to be practically exhausted and only the remaining thermal energy feeds the fading luminosity.

From this qualitative identification of observed types of stars with phases of stellar evolution let us turn now to more quantitative results. Data on the physical state and the internal structure are given in detailed numerical form in Tables 28.1 to 28.8 for a number of stars in various evolutionary phases. These data will form the basis of the following discussion.

Temperatures and Densities

An over-all view of the temperatures and densities occurring in typical stars is given by Fig. 28.1. This graph gives the temperature-density relation for four representative stars: two main-sequence stars (the sun and a B1 star of 10 solar masses), a red giant (1.3 solar masses) and a white dwarf (0.89 solar masses). Each star is represented by a solid curve. The dot at the upper end of the curve corresponds to the center of the star and the dot at the lower end approximately to its photosphere. The four intermediate dots on each curve divide the star into five shells of equal mass.

The complicated form at the bottom of the curves for the sun and for the red giant is a consequence of the hydrogen ionization zone which we have discussed in §11. The crosses on these two curves indicate the transition from the convective envelope to the radiative interior. The curves for the red giant and for the white dwarf each have a horizontal top portion which indicates an isothermal interior; the isothermal condition in the white dwarf is caused by the high electron conduction in the degenerate state, which does not permit appreciable temperature gradients; the isothermal condition in the red giant arises because with the hydrogen of the core exhausted there exists no energy flow which could cause a temperature gradient.

Comparison of the curves for the sun and for the B1 star in Fig. 28.1 shows that the upper main-sequence stars have higher temperatures and somewhat lower densities than the lower main-sequence stars. Comparison of the curve for the sun, representing an early evolutionary phase, with that for the red giant, representing an advanced evolutionary phase, indicates that as the evolution progresses the internal temperature tends

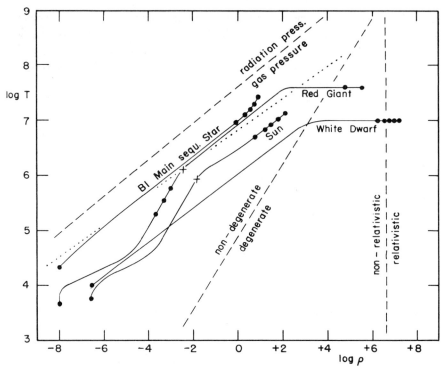

Fig. 28.1. Temperature-density diagram for four representative stars. Each line representing a star starts at the right with a point for the center and ends at the left with a point for the photosphere. The four intermediate points divide the star into five shells of equal mass. The dashed lines represent transitions in the equation of state. The dotted line separates the hot region in which electron scattering provides the main opacity, from the cooler region in which Kramers' opacity (bound-free and free-free transitions) dominates.

to increase and that the density of the core will increase by a large factor during these evolutionary phases.

If you ignore the outermost 20 percent of the mass of any star, that is the interval between the lowest two points on each curve in Fig. 28.1, and if you furthermore ignore the entire convective envelope of the red giant, then you find that the internal temperatures of the four representative stars shown in this figure range from about 4 million degrees to 40 million degrees. This is the main temperature range for the stellar interior. But both higher and lower temperatures may be encountered. Higher temperatures may nearly certainly be expected in the most advanced evolutionary phases, in which helium is transmuted into heavier elements. On the other hand, lower temperatures are found in the massive envelopes of red giants and may be expected even in the cores in three types of stars: stars in the early part of the pre-main-sequence contraction, the faintest main-sequence dwarfs, and old, faint, white dwarfs.

Detailed models do not yet exist, however, for either the especially hot or the especially cold stars.

The values of temperature and density in the stellar interior, as shown by Fig. 28.1, uniquely govern the equation of state, the atomic processes which provide the opacity, and the nuclear processes which provide the energy sources within a star. The results for these three gas characteristics run as follows.

Equation of State and Opacity

One of the dashed lines in Fig. 28.1 represents the degeneracy limit. Another represents the states in which the radiation pressure equals the gas pressure. These two lines (copied from Fig. 8.1) divide the temperature-density diagram into three parts, a diagonal middle strip in which the equation of state of an ideal gas holds, an upper left-hand corner in which radiation pressure dominates, and a right-hand portion in which the free electrons are in a degenerate state.

We see from Fig. 28.1 that most of the curves representing our sample stars fall in the middle strip, where the ideal gas law holds. Radiation pressure is important only in the interior of the most luminous main-sequence stars. Degeneracy rules in the white dwarfs, but occurs also in the cores of stars in advanced evolutionary phases, such as the red giants.

For the opacity we have to consider another limit which is represented in Fig. 28.1 by the dotted line (copied from Fig. 9.1). This limit separates the region in the temperature-density diagram where electron scattering dominates from that in which Kramers' opacity law holds. Here we take Kramers' law to represent the bound-free transitions of the heavier elements as well as the free-free transitions of hydrogen and helium; the former are more important in Population I stars and the latter in Population II stars where the abundance of the heavier elements is low.

For the opacity we have to consider also the degeneracy limit, since in the degenerate state electron conduction rather than radiative transfer is the main energy-transporting mechanism. Thus we again have the temperature-density diagram divided into three parts, the upper left-hand part with electron scattering, a central portion with Kramers' opacity and a right-hand part with electron conduction.

Fig. 28.1 indicates that Kramers' opacity is dominant in the outer portions of all our representative stars as well as throughout the interior of lower main-sequence stars such as the sun. In the interior of upper main-sequence stars, however, electron scattering provides the main opacity, and the same is true in a zone surrounding the isothermal core in red giants. Finally, electron conduction is important wherever degeneracy occurs, in white dwarfs and in the central cores of red giants.

The review of the equation of state and the opacity with the help of Fig. 28.1 emphasizes the variety of physical processes which have to

be taken into account when the equation of state and radiation transfer
are to be applied to the stellar interior. This variety of physical processes complicates the computations for the stellar interior. It does not,
however, introduce any indefiniteness since at any point in the temperature-density diagram relevant to the stellar interior the dominant physical
processes are uniquely determined and can by now be computed with fair
accuracy.

Nuclear Energy Sources

The energy radiated from the stars has its source nearly exclusively
in nuclear transmutations. Exceptions to this rule are stars in their pre-
main-sequence contraction and the white dwarfs. Stars in their early
contraction phase have as their source the gravitational energy freed by
the contraction; half of the gravitational energy is transformed into thermal energy while the other half is free for the radiation from the surface.
On the other hand, the white dwarfs shine most probably from the remainder of the thermal energy of the nuclei in their interior (the kinetic
energy of the free electrons is frozen by the degeneracy). These two
exceptional types of stars—characterizing the very beginning and the
very end of stellar life—provide only a minute fraction of the total energy
radiated from stars. In all other types of stars gravitational and thermal
energies play a minor role as energy sources compared with nuclear
processes—though gravitational contraction appears to play a decisive
role for the structural development during certain short phases of stellar
evolution.

Of all the nuclear processes which can provide energy sources two are
overwhelmingly the most important: the transmuattion of hydrogen into
helium and the transmutation of helium into heavier elements. The predominance of these two processes arises from two factors, first the large
mass defect of the proton and the alpha particle, and second the high
abundance of hydrogen in the initial composition of stars. Of these two
processes hydrogen burning provides a much bigger source than helium
burning because the energy released from one gram of hydrogen is about
ten times larger than that released from one gram of helium.

Hydrogen burning appears to provide the energy source for all the
main-sequence stars, the subdwarfs, the subgiants, and even most of the
red giants. Hydrogen burning can proceed by two processes, the proton-
proton reaction which is dominant at the lower temperatures in the fainter
stars, and the carbon cycle which is dominant at the higher temperatures
in the brighter stars. The sun happens to be near the dividing line of
the two processes; the proton-proton reaction provides most of the solar
energy but the carbon cycle appears to make a small contribution, too.

Helium burning requires higher temperatures than hydrogen burning and
therefore can occur only in the advanced evolutionary phases when a

star has exhausted the hydrogen in its core. Only one observed type of star has thus far been identified fairly certainly with a helium-burning phase, namely the top of the red-giant branch in globular clusters. It appears highly likely, however, that helium burning provides the main energy source in some, if not most, of the most advanced phases of stellar evolution that are as yet poorly identified.

We have finished our review of the main physical processes characteristic of the gases in a star. What now is the resulting over-all structure of a star?

Density Distribution

The structure of a star may best be symbolized by its density distribution throughout the interior from the center to the surface. Fig. 28.2 gives the density distribution (normalized to 1 at the center) for four stars: three main sequence stars (of 10, 1.0, and 0.6 solar masses) and a white dwarf. The striking feature of this figure is the similarity in the density distributions of these four very different types of stars. Their internal structure differs little from that of the "standard model" of two

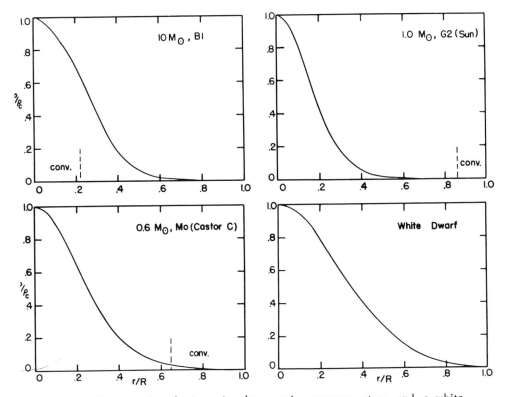

Fig. 28.2. Density distributions in three main-sequence stars and a white dwarf (data from Tables 28.1, 28.3, 28.4, and 28.8).

decades ago. In fact, the ratio of central to mean density for the three main-sequence models runs from 20 to 60, as compared with 54 for the "standard model."

Does this similarity in internal structure hold for all stars? No, only for stars with homogeneous composition, not for inhomogeneous stars. The four stellar models represented in Fig. 28.2 are all homogeneous in composition, and it is this feature which makes their structure so similar. But inhomogeneities in composition (such as those shown in Figs. 22.2, 22.3, and 23.1) are an essential feature in stellar evolution. They are caused by the nuclear transmutation in the inner portions of a star. They appear soon after the initial main-sequence state, last through all the subgiant and giant phases, and disappear only when all the nuclear fuel is exhausted and the star settles into the white dwarf state. It is these inhomogeneous phases which the stellar models of two decades ago could not well represent.

The density distribution in an inhomogeneous red giant, such as that represented in Table 28.7, cannot be shown in a graph like those of Fig. 28.2; the density drops much too steeply in the core, reaching one percent of its central value at less than one thousandth of the radius. Clearly the density distribution in a red giant is very different from that in main-sequence stars. The ratio of central to mean density for the red giant of Table 28.7 is about two billion, i.e. eight orders of magnitude larger than for main-sequence stars.

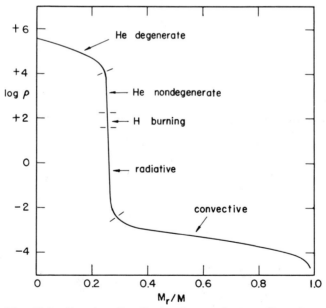

Fig. 28.3. Density distribution in a red giant (data from Table 28.7).

To give a picture of the denstiy distribution in a red giant, the logarithm of the density is plotted against the mass fraction in Fig. 28.3. This figure shows that the red giant consists of two sharply distinct parts, a degenerate helium core of high density and a convective hydrogen envelope of low density. Between these two parts there occurs a radiative transition zone which covers only a minute fraction of the stellar mass, but contains in its middle the hydrogen-burning shell.

While this internal structure which is characteristic for advanced phases of stellar evolution may look complicated, it is a straightforward consequence of the basic physical processes which govern stellar evolution.

TABLE 28.1

Model for star of 10 solar masses in initial main-sequence state.

$$\frac{M}{M_\odot} = 10, \quad \frac{L}{L_\odot} = 3000, \quad \frac{R}{R_\odot} = 3.63.$$

r/R	$\dfrac{M_r}{M}$	$\dfrac{L_r}{L}$	$\log P$	$\log T$	$\log \rho$	$\varkappa$	U	V	$n+1$
0.00	0.000	0.000	16.533	7.442	+0.892	conv.	3.000	0.000	2.500
0.02	0.000	0.010	16.530	7.440	0.890	”	2.995	0.013	2.500
0.04	0.002	0.061	16.522	7.437	0.885	”	2.982	0.051	2.500
0.06	0.006	0.170	16.508	7.432	0.877	”	2.959	0.114	2.500
0.08	0.013	0.331	16.489	7.424	0.865	”	2.927	0.203	2.500
0.10	0.025	0.511	16.464	7.414	0.850	”	2.887	0.318	2.500
0.12	0.042	0.678	16.433	7.402	0.832	”	2.838	0.460	2.500
0.14	0.065	0.809	16.397	7.387	0.811	”	2.779	0.629	2.500
0.16	0.093	0.899	16.356	7.371	0.786	”	2.713	0.825	2.500
0.18	0.128	0.952	16.308	7.351	0.757	”	2.638	1.051	2.500
0.20	0.168	0.980	16.254	7.330	0.724	”	2.556	1.307	2.500
0.22	0.214	0.993	16.194	7.306	0.689	”	2.465	1.595	2.500
0.24	0.264	1.000	16.128	7.279	0.649	0.542	2.368	1.918	2.580
0.26	0.317	1.000	16.055	7.252	0.604	0.555	2.256	2.273	2.780
0.28	0.374	1.000	15.976	7.225	0.554	0.567	2.130	2.650	2.961
0.30	0.431	1.000	15.891	7.197	0.497	0.580	1.994	3.045	3.124
0.32	0.488	1.000	15.800	7.168	0.435	0.594	1.852	3.456	3.269
0.34	0.543	1.000	15.703	7.139	0.368	0.607	1.709	3.875	3.397
0.36	0.596	1.000	15.602	7.110	0.296	0.619	1.566	4.303	3.509
0.38	0.648	1.000	15.496	7.080	0.220	0.631	1.426	4.736	3.606
0.40	0.694	1.000	15.386	7.050	0.140	0.643	1.290	5.174	3.692
0.42	0.736	1.000	15.272	7.019	+0.057	0.655	1.161	5.614	3.765
0.44	0.775	1.000	15.153	6.988	−0.030	0.666	1.039	6.061	3.829
0.46	0.809	1.000	15.032	6.957	−0.120	0.676	0.924	6.512	3.884
0.48	0.840	1.000	14.908	6.925	−0.212	0.688	0.818	6.968	3.931
0.50	0.867	1.000	14.780	6.892	−0.308	0.699	0.719	7.437	3.971
0.52	0.890	1.000	14.649	6.860	−0.406	0.708	0.629	7.918	4.006
0.54	0.910	1.000	14.515	6.826	−0.507	0.719	0.546	8.416	4.035
0.56	0.927	1.000	14.378	6.793	−0.610	0.728	0.472	8.936	4.060
0.58	0.941	1.000	14.239	6.758	−0.715	0.740	0.405	9.484	4.081
0.60	0.953	1.000	14.094	6.723	−0.825	0.748	0.344	10.07	4.100
0.62	0.963	1.000	13.946	6.687	−0.936	0.759	0.291	10.69	4.115
0.64	0.971	1.000	13.794	6.650	−1.052	0.770	0.243	11.37	4.127
0.66	0.978	1.000	13.639	6.612	−1.170	0.781	0.202	12.11	4.138
0.68	0.983	1.000	13.476	6.573	−1.294	0.791	0.165	12.94	4.147
0.70	0.988	1.000	13.307	6.532	−1.422	0.804	0.134	13.86	4.155
0.72	0.991	1.000	13.132	6.490	−1.555	0.816	0.107	14.89	4.161
0.74	0.994	1.000	12.948	6.446	−1.694	0.830	0.084	16.07	4.166
0.76	0.996	1.000	12.755	6.400	−1.842	0.841	0.065	17.44	4.171
0.78	0.997	1.000	12.549	6.350	−1.998	0.860	0.048	19.06	4.175
0.80	0.998	1.000	12.329	6.298	−2.166	0.873	0.035	20.99	4.178
0.82	0.999	1.000	12.092	6.241	−2.346	0.893	0.025	23.34	4.181
0.84	1.000	1.000	11.832	6.179	−2.544	0.914	0.017	26.28	4.184
0.86	1.000	1.000	11.546	6.111	−2.762	0.937	0.011	30.04	4.186
0.88	1.000	1.000	11.222	6.033	−3.009	0.967	0.007	35.06	4.189
0.90	1.000	1.000	10.850	5.946	−3.293	0.993	0.004	42.00	4.191
0.92	1.000	1.000	10.400	5.837	−3.636	1.04	0.002	52.61	4.192
0.94	1.000	1.000	9.834	5.702	−4.067	1.11	0.001	70.18	4.192
0.96	1.000	1.000	9.055	5.516	−4.661	1.20	0.000	105.2	4.192
0.98	1.000	1.000	7.746	5.204	−5.659	1.38	0.000	210.9	4.192
1.00	1.000	1.000	—	—	—	—	0.000	∞	4.192

TABLE 28.2   255

Model for star of 2.5 solar masses in initial main-sequence state.

$$\frac{M}{M_\odot} = 2.5, \quad \frac{L}{L_\odot} = 21.2, \quad \frac{R}{R_\odot} = 1.59.$$

r/R	$\dfrac{M_r}{M}$	$\dfrac{L_r}{L}$	$\log P$	$\log T$	$\log \rho$	$\varkappa$	U	V	$n+1$
0.00	0.000	0.000	17.166	7.297	$+1.684$	conv.	3.000	0.000	2.500
0.02	0.000	0.020	17.161	7.295	1.681	"	2.992	0.022	2.500
0.04	0.003	0.172	17.147	7.289	1.672	"	2.969	0.087	2.500
0.06	0.011	0.317	17.123	7.280	1.658	"	2.930	0.195	2.500
0.08	0.027	0.551	17.090	7.267	1.638	"	2.876	0.348	2.500
0.10	0.050	0.757	17.048	7.250	1.613	"	2.808	0.547	2.500
0.12	0.083	0.887	16.995	7.229	1.581	"	2.724	0.792	2.500
0.14	0.125	0.957	16.933	7.204	1.544	"	2.627	1.087	2.500
0.16	0.177	1.000	16.860	7.172	1.499	0.955	2.515	1.433	2.589
0.18	0.236	1.000	16.777	7.142	1.446	1.03	2.377	1.823	2.892
0.20	0.300	1.000	16.685	7.112	1.384	1.10	2.218	2.243	3.156
0.22	0.368	1.000	16.583	7.080	1.313	1.18	2.047	2.685	3.383
0.24	0.436	1.000	16.473	7.049	1.235	1.25	1.871	3.139	3.577
0.26	0.503	1.000	16.356	7.017	1.150	1.33	1.695	3.598	3.740
0.28	0.567	1.000	16.233	6.984	1.059	1.42	1.525	4.056	3.878
0.30	0.627	1.000	16.105	6.952	0.963	1.49	1.362	4.507	3.994
0.32	0.681	1.000	15.972	6.919	0.863	1.58	1.209	4.951	4.091
0.34	0.729	1.000	15.836	6.886	0.760	1.66	1.068	5.385	4.172
0.36	0.772	1.000	15.697	6.853	0.654	1.75	0.938	5.811	4.240
0.38	0.810	1.000	15.556	6.820	0.546	1.83	0.820	6.230	4.297
0.40	0.843	1.000	15.412	6.787	0.436	1.92	0.713	6.645	4.345
0.42	0.870	1.000	15.268	6.754	0.324	2.01	0.618	7.055	4.386
0.44	0.894	1.000	15.120	6.720	$+0.210$	2.11	0.533	7.470	4.420
0.46	0.914	1.000	14.972	6.687	$+0.096$	2.20	0.457	7.888	4.448
0.48	0.931	1.000	14.823	6.654	-0.020	2.29	0.391	8.313	4.472
0.50	0.944	1.000	14.672	6.620	-0.138	2.39	0.332	8.754	4.492
0.52	0.956	1.000	14.519	6.586	-0.257	2.50	0.281	9.212	4.508
0.54	0.965	1.000	14.364	6.551	-0.377	2.63	0.236	9.693	4.523
0.56	0.973	1.000	14.207	6.517	-0.500	2.73	0.197	10.20	4.534
0.58	0.979	1.000	14.048	6.482	-0.624	2.85	0.163	10.74	4.544
0.60	0.984	1.000	13.884	6.446	-0.751	2.99	0.134	11.34	4.553
0.62	0.988	1.000	13.719	6.410	-0.881	3.12	0.110	11.98	4.559
0.64	0.991	1.000	13.549	6.372	-1.014	3.28	0.089	12.68	4.565
0.66	0.993	1.000	13.376	6.334	-1.149	3.45	0.071	13.45	4.570
0.68	0.995	1.000	13.195	6.295	-1.290	3.61	0.056	14.33	4.574
0.70	0.996	1.000	13.009	6.254	-1.435	3.81	0.044	15.31	4.578
0.72	0.997	1.000	12.816	6.212	-1.586	4.02	0.033	16.42	4.581
0.74	0.998	1.000	12.613	6.168	-1.745	4.24	0.025	17.68	4.584
0.76	0.999	1.000	12.400	6.121	-1.912	4.52	0.019	19.18	4.586
0.78	0.999	1.000	12.173	6.072	-2.089	4.80	0.013	20.93	4.588
0.80	1.000	1.000	11.932	6.019	-2.278	5.16	0.009	23.04	4.590
0.82	1.000	1.000	11.672	5.963	-2.481	5.53	0.006	25.60	4.591
0.84	1.000	1.000	11.387	5.901	-2.704	6.00	0.004	28.82	4.593
0.86	1.000	1.000	11.073	5.832	-2.950	6.60	0.003	32.94	4.594
0.88	1.000	1.000	10.718	5.755	-3.228	7.29	0.001	38.44	4.596
0.90	1.000	1.000	10.310	5.666	-3.548	8.22	0.001	46.04	4.598
0.92	1.000	1.000	9.816	5.559	-3.934	9.48	0.000	57.69	4.598
0.94	1.000	1.000	9.196	5.424	-4.420	11.38	0.000	76.93	4.598
0.96	1.000	1.000	8.342	5.238	-5.089	14.71	0.000	124.1	4.598
0.98	1.000	1.000	6.908	4.926	-6.212	22.69	0.000	225.3	4.598
1.00	1.000	1.000	—	—	—	—	0.000	∞	4.598

TABLE 28.3.

Model for star of one solar mass in initial main-sequence state (initial sun).

$$\frac{M}{M_\odot} = 1, \ \frac{L}{L_\odot} = 0.578, \ \frac{R}{R_\odot} = 1.021.$$

$\dfrac{r}{R}$	$\dfrac{M_r}{M}$	$\dfrac{L_r}{L}$	$\log P$	$\log T$	$\log \rho$	$\varkappa$	U	V	$n+1$
0.00	0.000	0.000	17.130	7.093	+1.886	1.42	3.000	0.000	2.882
0.02	0.000	0.004	17.125	7.091	1.882	1.43	2.990	0.025	2.890
0.04	0.004	0.034	17.109	7.086	1.871	1.47	2.961	0.099	2.914
0.06	0.012	0.103	17.082	7.076	1.854	1.54	2.913	0.222	2.952
0.08	0.028	0.212	17.045	7.064	1.829	1.63	2.846	0.392	3.007
0.10	0.052	0.348	16.996	7.048	1.796	1.75	2.759	0.608	3.078
0.12	0.085	0.492	16.938	7.030	1.757	1.89	2.656	0.867	3.163
0.14	0.126	0.628	16.871	7.009	1.711	2.07	2.538	1.165	3.261
0.16	0.176	0.742	16.795	6.986	1.657	2.27	2.407	1.495	3.371
0.18	0.232	0.830	16.709	6.960	1.597	2.52	2.265	1.855	3.490
0.20	0.292	0.893	16.616	6.934	1.530	2.77	2.114	2.233	3.613
0.22	0.354	0.936	16.516	6.907	1.457	3.04	1.957	2.626	3.735
0.24	0.417	0.962	16.409	6.879	1.378	3.32	1.801	3.024	3.854
0.26	0.479	0.978	16.296	6.850	1.294	3.63	1.646	3.425	3.966
0.28	0.539	0.988	16.180	6.821	1.207	3.95	1.495	3.820	4.067
0.30	0.594	0.994	16.060	6.792	1.116	4.26	1.353	4.206	4.158
0.32	0.646	0.997	15.937	6.763	1.022	4.57	1.217	4.584	4.239
0.34	0.692	0.998	15.812	6.733	0.926	4.94	1.091	4.950	4.308
0.36	0.734	0.999	15.684	6.704	0.829	5.27	0.975	5.306	4.369
0.38	0.772	1.000	15.556	6.675	0.729	5.60	0.868	5.651	4.421
0.40	0.805	1.000	15.426	6.646	0.628	5.94	0.770	5.987	4.466
0.42	0.834	1.000	15.296	6.617	0.528	6.32	0.681	6.315	4.504
0.44	0.859	1.000	15.165	6.588	0.425	6.68	0.601	6.640	4.536
0.46	0.881	1.000	15.034	6.559	0.323	7.08	0.529	6.960	4.564
0.48	0.900	1.000	14.903	6.530	0.221	7.50	0.465	7.277	4.588
0.50	0.916	1.000	14.770	6.501	0.117	7.92	0.407	7.598	4.609
0.52	0.930	1.000	14.638	6.473	+0.014	8.30	0.355	7.920	4.625
0.54	0.942	1.000	14.506	6.444	−0.090	8.77	0.309	8.248	4.640
0.56	0.952	1.000	14.373	6.416	−0.195	9.16	0.268	8.585	4.652
0.58	0.960	1.000	14.240	6.387	−0.299	9.67	0.233	8.931	4.663
0.60	0.967	1.000	14.105	6.358	−0.405	10.18	0.200	9.293	4.671
0.62	0.973	1.000	13.970	6.329	−0.511	10.71	0.172	9.670	4.678
0.64	0.978	1.000	13.834	6.300	−0.618	11.24	0.147	10.06	4.684
0.66	0.982	1.000	13.697	6.271	−0.726	11.79	0.126	10.49	4.688
0.68	0.985	1.000	13.558	6.241	−0.835	12.43	0.106	10.94	4.689
0.70	0.988	1.000	13.417	6.211	−0.946	13.07	0.090	11.41	4.689
0.72	0.990	1.000	13.275	6.181	−1.058	13.72	0.075	11.93	4.685
0.74	0.992	1.000	13.130	6.150	−1.172	14.47	0.063	12.49	4.673
0.76	0.994	1.000	12.982	6.118	−1.288	15.32	0.052	13.10	4.650
0.78	0.995	1.000	12.830	6.085	−1.407	16.28	0.042	13.79	4.604
0.80	0.996	1.000	12.674	6.051	−1.529	17.34	0.035	14.51	4.521
0.82	0.997	1.000	12.514	6.015	−1.653	18.71	0.028	15.45	4.339
0.84	0.998	1.000	12.346	5.975	−1.781	20.70	0.022	16.55	3.983
0.86	0.998	1.000	12.170	5.927	−1.908	24.48	0.018	18.07	3.232
0.88	0.999	1.000	11.979	5.856	−2.029	Conv.	0.015	20.83	2.500
0.90	0.999	1.000	11.758	5.767	−2.161	''	0.012	24.97	2.500
0.92	1.000	1.000	11.490	5.660	−2.322	''	0.009	31.26	2.500
0.94	1.000	1.000	11.156	5.526	−2.522	''	0.005	41.60	2.500
0.96	1.000	1.000	10.691	5.340	−2.801	''	0.003	62.33	2.500
0.98	1.000	1.000	9.922	5.031	−3.261	''	0.000	125.0	2.500
1.00	1.000	1.000	—	—	—	''	0.000	∞	2.500

TABLE 28.4   257

Model for star of 0.6 solar masses in initial main-sequence state (Castor C).

$$\frac{M}{M_\odot} = 0.603, \frac{L}{L_\odot} = 0.565, \frac{R}{R_\odot} = 0.644$$

$\dfrac{r}{R}$	$\dfrac{M_r}{M}$	$\dfrac{L_r}{L}$	$\log P$	$\log T$	$\log \rho$	$\varkappa$	U	V	$n+1$
0.00	0.000	0.000	16.871	6.906	+1.813	2.63	3.000	0.000	3.140
0.02	0.000	0.002	16.869	6.905	+1.812	2.65	2.994	0.013	3.145
0.04	0.001	0.012	16.860	6.902	+1.806	2.69	2.979	0.051	3.157
0.06	0.004	0.040	16.846	6.898	+1.796	2.75	2.953	0.114	3.180
0.08	0.010	0.086	16.827	6.892	+1.783	2.84	2.917	0.202	3.212
0.10	0.019	0.153	16.802	6.884	+1.766	2.97	2.870	0.314	3.252
0.12	0.033	0.237	16.772	6.875	+1.745	3.12	2.815	0.450	3.300
0.14	0.051	0.332	16.737	6.865	+1.721	3.29	2.750	0.607	3.358
0.16	0.071	0.433	16.697	6.853	+1.693	3.51	2.676	0.784	3.424
0.18	0.097	0.531	16.652	6.840	+1.661	3.75	2.595	0.980	3.498
0.20	0.127	0.623	16.603	6.826	+1.625	4.03	2.507	1.191	3.579
0.22	0.161	0.704	16.549	6.811	+1.586	4.35	2.413	1.418	3.666
0.24	0.197	0.773	16.491	6.795	+1.544	4.71	2.313	1.654	3.758
0.26	0.236	0.830	16.429	6.779	+1.498	5.09	2.210	1.900	3.852
0.28	0.277	0.876	16.364	6.762	+1.450	5.52	2.104	2.151	3.948
0.30	0.319	0.910	16.296	6.745	+1.399	5.97	1.997	2.405	4.043
0.32	0.362	0.936	16.225	6.728	+1.345	6.43	1.890	2.660	4.138
0.34	0.405	0.955	16.152	6.711	+1.289	6.92	1.784	2.913	4.227
0.36	0.447	0.969	16.076	6.693	+1.231	7.48	1.679	3.163	4.310
0.38	0.488	0.979	15.999	6.675	+1.172	8.08	1.577	3.409	4.381
0.40	0.528	0.986	15.920	6.657	+1.111	8.71	1.478	3.649	4.450
0.42	0.566	0.991	15.841	6.640	+1.050	9.31	1.385	3.880	4.512
0.44	0.603	0.994	15.760	6.622	+0.986	10.00	1.293	4.109	4.559
0.46	0.637	0.996	15.678	6.604	+0.922	10.74	1.208	4.328	4.590
0.48	0.669	0.998	15.597	6.586	+0.859	11.55	1.127	4.540	4.605
0.50	0.700	0.999	15.515	6.568	+0.794	12.39	1.049	4.749	4.598
0.52	0.728	1.000	15.432	6.550	+0.730	13.30	0.978	4.953	4.561
0.54	0.755	1.000	15.349	6.532	+0.665	14.27	0.911	5.155	4.491
0.56	0.779	1.000	15.266	6.513	+0.601	15.45	0.849	5.358	4.375
0.58	0.802	1.000	15.183	6.494	+0.537	16.73	0.791	5.568	4.198
0.60	0.823	1.000	15.100	6.473	+0.474	18.43	0.737	5.792	3.940
0.62	0.843	1.000	15.015	6.451	+0.412	20.48	0.689	6.039	3.577
0.64	0.861	1.000	14.930	6.426	+0.353	23.41	0.648	6.340	3.058
0.66	0.877	1.000	14.841	6.393	+0.297	conv.	0.613	6.756	2.500
0.68	0.893	1.000	14.750	6.356	+0.242	,,	0.581	7.260	2.500
0.70	0.908	1.000	14.655	6.318	+0.185	,,	0.547	7.822	2.500
0.72	0.921	1.000	14.556	6.279	+0.126	,,	0.511	8.460	2.500
0.74	0.934	1.000	14.451	6.237	+0.063	,,	0.473	9.188	2.500
0.76	0.945	1.000	14.340	6.192	−0.004	,,	0.434	10.03	2.500
0.78	0.955	1.000	14.221	6.145	−0.076	,,	0.394	11.02	2.500
0.80	0.964	1.000	15.093	6.094	−0.152	,,	0.353	12.20	2.500
0.82	0.972	1.000	13.956	6.038	−0.235	,,	0.312	13.63	2.500
0.84	0.979	1.000	13.804	5.978	−0.326	,,	0.270	15.41	2.500
0.86	0.985	1.000	13.636	5.910	−0.427	,,	0.228	17.67	2.500
0.88	0.990	1.000	13.445	5.834	−0.541	,,	0.187	20.69	2.500
0.90	0.993	1.000	13.225	5.746	−0.673	,,	0.147	24.87	2.500
0.92	0.996	1.000	12.958	5.640	−0.833	,,	0.108	31.18	2.500
0.94	0.998	1.000	12.625	5.506	−1.033	,,	0.073	41.55	2.500
0.96	0.999	1.000	12.160	5.320	−1.312	,,	0.041	62.30	2.500
0.98	1.000	1.000	11.387	5.011	−1.776	,,	0.015	125.0	2.500
1.00	1.000	1.000	—	—	—	,,	0.000	∞	2.500

TABLE 28.5

Model for star of 10 solar masses in evolutionary phase when hydrogen
content at center is reduced to 6 percent.

$$\frac{M}{M_\odot} = 10, \quad \frac{L}{M_\odot} = 5220, \quad \frac{R}{R_\odot} = 6.09.$$

r/R	$\dfrac{M_r}{M}$	$\dfrac{L_r}{L}$	X	$\log P$	$\log T$	$\log \rho$	$\varkappa$	U	V	$n+1$
0.00	0.000	0.000	0.061	16.844	7.545	+1.075	conv.	3.000	0.000	2.500
0.02	0.001	0.131	0.061	16.824	7.537	+1.063	"	2.967	0.092	2.500
0.04	0.011	0.582	0.061	16.764	7.513	+1.027	"	2.869	0.371	2.500
0.06	0.035	0.974	0.061	16.662	7.473	+0.966	"	2.707	0.844	2.500
0.08	0.075	1.000	0.081	16.506	7.415	+0.868	0.417	2.419	1.486	2.520
0.10	0.119	1.000	0.325	16.223	7.357	+0.648	0.415	1.799	1.602	2.680
0.12	0.160	1.000	0.531	15.998	7.310	+0.475	0.414	1.566	1.665	2.724
0.14	0.202	1.000	0.720	15.826	7.268	+0.347	0.416	1.444	1.708	2.716
0.16	0.244	1.000	0.893	15.670	7.231	+0.231	0.418	1.368	1.738	2.665
0.18	0.288	1.000	0.900	15.575	7.196	+0.172	0.426	1.443	1.984	2.788
0.20	0.336	1.000	0.900	15.478	7.162	+0.111	0.434	1.473	2.260	2.917
0.22	0.387	1.000	0.900	15.379	7.129	+0.046	0.442	1.467	2.560	3.045
0.24	0.439	1.000	0.900	15.276	7.096	−0.022	0.451	1.433	2.880	3.168
0.26	0.491	1.000	0.900	15.170	7.063	−0.095	0.459	1.378	3.216	3.284
0.28	0.543	1.000	0.900	15.061	7.030	−0.171	0.468	1.309	3.562	3.390
0.30	0.593	1.000	0.900	14.949	6.998	−0.249	0.476	1.230	3.916	3.486
0.32	0.640	1.000	0.900	14.834	6.965	−0.331	0.485	1.145	4.277	3.572
0.34	0.684	1.000	0.900	14.717	6.933	−0.416	0.492	1.058	4.640	3.649
0.36	0.725	1.000	0.900	14.597	6.900	−0.503	0.501	0.970	5.008	3.717
0.38	0.763	1.000	0.900	14.476	6.868	−0.592	0.508	0.884	5.378	3.777
0.40	0.796	1.000	0.900	14.351	6.835	−0.683	0.516	0.800	5.753	3.830
0.42	0.826	1.000	0.900	14.226	6.802	−0.776	0.524	0.721	6.131	3.876
0.44	0.853	1.000	0.900	14.098	6.770	−0.871	0.530	0.645	6.517	3.917
0.46	0.876	1.000	0.900	13.968	6.737	−0.968	0.537	0.574	6.910	3.952
0.48	0.897	1.000	0.900	13.837	6.704	−1.066	0.544	0.509	7.312	3.983
0.50	0.914	1.000	0.900	13.704	6.670	−1.166	0.552	0.448	7.731	4.009
0.52	0.929	1.000	0.900	13.568	6.636	−1.267	0.560	0.392	8.166	4.032
0.54	0.942	1.000	0.900	13.431	6.602	−1.371	0.566	0.341	8.622	4.052
0.56	0.953	1.000	0.900	13.291	6.568	−1.477	0.572	0.295	9.104	4.069
0.58	0.962	1.000	0.900	13.148	6.533	−1.584	0.580	0.254	9.618	4.083
0.60	0.970	1.000	0.900	13.002	6.497	−1.695	0.587	0.216	10.17	4.096
0.62	0.977	1.000	0.900	12.854	6.461	−1.807	0.594	0.183	10.77	4.106
0.64	0.982	1.000	0.900	12.700	6.424	−1.923	0.602	0.153	11.43	4.116
0.66	0.986	1.000	0.900	12.544	6.386	−2.042	0.609	0.127	12.14	4.123
0.68	0.989	1.000	0.900	12.382	6.346	−2.164	0.619	0.104	12.94	4.130
0.70	0.992	1.000	0.900	12.213	6.306	−2.292	0.626	0.085	13.85	4.136
0.72	0.994	1.000	0.900	12.038	6.263	−2.425	0.637	0.068	14.87	4.141
0.74	0.996	1.000	0.900	11.854	6.219	−2.565	0.646	0.053	16.04	4.145
0.76	0.997	1.000	0.900	11.661	6.172	−2.712	0.657	0.041	17.41	4.149
0.78	0.998	1.000	0.900	11.455	6.123	−2.868	0.668	0.031	19.00	4.153
0.80	0.999	1.000	0.900	11.236	6.070	−3.034	0.682	0.023	20.93	4.156
0.82	0.999	1.000	0.900	11.000	6.013	−3.214	0.696	0.016	23.26	4.159
0.84	1.000	1.000	0.900	10.741	5.951	−3.411	0.711	0.011	26.20	4.161
0.86	1.000	1.000	0.900	10.456	5.883	−3.629	0.728	0.007	29.94	4.165
0.88	1.000	1.000	0.900	10.133	5.805	−3.875	0.750	0.004	34.95	4.168
0.90	1.000	1.000	0.900	9.762	5.716	−4.157	0.777	0.003	41.86	4.171
0.92	1.000	1.000	0.900	9.313	5.609	−4.499	0.809	0.001	52.45	4.172
0.94	1.000	1.000	0.900	8.750	5.474	−4.928	0.856	0.000	69.94	4.172
0.96	1.000	1.000	0.900	7.973	5.287	−5.520	0.931	0.000	104.9	4.172
0.98	1.000	1.000	0.900	6.669	4.975	−6.513	1.08	0.000	210.2	4.172
1.00	1.000	1.000	0.900	−	−	−	−	0.000	∞	4.172

TABLE 28.6 259

Model for star of one solar mass in evolutionary phase when hydrogen content at center is reduced to 50 percent (present sun, improved model by R. Weymann, *Ap.J. 126*, 208, 1957).

$$\frac{M}{M_\odot} = 1, \quad \frac{L}{L_\odot} = 1, \quad \frac{R}{R_\odot} = 1.$$

r/R	$\dfrac{M_r}{M}$	$\dfrac{L_r}{L}$	X	$\log P$	$\log T$	$\log \rho$	$\varkappa$	U	V	$n+1$
0.00	0.000	0.000	0.494	17.351	7.165	+2.128	1.07	3.000	0.000	3.264
0.02	0.001	0.006	0.498	17.335	7.162	+2.113	1.07	2.976	0.043	3.281
0.04	0.006	0.042	0.520	17.307	7.154	+2.084	1.10	2.903	0.166	3.284
0.06	0.018	0.124	0.545	17.265	7.141	+2.046	1.16	2.802	0.350	3.312
0.08	0.040	0.244	0.571	17.205	7.123	+1.995	1.24	2.680	0.597	3.350
0.10	0.073	0.396	0.611	17.135	7.102	+1.932	1.34	2.534	0.873	3.401
0.12	0.113	0.538	0.643	17.058	7.080	+1.867	1.45	2.398	1.159	3.448
0.14	0.162	0.668	0.670	16.970	7.054	+1.796	1.60	2.257	1.480	3.513
0.16	0.217	0.774	0.694	16.874	7.027	+1.721	1.76	2.115	1.748	3.592
0.18	0.276	0.854	0.714	16.774	7.000	+1.642	1.93	1.973	2.160	3.681
0.20	0.337	0.909	0.723	16.667	6.971	+1.561	2.12	1.838	2.524	3.776
0.22	0.399	0.945	0.728	16.554	6.942	+1.476	2.32	1.702	2.896	3.875
0.24	0.460	0.968	0.733	16.438	6.912	+1.389	2.55	1.565	3.265	3.972
0.26	0.519	0.981	0.737	16.319	6.882	+1.298	2.78	1.431	3.630	4.064
0.28	0.574	0.989	0.741	16.196	6.852	+1.205	3.02	1.303	3.988	4.149
0.30	0.626	0.994	0.744	16.072	6.823	+1.109	3.23	1.180	4.332	4.226
0.32	0.672	0.997	0.744	15.944	6.793	+1.011	3.47	1.066	4.676	4.293
0.34	0.716	0.998	0.744	15.816	6.763	+0.913	3.73	0.959	5.009	4.353
0.36	0.753	0.999	0.744	15.690	6.734	+0.816	3.99	0.862	5.327	4.403
0.38	0.788	1.000	0.744	15.562	6.705	+0.717	4.25	0.771	5.640	4.449
0.40	0.818	1.000	0.744	15.432	6.676	+0.616	4.51	0.687	5.944	4.488
0.42	0.844	1.000	0.744	15.302	6.648	+0.514	4.74	0.613	6.243	4.520
0.44	0.867	1.000	0.744	15.174	6.619	+0.415	5.04	0.543	6.534	4.549
0.46	0.887	1.000	0.744	15.045	6.591	+0.314	5.31	0.482	6.825	4.573
0.48	0.904	1.000	0.744	14.917	6.563	+0.214	5.60	0.426	7.111	4.595
0.50	0.919	1.000	0.744	14.788	6.535	+0.113	5.89	0.375	7.401	4.613
0.52	0.932	1.000	0.744	14.660	6.507	+0.013	6.21	0.330	7.692	4.628
0.54	0.943	1.000	0.744	14.531	6.480	−0.089	6.47	0.290	7.989	4.641
0.56	0.953	1.000	0.744	14.403	6.452	−0.189	6.83	0.254	8.291	4.653
0.58	0.961	1.000	0.744	14.274	6.424	−0.290	7.19	0.222	8.602	4.662
0.60	0.967	1.000	0.744	14.144	6.397	−0.393	7.48	0.193	8.926	4.669
0.62	0.973	1.000	0.744	14.015	6.369	−0.494	7.87	0.167	9.261	4.675
0.64	0.979	1.000	0.744	13.885	6.341	−0.596	8.27	0.145	9.612	4.679
0.66	0.982	1.000	0.744	13.755	6.313	−0.698	8.69	0.125	9.982	4.679
0.68	0.985	1.000	0.744	13.622	6.285	−0.803	9.08	0.107	10.38	4.676
0.70	0.988	1.000	0.744	13.489	6.256	−0.907	9.59	0.092	10.79	4.667
0.72	0.989	1.000	0.744	13.355	6.228	−1.013	10.01	0.078	11.24	4.648
0.74	0.992	1.000	0.744	13.218	6.198	−1.120	10.60	0.066	11.73	4.611
0.76	0.994	1.000	0.744	13.079	6.168	−1.229	11.17	0.056	12.27	4.547
0.78	0.995	1.000	0.744	12.937	6.136	−1.339	11.96	0.047	12.88	4.427
0.80	0.996	1.000	0.744	12.792	6.103	−1.451	12.86	0.039	13.57	4.199
0.82	0.997	1.000	0.744	12.642	6.065	−1.563	14.40	0.032	14.47	3.790
0.84	0.998	1.000	0.744	12.484	6.017	−1.673	17.53	0.027	15.77	2.905
0.86	0.998	1.000	0.744	12.312	5.947	−1.775	conv.	0.023	17.84	2.500
0.88	0.999	1.000	0.744	12.119	5.870	−1.891	''	0.018	20.83	2.500
0.90	0.999	1.000	0.744	11.898	5.782	−2.024	''	0.015	24.97	2.500
0.92	1.000	1.000	0.744	11.631	5.675	−2.184	''	0.011	31.25	2.500
0.94	1.000	1.000	0.744	11.297	5.541	−2.384	''	0.006	41.60	2.500
0.96	1.000	1.000	0.744	10.832	5.355	−2.663	''	0.003	62.33	2.500
0.98	1.000	1.000	0.744	10.063	5.046	−3.123	''	0.000	125.0	2.500
1.00	1.000	1.000	0.744	—	—	—	''	0.000	∞	2.500

TABLE 28.7.

Model for red giant of 1.3 solar masses in evolutionary phase when hydrogen is exhausted in inner 26 percent of mass (improved model by Schwarzschild and Selberg, unpublished).

$$\frac{M}{M_\odot} = 1.3, \quad \frac{L}{L_\odot} = 226, \quad \frac{R}{R_\odot} = 21.$$

r/R	M_r/M	L_r/L	$\log P$	$\log T$	$\log \rho$	$\varkappa$
0.00000	0.000	0.00	21.74	7.60	+5.54	degen.
0.00005	0.001	0.00	21.74	7.60	+5.53	"
0.00010	0.003	0.00	21.71	7.60	+5.52	"
0.00015	0.007	0.00	21.68	7.60	+5.50	"
0.00020	0.014	0.00	21.63	7.60	+5.46	"
0.00025	0.025	0.00	21.56	7.60	+5.43	"
0.00030	0.039	0.00	21.49	7.60	+5.38	"
0.00035	0.057	0.00	21.40	7.60	+5.32	"
0.00040	0.078	0.00	21.30	7.60	+5.26	"
0.00045	0.101	0.00	21.18	7.60	+5.19	"
0.00050	0.126	0.00	21.04	7.60	+5.10	"
0.00055	0.150	0.00	20.90	7.60	+5.10	"
0.00060	0.174	0.00	20.72	7.60	+4.90	"
0.00065	0.195	0.00	20.53	7.60	+4.78	"
0.00070	0.214	0.00	20.31	7.60	+4.64	"
0.00075	0.229	0.00	20.07	7.60	+4.48	"
0.00080	0.241	0.00	19.80	7.60	+4.29	"
0.00085	0.249	0.00	19.45	7.60	+4.05	1.10
0.00090	0.254	0.00	19.12	7.60	+3.73	0.78
0.00095	0.257	0.00	18.79	7.60	+3.41	0.45
0.0010	0.258	0.00	18.51	7.60	+3.21	0.32
0.0011	0.259	0.00	18.01	7.60	+2.63	0.23
0.0012	0.260	0.10	17.63	7.60	+1.84	0.37
0.0013	0.260	0.67	17.49	7.58	+1.72	0.37
0.0014	0.260	0.91	17.36	7.56	+1.61	0.37

TABLE 28.7 (continued)

r/R	M_r/M	L_r/L	$\log P$	$\log T$	$\log \rho$	$\varkappa$
0.0015	0.260	0.97	17.24	7.53	+1.53	0.37
0.0020	0.261	1.00	16.77	7.41	+1.16	0.37
0.0030	0.262	1.00	16.10	7.24	+0.66	0.38
0.0040	0.263	1.00	15.63	7.13	+0.32	0.38
0.0050	0.263	1.00	15.23	7.04	+0.05	0.38
0.006	0.263	1.00	15.00	6.97	−0.16	0.38
0.008	0.264	1.00	14.56	6.86	−0.49	0.38
0.010	0.265	1.00	14.24	6.78	−0.73	0.39
0.015	0.266	1.00	13.68	6.64	−1.15	0.39
0.020	0.267	1.00	13.31	6.55	−1.43	0.39
0.025	0.269	1.00	13.04	6.48	−1.63	0.40
0.030	0.270	1.00	12.83	6.43	−1.78	0.40
0.035	0.271	1.00	12.67	6.38	−1.90	0.40
0.040	0.272	1.00	12.53	6.35	−2.00	0.41
0.050	0.275	1.00	12.32	6.28	−2.16	0.41
0.06	0.278	1.00	12.12	6.23	−2.27	0.42
0.07	0.281	1.00	12.02	6.19	−2.36	0.43
0.08	0.285	1.00	11.90	6.14	−2.43	conv.
0.09	0.289	1.00	11.80	6.10	−2.49	"
0.10	0.294	1.00	11.77	6.07	−2.54	"
0.15	0.323	1.00	11.38	5.94	−2.74	"
0.20	0.360	1.00	11.15	5.84	−2.88	"
0.25	0.407	1.00	10.95	5.76	−3.00	"
0.30	0.460	1.00	10.77	5.69	−3.11	"
0.35	0.518	1.00	10.60	5.62	−3.21	"
0.40	0.580	1.00	10.43	5.56	−3.31	"
0.45	0.642	1.00	10.27	5.49	−3.41	"
0.5	0.704	1.00	10.10	5.42	−3.52	"
0.6	0.817	1.00	9.72	5.28	−3.74	"
0.7	0.908	1.00	9.29	5.11	−4.00	"
0.8	0.969	1.00	8.72	4.87	−4.34	"
0.9	1.000	1.00	7.85	4.52	−4.87	"
1.0	1.000	1.00	−	−	−	"

TABLE 28.8

Model for white dwarf of 0.9 solar masses, (Chandrasekhar, *Stellar Structure*, p. 496.)

$$\frac{M}{M_\odot} = 0.886, \frac{R}{R_\odot} = 0.00924.$$

$\dfrac{r}{R}$	$\dfrac{M_r}{M}$	$\log P$	$\log \rho$	U	V
0.00	0.000	24.205	7.196	3.000	0.000
0.02	0.000	24.202	7.194	2.996	0.008
0.04	0.001	24.198	7.191	2.989	0.029
0.06	0.002	24.191	7.186	2.974	0.064
0.08	0.005	24.181	7.179	2.953	0.113
0.10	0.010	24.166	7.169	2.927	0.176
0.12	0.016	24.149	7.157	2.896	0.252
0.14	0.025	24.130	7.144	2.859	0.342
0.16	0.037	24.108	7.128	2.816	0.444
0.18	0.051	24.082	7.110	2.770	0.557
0.20	0.069	24.053	7.090	2.720	0.683
0.22	0.089	24.023	7.069	2.665	0.821
0.24	0.112	23.989	7.046	2.606	0.970
0.26	0.137	23.952	7.020	2.545	1.129
0.28	0.166	23.914	6.994	2.480	1.298
0.30	0.196	23.872	6.965	2.412	1.477
0.32	0.229	23.828	6.935	2.342	1.667
0.34	0.263	23.782	6.903	2.270	1.866
0.36	0.299	23.733	6.870	2.196	2.075
0.38	0.336	23.682	6.835	2.120	2.294
0.40	0.373	23.628	6.798	2.044	2.522
0.42	0.412	23.573	6.761	1.966	2.761
0.44	0.450	23.514	6.721	1.888	3.011
0.46	0.489	23.453	6.680	1.809	3.273
0.48	0.527	23.390	6.638	1.729	3.546
0.50	0.565	23.325	6.594	1.649	3.834
0.52	0.601	23.257	6.549	1.569	4.137
0.54	0.637	23.186	6.502	1.489	4.456
0.56	0.672	23.114	6.454	1.409	4.795
0.58	0.705	23.037	6.403	1.330	5.155
0.60	0.736	22.959	6.352	1.252	5.541
0.62	0.766	22.877	6.298	1.174	5.955
0.64	0.794	22.792	6.243	1.096	6.404
0.66	0.820	22.703	6.185	1.019	6.893
0.68	0.845	22.610	6.125	0.943	7.430
0.70	0.860	22.513	6.063	0.868	8.027
0.72	0.888	22.411	5.998	0.793	8.696
0.74	0.906	22.303	5.929	0.720	9.453
0.76	0.923	22.189	5.857	0.648	10.32
0.78	0.938	22.067	5.780	0.577	11.33
0.80	0.951	21.937	5.698	0.509	12.51
0.82	0.962	21.797	5.611	0.442	13.96
0.84	0.972	21.641	5.514	0.376	15.76
0.86	0.980	21.467	5.405	0.313	18.04
0.88	0.986	21.273	5.286	0.253	21.05
0.90	0.991	21.047	5.146	0.196	25.28
0.92	0.995	20.778	4.982	0.143	31.56
0.94	0.998	20.440	4.776	0.094	42.02
0.96	0.999	19.972	4.492	0.052	62.87
0.98	1.000	19.164	4.020	0.018	125.5
1.00	1.000	—	—	0.000	∞

29. *Evolutionary Tracks in the Hertzsprung-Russell Diagram*

The physical laws and equilibrium conditions which we have assembled in Chapter II determine uniquely the physical state and structure for a star of given mass and initial composition from phase to phase throughout the star's evolution. Among the various physical characteristics of a star which are thus determined as functions of time there are two of special importance, the luminosity and the radius. Their special importance arises from the fact that—as we have seen in §1—they can be observed with good accuracy for many stars.

Stellar luminosities and radii can be represented in the Hertzsprung-Russell diagram as shown in Fig. 1.5. Hence a given star must follow a uniquely determined track in the Hertzsprung-Russell diagram during its evolution. These evolutionary tracks have become the most powerful tool we have at present for checking the theory of stellar evolution with observed data.

Let us then assemble and review the evolutionary tracks we have derived in the preceding chapters, first for massive stars, heavier than say two solar masses, and then for medium-weight stars, with masses between approximately one and two solar masses. We shall ingore here the small stars with masses less than one solar mass, since their evolution is so slow that none of them can have evolved noticeably beyond the initial main-sequence state during the lifetime of our galaxy.

Evolutionary Tracks of Massive Stars

In the Hertzsprung-Russell diagram of Fig. 29.1 the evolutionary tracks for three stars heavier than two solar masses are shown. With an eye on one of these tracks let us follow the major events in the life of such a star.

The star makes its appearance in the observable part of the Hertzsprung-Russell diagram during its original contraction from the interstellar gas. It appears in the right-hand part of the diagram at a point corresponding to large radius and low surface temperature. As the contraction proceeds the star moves corresponding to decreasing radius. In the meantime the luminosity increases somewhat, because of a reduction in the internal opacity. When the central temperature, which is steadily growing during the contraction, becomes high enough for the nuclear hydrogen burning to reach a significant rate, the contraction slows down, the gravitational energy release is replaced by nuclear energy release, and the internal structure of the star undergoes a change which causes a slight but rather abrupt reduction in the luminosity. Now the star settles down in its initial main-sequence state.

During the following phases evolution proceeds much more slowly than during the preceding contraction phases, owing to the large store of

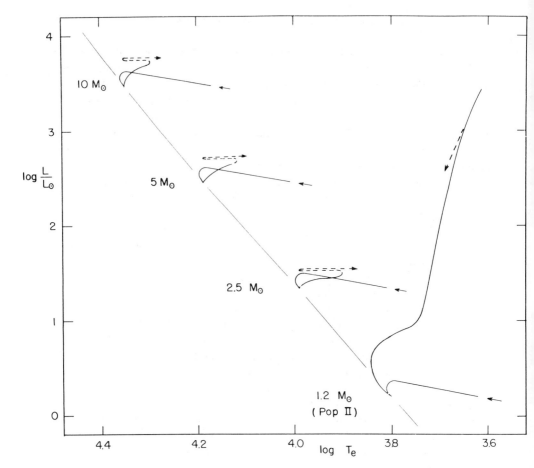

Fig. 29.1. Evolution tracks in the Hertzsprung-Russell diagram (data taken from Figs. 19.1, 22.4, and 24.3).

nuclear fuel. As a matter of fact, the star is most likely to spend the major part of its entire active life (excluding the final white-dwarf state) in the phases immediately following the initial main-sequence state. During these phases the hydrogen content in the core steadily diminishes, as shown in Figs. 22.2 and 22.3, while the hydrogen content of the outer parts of the star is unaffected by the nuclear processes. This growing inhomogeneity in composition produces a slow change in the structure of the star and causes the luminosity and the radius to increase somewhat. It is during these long phases of hydrogen burning in the core that the vast majority of the massive stars are observed.

After the hydrogen content of the core has dropped to about one percent, the depletion of the fuel—now proceeding percentage-wise fairly fast— has to be compensated by an increase in temperature to keep the energy production in balance with the luminosity. The required temperature increase is provided by a contraction of the whole star. Thus the evolutionary track in the Hertzsprung-Russell diagram takes a sudden turn to the left. This secondary contraction phase gets soon terminated when the hydrogen content of the core is exhausted and the hydrogen burning shifts outward into a shell surrounding the now inert core.

The subsequent evolutionary phases have not yet been computed through in detail. But qualitative investigations suggest the following picture. While the hydrogen-burning shell steadily moves further out and the helium core continues to grow in mass, this core cannot settle down into an isothermal state since for a massive star an isothermal core cannot provide the necessary pressure to support the outer layers. In consequence the helium core must continue to contract until it reaches the temperature at which helium burning commences. The combination of a helium core heated first by contraction and later by helium burning, a hydrogen-burning shell, and a hydrogen-rich envelope produces a stellar structure very different from that of the main-sequence phases. In particular, it produces a rapidly expanding envelope so that the star on its track in the Hertzsprung-Russell diagram moves to the right, rapidly crossing the "Hertzsprung gap," which separates the main-sequence stars from the red giants.

The fate of a massive star during and after the red-giant phase is still quite unclear. One dominating question arises: does mass ejection play a decisive role in the red-giant phase? We shall return to this question after we have reviewed the evolution of less massive stars.

Evolutionary Tracks of Medium-Weight Stars

The Hertzsprung-Russell diagram of Fig. 29.1 shows the evolutionary track of one medium-weight star (1.2 solar masses). This track, just as those for the more massive stars, begins with the original contraction phase, which leads the star fast to its initial main-sequence state. The following phases are only roughly sketched in our figure. To start with they proceed in much the same way as the corresponding phases in the more massive stars; the burning in the core depletes the hydrogen content, as shown in Fig. 23.1, and eventually the hydrogen in the central region is exhausted, an inert helium core forms, and the hydrogen burning occurs further out in a shell surrounding the core.

At this state a significant difference appears between the massive and the medium-weight stars. When the hydrogen is exhausted in the core, the central density in a medium-weight star is so high that the electron gas is degenerate there. With the higher degenerate pressure an isothermal core turns out to be capable of providing the necessary support for the outer parts of the star. In consequence the helium core of a medium-weight star will not immediately undergo a violent contraction, and hence the commencement of helium burning will be postponed to a much later phase.

Thus, after hydrogen exhaustion at the center, a medium-weight star will consist of a partially degenerate isothermal helium core which steadily grows in mass, a narrow hydrogen-burning shell which provides the entire energy, and an envelope with the original hydrogen-rich composition. Such a structure—much like the corresponding structure of massive stars with non-degenerate but contracting cores—produces a steady increase in the extent of the envelope and thus leads the evolutionary track to the right. This envelope expansion is slowed down, however, by the appearance of a deep convection zone characteristic of stellar envelopes with low surface temperatures.

As the helium core grows in mass and the envelope expands, the luminosity steadily increases, a consequence of the continuous increase of the average molecular weight for the star as a whole. Thus the evolutionary track in the Hertzsprung-Russell diagram will lead to the right and upwards, and the star will become successively a subgiant, a giant, and a supergiant.

When the mass of the helium core has reached about forty percent of the total stellar mass, the luminosity has increased by about a factor of a thousand compared with the luminosity of the initial main-sequence state. This very high luminosity requires a very high rate of hydrogen transmutation, which in turn causes a rapid change in the structure of the star. These structural changes produce a compression of the core which steadily accelerates, so that eventually in spite of the degeneracy the core heats up to the temperature critical for helium burning. The starting of the helium burning appears to terminate the red-giant branch of the evolutionary track of a medium-weight star.

What happens next in the evolution of a medium-weight star we do not know. It appears fairly likely that the envelope will contract again. It is completely uncertain, however, whether the central helium burning could cause convection extensive enough to mix the hydrogen from the outer parts of the star into the core and thus profoundly influence the subsequent evolutionary phases. Furthermore, the same dominant question which went unanswered for the massive stars faces us again for the medium-weight stars: does the star suffer a substantial loss of mass by ejection from the surface in those phases of its evolution when it is a red giant or supergiant with a greatly extended envelope?

Mass Ejection

Before we proceed with the main point of this section, the test of the theory of stellar evolution with the help of the Hertzsprung-Russell diagram, we should interject here a short but critical inspection of one major assumption which we have made implicitly throughout this book thus far, the assumption that the mass of the star remains substantially constant throughout its evolution. This assumption is far from trivial. As a matter of fact, both mass growth by accretion and mass diminution by surface ejection have been investigated extensively.

Regarding mass growth, the theory of accretion has shown that substantial increases of a stellar mass by accretion from an interstellar cloud will occur only if the relative velocity of star and cloud is very small, much smaller than the average relative velocities. Substantial accretion appears therefore most likely quite a rare phenomenon.

In contrast, a substantial mass loss by ejection from the surface may well happen to a large class of stars in a critical phase of their evolution. These critical phases do not include the main-sequence phases, at least not for the majority of stars. The present rate of mass loss of the sun, for example, is entirely negligible, and the same appears likely for most of the main-sequence stars, with the exception of some of the extremely massive stars of spectral types O and early B. These exceptional cases show spectroscopic evidence of significant mass ejection, possibly caused or at least aided by their strong radiation pressure.

In contrast with the main-sequence phases several indications suggest that mass ejection in the red-giant phases may occur with such speed as to greatly influence evolution. The most direct evidence comes from the spectroscopic observation of one double star, α Herculis. In this binary one component is a red supergiant while the spectrum of the other component shows absorption features most likely explained by circumstellar matter ejected from the red giant to large distances. Additional less direct evidence comes from the investigation of galactic clusters such as the Hyades or Praesepe. In these clusters the number of red giants is so low and their luminosity so moderate that it is hard to see how, in the red-giant phases, more than just a few percent of the total hydrogen store of a star could be burned. The addition of this small amount of hydrogen consumption to that occurring in the preceding main-sequence phases gives a total hydrogen consumption during the entire life of a star of probably less than 20 percent. It would thus appear likely that these stars do not end their active life by running out of hydrogen fuel, but instead perhaps by ejecting more than half of their mass (and hence most of their remaining hydrogen fuel store) back into interstellar space. Additional statistical evidence for this mode of ending a star's active life will come up in §30.

On the other hand, from theoretical considerations does it seem likely that a red giant is at all capable of ejecting a large fraction of its mass? Yes—for example, the following mechanism might cause such large-scale ejection. The average velocity of the macroscopic turbulence in the photosphere of red giants is known from spectroscopic investigations to approach sound velocity. Such violent turbulence will act as a strong source of acoustic noise. Some of these noise waves will progress upwards and will represent a mechanical energy flux from the photosphere into the chromosphere and corona of the star. The order of magnitude of this energy flux can be estimated, and turns out quite sufficient to provide the necessary rate of mass ejection, even if only a small percent of it is transformed into kinetic energy for escape.

Altogether, we have to conclude that according to various indications it seems possible, if not even likely, that many stars suffer in the red-giant phase a substantial mass loss, and that this mass-ejection mechanism will have to be investigated before we can follow these stars securely through their final evolutionary phases.

Time Lines in the Hertzsprung-Russell Diagram

We are finally ready for the key test of the theory of stellar evolution, the comparison of the theoretical evolutionary tracks shown in Fig. 29.1 with the observed Hertzsprung-Russell diagrams of clusters shown in Figs. 1.1 to 1.5—with a proper caution regarding all red-giant phases because of the uncertainty of possible mass losses.

The line on which the stars of a cluster are found to fall in the Hertzsprung-Russell diagram cannot be an evolutionary track. On the contrary, it must represent a time line, that is a line on which all stars of various masses must fall at a definite time after the birth of the stars. This interpretation of the line of the cluster as a time line presupposes that all the stars of one cluster were born essentially at the same moment, possessed essentially the same initial composition, and differ only in their mass, a presumption we have discussed already in §1. Therefore, to make the comparison between theory and observations possilbe, we have first to derive theoretical time lines from the theoretical evolutionary tracks.

The derivation of the theoretical time lines follows directly from the model construction work described in the preceding chapters where we have not only determined evolutionary model sequences but have also computed the age at which a given star reaches a particular model in a given sequence. These ages are related to the luminosity of the star and also to the amount of hydrogen burned according to Eq.22.10.

The results of these computations are shown in Fig. 29.2 for stars ranging in mass from 2.5 to 5.6 solar masses. In this figure are drawn the evolutionary tracks for eight stars of different masses for those evolutionary phases in which these stars spend probably more than three quarters of their active life, starting from the initial main-sequence phase (onset of hydrogen burning) to the phase of near-exhaustion of the hydrogen in the core (just preceding the short secondary contraction and the fast subsequent expansion into the red-giant phase). On each evolutionary track are marked those points which the star will reach at an age of 100, 200, 400 and 800 million years respectively. Finally, the points corresponding to a fixed age are connected by a line across the evolutionary tracks, the time line corresponding to the age considered.

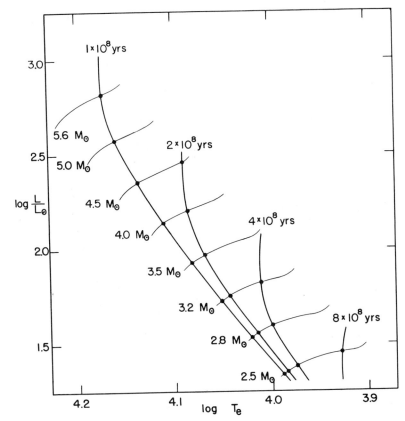

Fig. 29.2. Time lines in Hertzsprung-Russell diagram for young clusters (data from Fig. 22.5).

Comparison of the time lines of Fig. 29.2 with the Hertzsprung-Russell diagrams of galactic clusters shown in Fig. 1.5 (in which we here ignore the red giants) shows an agreement qualitatively striking and quantitatively satisfactory—a most encouraging first result of our key test.

An equally happy result is found when the theoretical time lines are compared with the observed Herzsprung-Russell diagrams of globular clusters. To make this comparison possible, approximate evolutionary tracks are given in Fig. 29.3 for three stars ranging from 1.15 to 1.25 solar masses. These tracks cover the phases from the initial main-sequence to the final red-giant phase in which helium burning is suspected to commence at the center. On each of the three tracks the point is marked which corresponds to a stellar age of six billion years and a line is drawn through these three points across the three evolutionary tracks, thus forming the six-billion-year time line. This time line is, b⟩ general character, in excellent agreement with the corresponding line in the observed Hertzsprung-Russell diagram of the globular cluster M3,

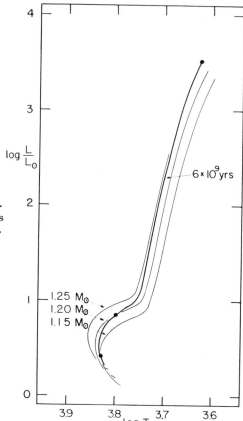

Fig. 29.3. Time line in Hertzsprung-
Russell Diagram for old clusters
(data—very approximate—from Fig.
29.1).

shown in Figs. 1.4 and 1.5. Even quantitatively the agreement is satisfactory if one remembers the approximate nature of the theoretical evolutionary tracks, in particular the uncertainty caused by the possible
mass loss during the red-giant phases and the inaccuracy of the radii of
the red-giant models (caused by the unknown efficiency of the convective
energy transport in the surface layers, discussed in §11).

One more item can be checked, at least qualitatively. According to
the theoretical evolutionary tracks in Fig. 29.1, the faintest dwarfs in
the youngest clusters should still be in the pre-main-sequence contraction
phase, and hence should lie to the right of or above the initial main
sequence. That this is in fact the case is shown by the Hertzsprung-
Russell diagram of the very young cluster NGC 2264 shown in the upper
graph of Fig. 1.3.

In summary, it appears that the observed Hertzsprung-Russell diagrams
of clusters have provided a significant and positive check on the theory
of stellar evolution, at least as far as the initial main-sequence state
and the subsequent early evolutionary phases are concerned. This

positive outcome of our key test is worth great emphasis. But equal emphasis needs to be put on our present bad lack of understanding of the late evolutionary phases of a star which, though comprising only a small fraction of the total active stellar life, probably include all the more sparkling events, particularly those in which the heavy elements are formed.

Age and Composition of Clusters

The comparison of the time lines derived from theory and the observed Hertzsprung-Russell diagrams of clusters not only proves valuable as a check on the theory but also provides two sets of fundamental data, the first regarding the ages of the clusters and the second regarding the difference in initial composition between the stellar populations.

The method of determining the age of a cluster follows directly from our preceding discussion of the theoretical time lines. All that is necessary to do is to fit the observed Hertzsprung-Russell diagram of a cluster into a family of theoretical time lines such as that of Fig. 29.2, determine by interpolation that time line which fits the observed diagram, and read off the age corresponding to this time line.

If one applies this method to typical galactic clusters one finds, for example, for the age of the Pleiades about 100 million years and of the Hyades as well as Praesepe about one billion years. With very few exceptions, such as M67 (shown in the lower graph of Fig. 1.3), the galactic clusters are found to have ages less than two billion years; some galactic clusters with extremely bright stars, such as NGC 2264 (shown in the upper graph of Fig. 1.3), turn out to be even younger than 10 million years. This short time scale for the very youngest clusters is in agreement with the ages determined by the observed expansion rate for some of these clusters.

On the other hand, when one fits the Hertzsprung-Russell diagram of a globular cluster such as M3 (shown in Fig. 1.4) into a family of theoretical time lines—the most relevant of which is shown in Fig. 29.3—one finds an age of approximately 6 billion years, even a little longer than the 4.5 billion years estimated for the sun's age from geophysical evidence. This age determination for the globular clusters, however, must still be considered with appreciable caution; it may well be wrong by 30 percent. This distressing uncertainty is caused in part by the present uncertainty of the transformation from apparent to absolute magnitude of the observed Hertzsprung-Russell diagrams of globular clusters and in part by the present lack of accurate evolutionary model sequences for a variety of stellar masses and initial compositions.

But in spite of these inaccuracies in the present age determinations for clusters it now seems certain that the majority of the galactic clusters, representing the young Population I stars, are less than one billion years old, while the globular clusters, characteristic for the extreme

Population II stars, are more than three billion years old. This approximate quantitative determination of the age difference between the extreme populations is perhaps the strongest argument for interpreting the sequence of stellar populations represented in Table 3.1 as an age sequence.

In addition to the ages of the clusters one other datum can be derived from the theoretical evolutionary tracks, namely a confirmation of the difference in initial composition between the stellar populations which previously had been derived from spectroscopic observations. The difference in the theoretical evolutionary tracks for medium-weight stars of different heavy-element abundances is shown in Fig. 24.3. If one compares this graph (with its decrease in effective temperature of the red-giant branch with increasing metal content) with the Hertzsprung-Russell diagrams for the old galactic cluster M67 and the typical globular cluster M3 (Fig. 1.5), one finds the spectroscopic observation confirmed that Population II has a low heavy-element abundance compared with Population I. This qualitative confirmation of the composition difference between stellar populations should be fairly secure since it is based on differential effects, even though the theoretical evolutionary tracks for the red-giant phases used here are much less certain than those for the early evolutionary phases used in the age determinations.

The determination of the ages of stellar clusters and the confirmation of the composition differences between stellar populations appear to be the first fruits of the theory of stellar evolution which provide some of the basic data on which a theory of the evolution of our galaxy as a whole may be built.

30. Vital Statistics of the Stars

In the preceding section we have applied the theory of stellar evolution to individual stars and special families of stars as represented by stellar clusters. In this final section we shall apply the theory to the statistics of the entire assembly of stars in the solar neighborhood and to some extent even to the whole galaxy.

Clearly the theory of stellar evolution is as yet too inaccurate and too incomplete to give secure and definite answers to most of the central questions in this wide application. Accordingly the main point of this section is not so much to describe the present tentative answers to these questions as to bring out the main problems of the vital statistics of the stars for which the theory of stellar evolution may provide definite answers in the forseeable future.

Before we start with these theoretical problems let us review in brief the statistical data derived from observations of the stars in our immediate galactic surroundings.

Galactic Densities and Luminosity Functions

Table 30.1 gives a very condensed and approximate summary of the statistical results obtained from the observations of the stars in the solar neighborhood. This table lists the major types of objects in the solar neighborhood, including the interstellar gas, and gives for each type an estimated mass density, in units of solar masses per cubic parsec. The main-sequence stars are here divided into two types, the bright ones and the faint ones, with the division line placed at approximately that luminosity which divides those with lifetimes shorter than 5 billion years from those with lifetimes longer than this value.

Table 30.1 shows that the bright stars, be they main-sequence stars or red giants, contribute little to the total mass density, that the interstellar gas represents about 20 percent of the total mass, and that the main contribution, approximately 70 percent, rests in the most unstirring stellar type, the faint main-sequence stars which evolve so slowly that little has happened to them since they were born.

Even for our rough survey purposes the summary of Table 30.1 is not quite sufficient since it groups together in each type stars of widely varying luminosities, that is stars with great differences in speed of evolution and hence in vital statistics. More detailed information regarding the frequency distribution of stellar luminosities is given by the luminosity function, which represents the number of stars per cubic parsec in the solar neighborhood per magnitude interval as a function of the visual absolute magnitude. The choice of the visual absolute magnitude as the independent variable for the luminosity function is purely a matter of convenience in handling the observational data. Theoretically the basic quantity for which we should investigate the frequency distribution is the stellar mass. The transformation from the visual absolute magnitude to the mass is an unnecessary complication, however, since these two quantities are uniquely related to each other, at least for the bulk of the stars, those of the main sequence.

TABLE 30.1

Approximate mass densities in the solar neighborhood. (Lectures by J. Oort in Princeton, 1952, unpublished.)

Objects	Density ($M_\odot$/cu. ps.)
Interstellar gas	0.012
Bright main-sequ. stars ($M_{vis} < +3$)	0.002
Faint main-sequ. stars ($M_{vis} > +3$)	0.04
Red giants	0.001
White dwarfs	0.005:
Total	0.06

Table 30.2 gives not only the luminosity function $\phi_{Tot.}$ for all stars, but also the luminosity functions ϕ_{MS} and ϕ_{RG} for the main-sequence

TABLE 30.2

Luminosity functions. (van Rhijn, Groningen Pub., No. 47, 1936; Trumpler and Weaver, *Statistical Astronomy*, p. 409, Univ. of California Press, 1953.)

M_{vis}	$\log \phi_{Tot}$	$\dfrac{\phi_{MS}}{\phi_{Tot}}$	$\log \phi_{MS}$	$\log \phi_{RG}$
−6	1.63-10	0.45	1.28-10	1.37-10
−5	2.77	0.50	2.47	2.47
−4	3.58	0.55	3.32	3.23
−3	4.12	0.60	3.90	3.72
−2	4.71	0.65	4.52	4.25
−1	5.32	0.60	5.10	4.92
0	5.98	0.50	5.68	5.68
+1	6.59	0.50	6.29	6.29
+2	6.71	0.80	6.61	6.01
+3	6.98	0.90	6.93	5.98
+4	7.29	1.00	7.29	—
+5	7.40	1.00	7.40	—
+6	7.45	1.00	7.45	—

stars and for the red giants respectively. The division of all the stars into main-sequence stars and red giants can be accomplished for the brighter magnitudes in a fairly well-defined manner by the use of the Hertzsprung gap, which separates the red giants from the upper main sequence. In the F stars, however, where the subgiants merge with the main sequence, the division is somewhat arbitrary, and in consequence the two separate luminosity functions are slightly less well defined in this region.

As a first preliminary application of the luminosity functions let us compute the total luminosity emitted by all the main-sequence stars, as well as that emitted by all the red giants, in one cubic parsec. To derive these total luminosities we multiply the respective luminosity functions by the luminosity (in units of the solar luminosity) and integrate over the visual absolute magnitude. We thus obtain

$$\int_{-\infty}^{+\infty} \phi_{MS} \cdot \frac{L}{L_{\odot}} \cdot dM_{vis} = 0.13 \, \frac{L_{\odot}}{\text{cu. ps.}}, \qquad (30.1)$$

$$\int_{-\infty}^{+\infty} \phi_{RG} \cdot \frac{L}{L_{\odot}} \cdot dM_{vis} = 0.017 \, \frac{L_{\odot}}{\text{cu. ps.}}. \qquad (30.2)$$

To see which stars contribute most to these total luminosities the integrands of Eqs.(30.1) and (30.2) are shown in Fig. 30.1.

The left-hand graph of Fig. 30.1 shows the situation for the main-sequence stars. Here the major portion of the total luminosity is provided by the B and O stars. Unfortunately the size of this major contribution is very uncertain owing to the great uncertainty in the bolometric correction. This correction is needed for the transformation from the observed visual luminosities to the over-all bolometric luminosities, which for the hot O and B stars fall largely in the unobserved ultra-violet. Accordingly, the bolometric corrections which are listed in Table 30.3 for the main-sequence stars are very uncertain at the brightset luminosities and the curve in Fig. 30.1 is rather a pure extrapolation for visual absolute magnitudes brighter than -2.

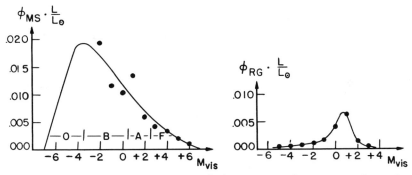

Fig. 30.1. Total luminosity emitted from one cubic parsec by main-sequence stars (left) and by red giants (right).

The right-hand graph of Fig. 30.1 shows a simple situation for the red giants: the ordinary red giants with visual absolute magnitudes around $+1$ provide most of the total luminosity, but also the supergiants (nega-tive visual absolute magnitudes) contribute a substantial fraction. For the red giants the bolometric correction does not present a serious problem. Since the majority of them are of spectral types G and K the constant bolometric correction of -0.2, which was used for the computa-tion of Eq. (30.2), should be sufficiently accurate for our purposes.

In spite of the serious uncertainty in the main-sequence luminosity function, two conclusions seem safe. First, stars as faint as or fainter than the sun contribute virtually nothing to the total luminosity emitted per cubic parsec; their very large number does not make up for their very low luminosity. Second, the main-sequence stars emit a total luminosity exceeding that emitted by the red giants by a factor of at least 5.

These conclusions have a direct bearing on the statistics of stellar evolution since the total luminosity is a direct measure of the total rate of nuclear fuel consumption. It thus follows, first, that the amount of nuclear fuel burned by the stars less massive than the sun is negligible compared to that burned by the stars heavier than the sun, and second, that the heavier stars burn much more fuel in their main-sequence phase than in their red-giant phase. This second deduction lends further weight to the suspicion that the heavier stars spend most of their active life in the main-sequence phases and a much shorter time in the subse-quent red-giant phases, a suspicion which we have discussed already in the preceding section.

Birth-Rate Function

As our second and principal application of the luminosity function let us derive the central item of the vital statistics of the stars, their birth rate. Let us define a birth-rate function ψ as the number of star births per cubic parsec per year per magnitude interval as a function of the visual absolute magnitude. To derive this function we have to consider separately the faint and the bright main-sequence stars. For the faint stars the time spent in the main-sequence phases τ_{MS}, will be longer than the relevant time of galactic development, say 5 billion years. Accordingly, their average birth rate during this interval can be computed by simply dividing the present number of main-sequence stars, ϕ_{MS}, by 5 billion years. For the bright stars this procedure would give erroneous results because a majority of the bright stars born during the past 5 billion years will have passed the main sequence phases in their evolution long ago and will by now be white dwarfs. All the bright stars now on the main sequence must have been born in a time interval equal to their life time on the main sequence, τ_{MS}. Accordingly, to obtain their birth rate one has to divide their present number, ϕ_{MS}, not by 5 billion years but by their main-sequence lifetime. Thus the birth-rate function for stars of all brightnesses can be derived by the formulae

$$\psi = \frac{\phi_{MS}}{5 \times 10^9} \text{ if } \tau_{MS} > 5 \times 10^9 \text{ yrs.,}$$

$$\psi = \frac{\phi_{MS}}{\tau_{MS}} \quad \text{ if } \tau_{MS} < 5 \times 10^9 \text{ yrs.}$$

(30.3)

Eqs. (30.3) give the birth rate as a time average, over a time interval that is short and recent for the brightest stars but very long for the fainter stars. Since the birth rate may well have varied systematically during this long time interval, we shall have to be much more cautious in the use of the birth-rate function for the fainter stars than in the use of that for the brighter stars.

Before we can apply the second of Eqs. (30.3) we have to determine the main-sequence lifetimes for the bright stars. During the main-sequence phases a star burns a certain fraction ΔX_{MS} of its hydrogen content. Multiplying this fraction by the total mass of the star gives the mass of hydrogen burnt. Multiplying this mass by the transmutation factor E^*_{cc} defined in Eq. (10.21) gives the total energy released by the hydrogen burning, and, finally, dividing by the luminosity gives the time the star spends on the main sequence:

$$\tau_{MS} = \frac{\overline{\Delta X}_{MS} \cdot M \cdot E^*_{cc}}{L} .$$

(30.4)

According to the computations of §22, the fraction of hydrogen burned during the main-sequence phases is about 13 percent, i.e. $\Delta \overline{X}_{MS} = 0.13$, and this value appears to be fairly constant over a rather wide range of stellar masses. With this numerical value Eq.(30.4) takes the convenient form

$$\log \tau_{MS} \text{ (yrs).} = 10.11 + \log \frac{M}{M_\odot} - \log \frac{L}{L_\odot} . \qquad (30.5)$$

If we introduce into this equation the masses and luminosities of main-sequence stars listed in Table 30.3 we obtain the lifetimes given in Table 30.4.

We now have all the data necessary to compute the complete birth-rate function with the help of Eqs.(30.3). The results of the computations are given in the last column of Table 30.4.

The birth-rate function thus derived is shown in Fig. 30.2, in comparison with the observed main-sequence luminosity function. The slope

TABLE 30.3

Luminosities and masses for main-sequence stars. (Spectral types from Trumpler and Weaver, *Statistical Astronomy*, p. 359 and p. 409; colors from Johnson and Morgan, *Ap.J. 117*, p. 313, 1953; bolometric corrections from Table 1.1; masses from Fig. 2.1.)

M_{vis}	Sp. T.	B−V	B. C.	M_{bol}	$\log \frac{L}{L_\odot}$	$\log \frac{M}{M_\odot}$
−2	B2	−0.24	−2.8:	−4.8:	+3.77:	+1.26
−1	B5	−0.16	−1.8	−2.8	+2.97	+0.86
0	B8	−0.09	−1.19	−1.2	+2.33	+0.60
+1	A0	0.00	−0.72	+0.3	+1.84	+0.40
+2	A5	+0.15	−0.29	+1.7	+1.17	+0.24
+3	F1	+0.33	−0.07	+2.9	+0.69	+0.14
+4	F7	+0.50	0.00	+4.0	+0.25	+0.04
+5	G2	+0.63	−0.05	+4.9	−0.11	−0.04
+6	K0	+0.82	−0.20	+5.8	−0.47	−0.10

TABLE 30.4

Lifetime of main-sequence stars, τ_{MS} (in yrs.) and the birth rate function ψ (in no. of births per cubic parsec per year).

M_{vis}	Sp. T.	$\log \tau_{MS}$	$\log \psi$
−2	B2	7.60	6.92-20
−1	B5	8.00	7.10
0	B8	8.38	7.30
+1	A0	8.67	7.62
+2	A5	9.18	7.43
+3	F1	9.56	7.37
+4	F7	[9.90]	7.59
+5	G2	[10.18]	7.70
+6	K0	[10.48]	7.75

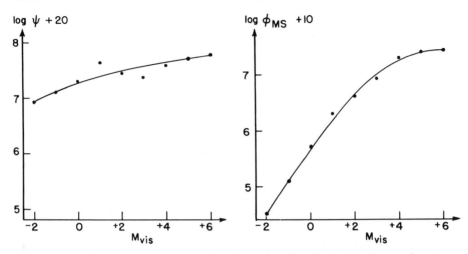

Fig. 30.2. The birth-rate function (left) compared with the luminosity function for the main-sequence stars (right).

of the birth-rate function is surprisingly shallow; relatively faint stars like the sun ($M_{vis} \approx +5$) have a birth rate only about four times higher than bright stars like typical B stars ($M_{vis} \approx -1$). In contrast, the observed luminosity function has a very steep slope for stars brighter than the sun. We can now explain this observed scarcity of the brightest stars, not so much by a low birth rate on their part, but rather by the very short time which they spend in the main-sequence phases, which automatically reduces the number of bright stars existing at any one time.

The derivation of the birth-rate function is based clearly on a highly simplified picture of the statistics of stellar origin and evolution. Under the circumstances it is particularly valuable that a check on the birth-rate function is provided by the stellar clusters—which have given us again and again major clues and checks in these investigations. If we consider a very young galactic cluster, and if we assume that the star-forming process in the cluster is similar to that in the general neighborhood of the sun, then we should expect the present frequency distribution in visual absolute magnitudes of the cluster stars to be directly proportional to the birth-rate function, as long as we exclude from our considerations those extremely bright stars which may have left the main-sequence phases by now in spite of the short age of a young cluster. The available data for the luminosity functions in young clusters are still rather limited, but as far as they go they indicate a frequency distribution of stellar magnitudes in these clusters which is satisfactorily close to the derived birth-rate function and has definitely a much shallower slope than the observed luminosity function of the solar neighborhood. Thus the first check on the attempts of deriving the vital statistics for the stars has turned out positive.

With the help of the birth-rate function we can immediately gain an approximate answer to the question: what fraction of the interstellar gas that goes into star formation is forming bright stars (brighter than, say, $+3$ in visual absolute magnitude). If we multiply the birth-rate function of Table 30.4 by the masses of Table 30.3 and integrate over all visual absolute magnitudes brighter than $+3$ we obtain

$$\int_{-\infty}^{+3} \psi \cdot \frac{M}{M_\odot} \cdot dM \bigg|_{vis} = 9.10^{-12} \frac{M_\odot}{cu.\ ps.\ yr.}\ . \tag{30.6}$$

Furthermore, if we multiply this rate of mass consumption for the formation of bright stars by the relevant time interval we find

$$0.05 \frac{M_\odot}{cu.\ ps.}\ \text{in } 5 \times 10^9 \text{ yrs.} \tag{30.7}$$

This value represents the total mass per cubic parsec which has gone into the formation of bright stars during the past 5 billion years. This value is much larger than the value given in Table 30.1 for the presently existing bright main-sequence stars, as was to be expected since the vast majority of the bright stars formed during the past 5 billion years have long ago passed beyond the main-sequence phases.

On the other hand, the value we have just found for the mass that has gone into the formation of bright stars is just about equal to the value given in Table 30.1 for the mass of the present faint main-sequence stars. This latter value must of course be equal to the total mass that has gone into the formation of faint stars all through the past 5 billion years, since practically none of these faint stars can have evolved beyond the main-sequence state. Even though the close coincidence of these two numerical values is purely fortuitous—both of them being very uncertain, especially the one given by Eq.(30.7)—we may conclude that by order of magnitude an equal amount of interstellar matter goes, in a given time interval and space volume, into the formation of bright stars as goes into the formation of faint stars. We shall make use of this deduction in the last part of this section.

We must not omit reminding ourselves of one serious limitation in the whole preceding discussion. All the observed statistical data which we have used refer to the solar neighborhood. This solar neighborhood is at best representative of one specific short portion of one specific spiral arm in our galaxy, and can thus hardly be taken as representative of the whole galaxy. In particular, the vast majority of stars in the solar neighborhood belong to Population I, and our results therefore refer to the vital statistics of this population only. On the other hand, for the galaxy as a whole Population II most likely represents a substantial fraction of the total mass, and the vital statistics of this population, of which thus far we know very little, must have a decisive influence on the over-all galactic development. Specifically, the question whether the result we have just derived for Population I (that is, that a large fraction of the mass goes into the formation of the brighter stars) held also for Population II during its formation time is a major unsolved problem at present and a serious stumbling block for the theory of the over-all galactic evolution.

Death Rate

For the brighter stars, which have life times shorter than say 5 billion years we may expect the average death rate to be in approximate balance with the birth rate. Consequently we can compute the total number of star deaths per cubic parsec per year by simply integrating the birth-rate function of all the brighter stars down to that value of the visual absolute magnitude at which the total stellar lifetime is approximately equal to 5 billion years.

Earlier in this section we have found arguments indicating that the brighter stars spend most of their active life in the main-sequence phases. Consequently we may assume that the total lifetime of these stars will exceed by only a little the lifetime for the main-sequence phases given in Table 30.4. Accordingly, we find from the table that to obtain the total death rate we should integrate the birth-rate function down to a visual absolute magnitude of about $+3$. Thus we find

$$\int_{-\infty}^{+3} \psi \cdot dM_{vis} = 1.3 \times 10^{-12} \frac{\text{deaths}}{\text{cu. ps. yr.}} . \tag{30.8}$$

If we multiply this death rate by 5 billion years we obtain for the total number of stars that died in the past

$$0.006 \frac{\text{deaths}}{\text{cu. ps.}} \text{ in } 5 \times 10^{9} \text{ yrs.} \tag{30.9}$$

Obviously we should compare this total number of star deaths with the number of observed white dwarfs, which we feel sure represent the very stars that died. The observational determination of the number of white dwarfs per cubic parsec is as yet difficult and uncertain. Present estimates give approximately

$$0.005 \frac{\text{white dwarfs}}{\text{cu. ps.}} . \tag{30.10}$$

The close coincidence of this number with that for the past star deaths given in Eq. (30.9) is entirely accidental; each of the two numbers may easily be wrong by a factor 2 and possibly by a rather larger factor. It is, nevertheless, most encouraging to find that our statistical data do not at all disagree with the hypothesis that the white dwarfs now existing might all be the consequence of previous star deaths.

Again we should remind ourselves that our vital statistics thus far all refer to Population I stars, but that their uncertainties are so great that it is far from excluded that a substantial fraction of the observed white dwarfs have come into being by deaths of Population II stars.

We should discuss one more phenomenon that is thought to be associated with star deaths, the supernovae. It is estimated that in a galaxy supernovae occur at a rate of 1 in approximately 500 years. If we assume that this rate is representative for the past 5 billion years, and if we reduce the rate per galaxy into a rate per cubic parsec by using the ratio of the mass of the galaxy to that of a cubic parsec in the solar neighborhood (which is approximately 10^{12}), we find for the total number of past supernovae

$$0.00001 \frac{\text{supernovae}}{\text{cu. ps.}} \quad \text{in } 5 \times 10^9 \text{ yrs.} \tag{30.11}$$

If we compare this number with the total number of star deaths given in Eq.(30.9) we have to conclude that the vast majority of star deaths do not lead to the supernova phenomenon, but on the contrary that at the very most 1 percent of all stars go through the supernova process at the end of their active life. We do not know as yet whether in order to become a supernova the star must fulfill very narrow conditions regarding mass and composition, or whether the selection occurs according to a condition on a secondary characteristic, such as unusually low rotational momentum.

Irreversible Developments

Let us close this book by considering the problem of the over-all long-range development of matter in our galaxy—or, more precisely and more modestly, of the matter in the solar neighborhood. Specifically, two processes in this development appear of cosmological interest because of their one-way character: the transformation of interstellar matter into white dwarfs and the transmutation of hydrogen into the heavy elements. It must be clear from the preceding discussions that our knowledge of the evolution and vital statistics of the stars is far from sufficient to deduce a secure picture of these one-way processes. We can, however, draw a tentative picture of them which appears to fit the available theoretical and observational data and which, though far from certain, may be useful as a working hypothesis for future investigations. This picture runs as follows.

Let us suppose that the matter of the solar neighborhood started in the form of interstellar matter. Then stars condensed out of it, first the Population II stars and then the Population I stars. The rate of star formation may possibly have been very high during the initial phases of galactic development, but then presumably settled down to values approximating the birth-rate function given in Table 30.4.

According to the third line of Table 30.1 and Eq.(30.7) about half of the condensing mass goes into the formation of bright stars with relatively short lifetimes, and the other half into the formation of faint stars with long lives. A bright star spends most of its active life in the main-sequence phases, subsequently spends a limited time in the red-giant phases, and may then terminate its active life by ejecting a large portion of its mass (possibly as much as 85 percent) while the rest settles into a white dwarf. Thus most of the mass condensing into bright stars is again and again returned to the interstellar gas; only about 10 percent of all the mass in the solar neighborhood has as yet been permanently bound by transformation into white dwarfs at the death of bright stars (according to the fifth line of Table 30.1).

The matter that has at any time during galactic life condensed into faint stars still exists largely in the form of faint main-sequence stars because of their long main-sequence lifetimes. According to the third line of Table 30.1 this matter amounts to about two thirds of the total mass in the solar neighborhood. It is not known whether these faint stars, upon termination of their active life, will eject any substantial fraction of their mass. It seems plausible, however, that this fraction will not exceed, say, one third, since even without any ejection the masses of these faint stars fall below the limit permissible for white dwarfs. If the last assumption is right, we can conclude that about 50 percent of the mass of the solar neighborhood is by now permanently bound within stars (roughly 40 percent in the faint main-sequence stars and 10 percent in the white dwarfs) while in the interstellar gas, according to the first line of Table 30.1, there is left at present only about 20 percent of the total mass. This latter percentage may, however, be supplemented in the future by another 30 percent through ejection from dying faint stars.

Thus it seems likely that at the present time the permanant transformation of the interstellar matter into stars is not far from having reached the halfway mark.

How does it stand with the second one-way process, the transmutation of hydrogen into the heavy elements? Here the principal questions are: first, can we uphold the assumption that the galaxy initially had a composition of pure hydrogen, and second, how far has the transmutation into heavier elements proceeded by now?

The first step of the nuclear transmutations, that from hydrogen to helium, is the main source of stellar energy. Its over-all rate in the solar neighborhood can therefore be estimated by dividing the total luminosity of one cubic parsec, as given by the sum of Eqs.(30.1) and (30.2), by the total mass of the same volume, as given by the last line of Table 30.1. This ratio turns out to be about 2.5 times larger than the corresponding ratio for the sun. Since the sun has transformed approximately 5 percent of its mass into helium during the past 5 billion years we may conclude that all the stars in the solar neighborhood together have produced a 12 percent helium abundance during the past 5 billion years.

This amount of helium is sufficient to supply the material necessary for the formation of the existing white dwarfs. It is not sufficient, however, to account for the helium content of Population I stars such as the sun. Nor should it be, since this helium (amounting to roughly another 12 percent of the galactic mass) must have existed in the interstellar matter at the time of the very beginning of the formation of Population I stars. Thus, if we want to hold on to the assumption of a pure hydrogen composition of the initial galaxy, we have to burden the early born, fast living, bright Population II stars (whose number appears completely unknown) with the task of producing and ejecting all the helium now found in Population I, a heavy but not impossible task.

The second step of the nuclear transmutations, that from helium to the intermediate elements such as oxygen, does not appear to present any over-all problems different from those of the first step. We have seen that the bright stars, soon after having transformed all the hydrogen into helium in their cores, will raise their central temperatures by contraction to those values necessary for the transmutation of helium into the intermediate elements. Thus whenever helium is produced and ejected during the evolution of a bright star, the same is likely to be true for the intermediate elements to a lesser but sufficient degree.

The last step in the nuclear transmutations, that from the intermediate elements to the heavy ones such as iron, poses a new problem. This step requires such high temperatures that it appears unlikely that they are reached anywhere except in the special evolutionary phases just preceding or during the supernova phenomenon. If the present supernova rate given in Eq.(30.11) had had to provide the total iron abundance in Population I stars, which according to Table 4.1 amounts to approximately 0.1 percent, each supernova would have had to eject six solar masses of pure iron—an improbably large amount. In reality, however, this requirement does not exist since the heavy elements now contained in Population I stars must have been produced prior to the beginning of the formation of these stars. Thus we must burden the bright stars of Population II evolving early in galactic history also with this task, heavy again but not impossible.

Our present tentative answers to the principal questions regarding galactic composition may then be these. We may retain the assumption of an initial galactic composition of pure hydrogen; nothing can contradict this assumption until we gain some knowledge about the number and evolution of the bright stars early in galactic history. And regarding the present degree of transmutation into the heavier elements, some 25 percent of the whole galactic mass appears to have made at least the first step to helium.

VIII. SUMMARY AND REVIEW

Do we then live in an adult galaxy which has already bound half of its mass permanently into stars, and has consumed a fourth of all its fuel? Are we too late to witness the turbulent sparkle of galactic youth but still in time to watch stars in all their evolutionary phases before they settle into the permanence of the white dwarfs?

REFERENCES

(Note: *Ap.J.* = *Astrophysical Journal*; *A.J.* = *Astronomical Journal*; *M.N.* = *Monthly Notices of the Royal Astronomical Society*)

INTRODUCTION

Basic early text books
A. S. Eddington, *The Internal Constitution of the Stars*, Cambridge University Press, 1930. Classical Description of stars in radiative equilibrium. (Numerical results still affected by neglect of hydrogen and helium abundance. Nuclear energy sources not yet identified.)
S. Chandrasekhar, *Stellar Structure*, University of Chicago Press, Chicago, 1938. Exhaustive theory of homogeneous stellar models, including theory of white dwarfs. (First identifications of nuclear energy sources reviewed.)

Recent texts and summaries
W. H. McCrea, *Physics of the Sun and Stars* (Chapters 7 and 9), Hutchinson's University Library, London, 1950.
Otto Struve, *Stellar Evolution*, Princeton University Press, Princeton, 1950.
S. Chandrasekhar, "The Structure, the Composition, and the Source of Energy of the Stars," Chapter 14 of *Astrophysics*, ed. J. A. Hynek, McGraw-Hill, New York, 1951.
B. Strömgren, "The Sun as a Star," Chapter 2 of *The Sun*, ed. G. P. Kuiper, University of Chicago Press, Chicago, 1952.
B. Strömgren, "Evolution of Stars," *A.J.* 57, 65, 1952
L. H. Aller, *Astrophysics, Nuclear Transformations, Stellar Interiors, and Nebulae* (Chapters 1 and 2), Ronald Press, New York, 1954

Recent popular books
C. Payne-Gaposchkin, *Stars in the Making*, Harvard University Press, Cambridge, 1952
Fred Hoyle, *Frontiers of Astronomy*, Harper, New York, 1955

Texts on major topics omitted in this book
S. Rosseland, *Pulsation Theory of Variable Stars*, Clarendon Press, Oxford, 1949.
O. Struve, *Stellar Evolution* (Chapter III, "Origin and Development of Close Double Stars"), Princeton University Press, 1950
E. M. Burbidge, G. R. Burbidge, W. A. Fowler, F. Hoyle, "Synthesis of the Elements in Stars," *Review of Modern Physics*, in press, 1957

§ 1. LUMINOSITIES AND RADII

The Hertzsprung-Russell diagram
Recent photoelectric observations of H-R diagrams of clusters:
H. L. Johnson, A. R. Sandage, *Ap.J.* 124, 397, 1956
O. Heckmann, H. L. Johnson, *Ap.J.* 124, 477, 1956
M. F. Walker, *Ap.J.* 125, 636, 1957
For further references see A. R. Sandage, *Ap.J.* 125, 435, 1957.

Transformation to luminosities and radii
Determination of initial main sequence:
H. L. Johnson, W. A. Hiltner, *Ap.J.* 123, 267, 1956

A. R. Sandage, *Ap.J. 125*, 435, 1957
Bolometric corrections and effective temperatures:
G. Kuiper, *Ap.J. 88*, 429, 1938
P. C. Keenan, W. W. Morgan, Chapter 1 of *Astrophysics*, ed. J.A. Hynek, McGraw-Hill, New York, 1951

Subdwarfs and white dwarfs
Colors and magnitudes of subdwarfs:
N. G. Roman, A.J. *59*, 307, 1954
M. Schwarzschild, L. Searle, R. Howard, *Ap.J. 122*, 353, 1955
Colors and magnitudes of white dwarfs:
D. L. Harris III, *Ap.J. 124*, 665, 1956

§ 2. MASSES

Spectroscopic binaries
Recent summary:
Z. Kopal, *Transactions of the International Astronomical Union*, Vol. IX, in press, 1955

Visual binaries
Recent summaries:
P. van de Kamp, *A.J. 59*, 447, 1954
K. Aa. Strand, R. G. Hall, *Ap.J. 120*, 322, 1954

Mass-luminosity relation
Earlier summaries:
A. S. Eddington, *Internal Constitution of the Stars* (Chapter VII), Cambridge University Press, 1930
G. P. Kuiper, *Ap.J. 88*, 472, 1938
H. N. Russell, C. E. Moore, *Masses of the Stars*, University of Chicago Press, Chicago, 1939
R. M. Petrie, *Publications of the Dominion Astrophysical Observatory*, Vol. 8, 341, 1950

Masses of giants and white dwarfs
Major correction of classical value of mass of Capella:
O. Struve, S. M. Kung, *Ap.J. 117*, 1, 1953
For further references see Table 2.3.

§3. STELLAR POPULATIONS

Brightest stars in stellar systems
Introduction of concept of stellar populations:
W. Baade, *Ap.J. 100*, 137, 1944
Expanding associations:
J. Delhaye, A. Blaauw, *Bulletin of the Astronomical Institutes of the Netherlands 12*, No. 448, p. 72, 1953
A. Blaauw, *Ap.J. 123*, 408, 1956

Kinematical behavior
Original discussion of high-velocity stars:
J. Oort, *Publications of the Kapteyn Astronomical Laboratory at Groningen*, No. 40, 1926
Recent summaries of high-velocity stars:
G. Miczaika, *Astronomische Nachrichten 270*, 249, 1940
N. G. Roman, *Ap.J.*, Supplement No. 18, 1955
For velocity dispersions of various star types see:
L. Spitzer, Jr., M. Schwarzschild, *Ap.J. 114*, 385, 1951 (see especially Table 4)
Maintenance of turbulence in interstellar gas:
J. H. Oort, L. Spitzer, Jr., *Ap.J. 121*, 6, 1955
Stellar encounters with cloud complexes:

L. Spitzer, Jr., M. Schwarzschild, *Ap.J. 114*, 385, 1951
L. Spitzer, Jr., M. Schwarzschild, *Ap.J. 118*, 106, 1953

Spectroscopic differences
Differences found by spectral classification:
W. W. Morgan, P. C. Keenan, E. Kellman, *An Atlas of Stellar Spectra*, University
of Chicago Press, p. 35, 1943
P. C. Keenan, G. Keller, *Ap.J. 117*, 241, 1953
N. G. Roman, *Ap.J. 112*, 554, 1950
N. G. Roman, *Ap.J. 116*, 122, 1952
A. N. Vyssotsky, A. Skumanitch, *A.J. 58*, 96, 1953
Differences found by spectrophotometry:
D. Chalonge, *Symposium on Stellar Populations*, Vatican Academy, Rome, 1957
(in press)
B. Strömgren, *Symposium on Stellar Populations*, Vatican Academy, Rome, 1957
(in press)

§4. ABUNDANCES OF THE ELEMENTS

Composition of sun
For summary see:
L. H. Aller, *Astrophysics: The Atmospheres of the Sun and Stars*, (p. 327),
Ronald Press, New York, 1953
Recent discussion (too recent for use in text):
L. Goldberg, E. A. Muller, L. H. Aller, *A.J. 62*, 15, 1957

Composition differences between populations
Differences between high- and low-velocity stars:
W. Iwanowska, *Bulletin of the Astronomical Observatory of N. Copernicus University in Torun*, No. 9, p. 25, 1950
M. and B. Schwarzschild, *Ap.J. 112*, 248, 1950
M. Schwarzschild, L. Spitzer, Jr., R. Wildt, *Ap.J. 114*, 398, 1951
G. and E. Burbidge, *Ap.J. 124*, 116, 1956
L. Gratton, *Mémoires de la société royale des sciences de Liège*, 4th Ser., 14,
p. 419, 1954
M. and B. Schwarzschild, L. Searle, A. Meltzer, *Ap.J. 125*, 123, 1957
Differences between subdwarfs and ordinary dwarfs:
J. W. Chamberlain, L. H. Aller, *Ap.J. 114*, 52, 1951

§5. HYDROSTATIC AND THERMAL EQUILIBRIUM

See for example:
A. S. Eddington, *The Internal Constitution of the Stars*, Cambridge University
Press, 1930 (Chapter IV)

§6. RADIATIVE ENERGY TRANSPORT

See for example:
A. S. Eddington, *The Internal Constitution of the Stars*, Cambridge University
Press, 1930 (Chapter V)
S. Chandrasekhar, *Stellar Structure*, University of Chicago Press, 1938 (Chapter
V)

§7. CONVECTIVE ENERGY TRANSPORT

See for example:
L. Biermann, *Ergebnisse der exakten Naturwissenschaften 21*, 1, 1945

§8. EQUATION OF STATE

See for example:
S. Chandrasekhar, *Stellar Structure*, University of Chicago Press, 1938 (pp. 254–
261 for non-degenerate state; Chapter X for degenerate state.)

§9. OPACITY

Early opacity computations:
B. Strömgren, *Zeitschrift für Astrophysik 4*, 118, 1932 and 7, 222, 1933
P. M. Morse, *Ap.J. 92*, 27, 1940
Recent extensive opacity table:
G. Keller, R. E. Meyerott, *Ap.J. 122*, 32, 1955
Electron conduction in degenerate state:
R. E. Marshak, *Annals of the New York Academy of Science 41*, 49, 1941
L. Mestel, *Proceedings of the Cambridge Philosophical Society 46*, 331, 1950
Interpolation formula for pure H-He mixture:
A. Reiz, *Ap.J. 120*, 342, 1954

§10. NUCLEAR REACTIONS

Recent summary (too recent to be used in text):
E. M. Burbidge, G. R. Burbidge, W. A. Fowler, F. Hoyle, "Synthesis of the Elements in Stars," *Review of Modern Physics*, in press, 1957
Early identifications of nuclear energy sources:
C. F. v. Weizsäcker, *Physikalische Zeitschrift 38*, 176, 1937
H. Bethe, C. L. Critchfield, *Physical Review 54*, 248, 1938
H. Bethe, *Physical Review 55*, 434, 1939
Proton-proton reaction:
E. E. Salpeter, *Physical Review 88*, 547, 1952
E. A. Frieman, L. Motz, *Physical Review 89*, 648, 1953
Carbon cycle:
W. A. Fowler, *Mémoires de la société royale des sciences de Liège,* 4th Ser., 14, p. 88, 1954
W. A. Lamb, R. E. Hester, *Bulletin of the American Physical Society 2*, 181, 1957 (too recent to be used in text; indicates ten times lower rate for carbon cycle than given in Table 10.1)
Triple-alpha process:
E. E. Salpeter, *Ap.J. 115*, 326, 1952
Processes involving D, Li, Be, and B:
E. E. Salpeter, *Physical Review 97*, 1237, 1955
Neutron production from C^{13}:
A. G. W. Cameron, *Stellar Evolution, Nuclear Astrophysics and Nucleogenesis*, p. 79, Chalk River, Ontario, 1957
Reaction rates at high densities:
E. Schatzman, *Journal de physique et le radium (8) 9*, 46, 1948
G. Keller, *Ap.J. 118*, 142, 1952
E. E. Salpeter, *Australian Journal of Physics 7*, 373, 1954

§11. SURFACE LAYERS

Early investigations of deep convective envelopes:
L. Biermann, *Astronomische Nachrichten 257*, 269, 1935
T. G. Cowling, *Astronomische Nachrichten 258*, 133, 1936
Convective envelope of sun:
E. Vitense, *Zeitschrift für Astrophysik 32*, 135, 1953
Convective envelope for faint dwarf:
D. E. Osterbrock, *Ap.J. 118*, 529, 1953
Convective envelope for red giant:
P. Dumezil-Curien, *Annales d'astrophysique 17*, 197, 1954
Boundary conditions for red giants:
F. Hoyle, M. Schwarzschild, *Ap.J.*, Supplement No. 13, Section X, 1955

§12. THE OVER-ALL PROBLEM

See for example:
H. Vogt, *Aufbau und Entwicklung der Sterne*, Akadem. Verlagsgesellschaft Becker und Erler, Leipzig, 1943

REFERENCES

§13. TRANSFORMATIONS AND INVARIANTS

See for example:
S. Chandrasekhar, *Stellar Structure*, University of Chicago Press, 1938 (Chapter IV)

§14. NUMERICAL INTEGRATIONS

For tables of numerical integrations and for further references see:
R. Härm, M. Schwarzschild, *Ap.J.*, Supplement No. 10, 1955

§15. THE UPPER MAIN SEQUENCE

Early investigations of models with convective cores:
T. G. Cowling, *M.N. 96*, 42 (Appendix), 1935
L. R. Henrich, *Ap.J. 98*, 192, 1943 (electron scattering)
R. E. Williamson, G. F. Duff, *M.N. 109*, 46 and 55, 1949, (modified Kramers' opacity)
Recent models for upper main sequence:
R. S. Kushwaha, *Ap.J. 125*, 242, 1957.

§ 16. THE LOWER MAIN SEQUENCE

Proton-proton reaction in sun:
J. B. Oke, *Journal of the Royal Astronomical Society of Canada 44*, 135, 1950
I. Epstein, *Ap.J. 112*, 207, 1950
Extent of convection in sun:
B. Strömgren, Mat. Tidsskr. B, *Festskr. t. J. Nielsen*, p. 96, Copenhagen, 1950
Models for red dwarfs:
D. E. Osterbrock, *Ap.J. 118*, 529, 1952
Model for initial sun:
M. Schwarzschild, R. Howard, R. Härm, *Ap.J. 125*, 233, 1957

§ 17. THE SUBDWARFS

Models with neglibible heavy element content:
A. Reiz, *Ap.J. 120*, 342, 1957

§ 18. THE APSIDAL MOTION TEST

Original derivation of apsidal motion test:
H. N. Russell, *M. N. 88*, 641, 1928
Theory of apsidal motion:
T. G. Cowling, *M.N. 98*, 734, 1938
T. E. Sterne, *M.N. 99*, 451, 1939
Observations of apsidal motion:
See references in heading of Table 18.1.
Recent application of test to upper main sequence :
R. S. Kushwaha, *Ap.J. 125*, 242, 1957

§ 19. PRE-MAIN-SEQUENCE CONTRACTION

Nuclear processes during pre-main-sequence contraction:
E. E. Salpeter, *Mémoires de la société royale des sciences de Liège*, 4th Ser., 14, p. 116, 1954
General investigation of contracting models:
L. H. Thomas, *M.N. 91*, 122 and 619, 1931
Model sequence derived with electronic computer:
L. G. Henyey, R. LeLevier, R. D. Levée, *Publications of the Astronomical Society of the Pacific*, Vol. 67, No. 396, 1955
Model for homologously contracting star:
R. D. Levée, *Ap.J. 117*, 200, 1953

§ 20. SIMPLIFIED EXAMPLE OF EVOLUTION

Early investigations of inhomogeneous models:
E. Öpik, *Publications de l'observatoire astronomique de l'universite de Tartu*, Vol. 30, No. 3, 1938; Vol. 30, No. 4, 1939; and Vol. 31, No. 1, 1943
F. Hoyle, R. A. Lyttleton, *Proceedings of the Cambridge Philosophical Society* 35, 592, 1939; *M.N. 102*, 218, 1942; and *M.N. 109*, 614, 1949
M. Schönberg, S. Chandrasekhar, *Ap.J.* 96, 161, 1942
Convective instability at composition discontinuity:
F. Hoyle, R. A. Lyttleton, *M.N. 106*, 525, 1946
E. Öpik, *M.N. 110*, 559, 1950 (see especially p. 592)

§ 21. ROTATIONAL MIXING

Original qualitative theory:
A. S. Eddington, *M.N. 90*, 54, 1929
Detailed quantitative theory:
P. A. Sweet, *M.N. 110*, 548, 1950
Effect of inhomogeneity in composition:
L. Mestel, *M.N. 113*, 716, 1953

§22. EVOLUTION OF UPPER MAIN-SEQUENCE STARS

Early derivation of evolutionary model sequence with inhomogeneous intermediate zone:
R. J. Taylor, *Ap.J. 120*, 332, 1954
Recent investigation of upper main-sequence evolution:
R. S. Kushwaha, *Ap.J. 125*, 242, 1957

§ 23. EVOLUTION OF THE SUN

Early inhomogeneous model for the sun:
P. Ledoux, *Annales d'astrophysique 11*, 174, 1948
Recent inhomogeneous models for the sun:
M. Schwarzschild, R. Howard, R. Härm, *Ap.J. 125*, 233, 1957
R. Weymann, *Ap.J.*, in press, 1957 (too recent to be used in text, but used for Table 28.6)
Evolutionary model sequence for a lower main-sequence star:
C. B. Haselgrove, F. Hoyle, *M.N. 116*, 527, 1956

§ 24. GROWTH OF ISOTHERMAL CORE

Early investigations of shell source models:
C. L. Critchfield, G. Gamow, *Ap.J. 89*, 244, 1939
L. R. Henrich, S. Chandrasekhar, *Ap.J. 94*, 525, 1941
M. Schönberg, S. Chandrasekhar, *Ap.J. 96*, 161, 1942
G. Gamow, *Physical Review 65*, 20, 1944; and *67*, 120, 1945
G. Gamow, G. Keller, *Review of Modern Physics 17*, 125, 1945
Recent investigations of evolutionary model sequences with isothermal cores:
F. Hoyle, M. Schwarzschild, *Ap.J.*, Supplement No. 13, 1955
C. B. Haselgrove, F. Hoyle, *M.N. 116*, 527, 1956
C. Hayashi, *Progress of Theoretical Physics 17*, 737, 1957

§ 25. HEATING OF CORE BY CONTRACTION

Core contraction in massive stars:
A. R. Sandage, M. Schwarzschild, *Ap.J. 116*, 463, 1952
Core contraction in medium-weight stars:
F. Hoyle, M. Schwarzschild, *Ap.J.*, Supplement No. 13, 1955

§ 26. STRUCTURE OF WHITE DWARFS

Basic theory:
S. Chandrasekhar, *Stellar Structure,* University of Chicago Press, 1938
     Chapter XI)
Corrections to the degenerate equation of state :
M. Rudkjobing, *Danske Videnskabernes Selskab, Mat.-Fys. Medd.* 27, No. 5, 1952
E. Schatzman, *White Dwarfs,* North-Holland Publishing Co., Amsterdam, 1957

§ 27. THERMODYNAMICS OF WHITE DWARFS

Early investigation of composition of white dwarfs:
R. E. Marshak, *Ap.J.* 92, 321, 1940
Recent papers on thermal state and composition of white dwarts:
S. A. Kaplan, *Astronomical Journal of the Soviet Union* 27, 31, 1950
E. Schatzman, *Annales d'astrophysique* 15, 126, 1952
L. Mestel, *M.N.* 112, 583, 1952
E. J. Öpik, *Mémoires de la société royale des sciences de Liège,* 4th Ser., 14,
     p. 131, 1954
Pulsational instability of white dwarfs:
P. J. Ledoux, E. Sauvenier-Goffin, *Ap.J.* 111, 611, 1950
Recent monograph:
E. Schatzman, *White Dwarfs,* North-Holland Publishing Co., Amsterdam, 1957

§ 28. PHYSICAL STATE OF THE STELLAR INTERIOR

For relevant references see preceding sections.

§ 29. EVOLUTIONARY TRACKS IN THE HERTZSPRUNG-RUSSELL DIAGRAM

Spectroscopic observations of mass ejection:
A. J. Deutsch, *Ap.J.* 123, 210, 1956
Investigations of stellar evolution with mass ejection:
V. G. Fesenkov, *Transactions of the International Astronomical Union* 8, 702,
     1952
A. G. Massewitsch, *Astronomical Journal of the Soviet Union* 30, 508, 1953
For further references see preceding sections.

§ 30. VITAL STATISTICS OF THE STARS

Derivation of birth-rate function:
E. E. Salpeter, *Ap.J.* 121, 161, 1955
A. R. Sandage, *Ap.J.* 125, 422, 1957
Check of birth-rate function with galactic clusters:
A. R. Sandage, *Ap.J.* 125, 422, 1957
M. F. Walker, *Ap.J.* 125, 636, 1957

INDEX

absorption coefficient: definition, 37; heavier elements, 68; hydrogen and helium, 70; Rosseland mean, 65ff; temperature-density diagram, 72

absorption coefficient in stars: summary, 249ff; tables, 254

abundance of elements: 23ff; Castor C, 138; differences between stellar populations, 27, 272; sun, 25, 205ff; upper main-sequence stars, 130; white dwarfs, 240

abundance of heavier elements, effect on luminosity, 140

abundance of metals, effect on evolution track, 219

adiabatic temperature gradient, definition, 46

age of clusters, 271ff

age of expanding clusters, 19

apsidal motion, 146

apsidal motion coefficient: evolutionary models, 198; initial models, 155; observational results, 155;

beryllium, 26, 87, 156

binaries: apsidal motion, 147; spectroscopic, 13ff; visual, 14ff

birth-rate function, 276ff

bolometric correction; 7ff; dependence on color index, 9

bound-free transitions, 63; *see also* Kramers' law

boundary conditions, 89ff, 97

boundary value problem, 96

carbon stars, 28

carbon cycle: 76ff; red giants, 216; simplified evolution, 165; summary of occurrence of, 250; upper main sequence, 122

Castor C: model, 137ff; table of model, 257

chemical composition, *see* abundances of elements

circulation, in rotating star, 175ff

clusters: age, 271ff; galactic, 3ff; globular, 4ff; luminosity function, 278; theoretical Hertzsprung-Russell diagram, 269ff

color index, 1

composition, *see* abundance of elements

conduction: 37; red giants, 227; white dwarfs, 236

contraction, *see* gravitational contraction

convection: energy transport, 47ff; mixing, 51; mixing length, 133, 218; velocity, 48ff

convective core, for upper main sequence, 124, 186

convective envelope: 92; depth in sun, 205ff; effect in Hertzsprung-Russell diagram, 143ff; lower main sequence, 133; red giants, 213ff

convective, equilibrium, 47

cooling time, of white dwarfs, 243ff

death rate, 280ff

degeneracy: 56ff; limits, 60; partial, 60ff; red giants, 211ff; semi-relativistic, 61ff; summary of occurrence, 249; white dwarfs, 230ff

density, galactic, 273

density distribution: main-sequence stars, 251; red giants, 252; white dwarfs, 251

density in stars: summary, 247ff; tables, 254ff

deuterium, 75, 87, 156

discontinuity in composition, 165, 210, 222ff

distortion: by tidal force, 147ff; in rotating star, 175ff, 178ff

292

A CATALOGUE OF
SELECTED DOVER BOOKS
IN ALL FIELDS OF INTEREST

A CATALOGUE OF SELECTED DOVER
BOOKS IN ALL FIELDS OF INTEREST

CELESTIAL OBJECTS FOR COMMON TELESCOPES, T. W. Webb. The most used book in amateur astronomy: inestimable aid for locating and identifying nearly 4,000 celestial objects. Edited, updated by Margaret W. Mayall. 77 illustrations. Total of 645pp. 5⅜ x 8½.
20917-2, 20918-0 Pa., Two-vol. set $9.00

HISTORICAL STUDIES IN THE LANGUAGE OF CHEMISTRY, M. P. Crosland. The important part language has played in the development of chemistry from the symbolism of alchemy to the adoption of systematic nomenclature in 1892. ". . . wholeheartedly recommended,"—Science. 15 illustrations. 416pp. of text. 5⅝ x 8¼.
63702-6 Pa. $6.00

BURNHAM'S CELESTIAL HANDBOOK, Robert Burnham, Jr. Thorough, readable guide to the stars beyond our solar system. Exhaustive treatment, fully illustrated. Breakdown is alphabetical by constellation: Andromeda to Cetus in Vol. 1; Chamaeleon to Orion in Vol. 2; and Pavo to Vulpecula in Vol. 3. Hundreds of illustrations. Total of about 2000pp. 6⅛ x 9¼.
23567-X, 23568-8, 23673-0 Pa., Three-vol. set $27.85

THEORY OF WING SECTIONS: INCLUDING A SUMMARY OF AIR-FOIL DATA, Ira H. Abbott and A. E. von Doenhoff. Concise compilation of subatomic aerodynamic characteristics of modern NASA wing sections, plus description of theory. 350pp. of tables. 693pp. 5⅜ x 8½.
60586-8 Pa. $8.50

DE RE METALLICA, Georgius Agricola. Translated by Herbert C. Hoover and Lou H. Hoover. The famous Hoover translation of greatest treatise on technological chemistry, engineering, geology, mining of early modern times (1556). All 289 original woodcuts. 638pp. 6¾ x 11.
60006-8 Clothbd. $17.95

THE ORIGIN OF CONTINENTS AND OCEANS, Alfred Wegener. One of the most influential, most controversial books in science, the classic statement for continental drift. Full 1966 translation of Wegener's final (1929) version. 64 illustrations. 246pp. 5⅜ x 8½. 61708-4 Pa. $4.50

THE PRINCIPLES OF PSYCHOLOGY, William James. Famous long course complete, unabridged. Stream of thought, time perception, memory, experimental methods; great work decades ahead of its time. Still valid, useful; read in many classes. 94 figures. Total of 1391pp. 5⅜ x 8½.
20381-6, 20382-4 Pa., Two-vol. set $13.00

GEOMETRY, RELATIVITY AND THE FOURTH DIMENSION, Rudolf Rucker. Exposition of fourth dimension, means of visualization, concepts of relativity as Flatland characters continue adventures. Popular, easily followed yet accurate, profound. 141 illustrations. 133pp. 5⅜ x 8½.
23400-2 Pa. $2.75

THE ORIGIN OF LIFE, A. I. Oparin. Modern classic in biochemistry, the first rigorous examination of possible evolution of life from nitrocarbon compounds. Non-technical, easily followed. Total of 295pp. 5⅜ x 8½.
60213-3 Pa. $4.00

PLANETS, STARS AND GALAXIES, A. E. Fanning. Comprehensive introductory survey: the sun, solar system, stars, galaxies, universe, cosmology; quasars, radio stars, etc. 24pp. of photographs. 189pp. 5⅜ x 8½. (Available in U.S. only)
21680-2 Pa. $3.75

THE THIRTEEN BOOKS OF EUCLID'S ELEMENTS, translated with introduction and commentary by Sir Thomas L. Heath. Definitive edition. Textual and linguistic notes, mathematical analysis, 2500 years of critical commentary. Do not confuse with abridged school editions. Total of 1414pp. 5⅜ x 8½.   60088-2, 60089-0, 60090-4 Pa., Three-vol. set $18.50

Prices subject to change without notice.

Available at your book dealer or write for free catalogue to Dept. GI, Dover Publications, Inc., 180 Varick St., N.Y., N.Y. 10014. Dover publishes more than 175 books each year on science, elementary and advanced mathematics, biology, music, art, literary history, social sciences and other areas.